Disaster Theory

Disaster Theory
An Interdisciplinary Approach to Concepts and Causes

David Etkin

AMSTERDAM • BOSTON • HEIDELBERG • LONDON
NEW YORK • OXFORD • PARIS • SAN DIEGO
SAN FRANCISCO • SINGAPORE • SYDNEY • TOKYO

Butterworth-Heinemann is an imprint of Elsevier

Acquiring Editor: Sara Scott
Editorial Project Manager: Marisa LaFleur
Project Manager: Vijayaraj Purushothaman
Designer: Maria Inês Cruz

Butterworth-Heinemann is an imprint of Elsevier
The Boulevard, Langford Lane, Kidlington, Oxford OX5 1GB, UK
225 Wyman Street, Waltham, MA 02451, USA

Copyright © 2016 Elsevier Inc. All rights reserved.

No part of this publication may be reproduced or transmitted in any form or by any means, electronic or mechanical, including photocopying, recording, or any information storage and retrieval system, without permission in writing from the publisher. Details on how to seek permission, further information about the Publisher's permissions policies and our arrangements with organizations such as the Copyright Clearance Center and the Copyright Licensing Agency, can be found at our website: www.elsevier.com/permissions.

This book and the individual contributions contained in it are protected under copyright by the Publisher (other than as may be noted herein).

Notices
Knowledge and best practice in this field are constantly changing. As new research and experience broaden our understanding, changes in research methods, professional practices, or medical treatment may become necessary.

Practitioners and researchers must always rely on their own experience and knowledge in evaluating and using any information, methods, compounds, or experiments described herein. In using such information or methods they should be mindful of their own safety and the safety of others, including parties for whom they have a professional responsibility.

To the fullest extent of the law, neither the Publisher nor the authors, contributors, or editors, assume any liability for any injury and/or damage to persons or property as a matter of products liability, negligence or otherwise, or from any use or operation of any methods, products, instructions, or ideas contained in the material herein.

British Library Cataloguing-in-Publication Data
A catalogue record for this book is available from the British Library.

Library of Congress Cataloging-in-Publication Data
Application submitted

ISBN: 978-0-12-800227-8

For information on all Butterworth-Heinemann
publications visit our website at http://store.elsevier.com

Dedication

In memory of my parents Ben and Maya, and particularly to my father, the quintessential academic who at age 96 and two weeks before he passed away, sent me two corrections to my final chapter.

Contents

Digital Assets	xiii
Foreward, by Ian Burton	xv
Preface	xvii
Introduction	xix

1.	What is a Disaster?		1
	1.1	Why This Topic Matters	2
	1.2	Recommended Readings	3
	1.3	The Meaning of Disaster	3
	1.4	Summary	10
	1.5	Case Study: The 2003 Heat Wave in Europe	10
		Further Reading about the 2003 Heat Wave	18
		End Notes	19
2.	Disaster Data: A Global View of Economic and Life Loss		23
	with Susan McGregor		
	2.1	Why This Topic Matters	24
	2.2	Recommended Readings	25
	2.3	Introduction	26
	2.4	Measuring Loss	26
	2.5	Data Quality	30
	2.6	Databases	35
	2.7	Conclusions	41
	2.8	Case Study: Hurricane Hazel and Toronto	46
		Further Readings	50
		End Notes	50

3.	**Disaster Risk**		**53**
	3.1 Why This Topic Matters		55
	3.2 Recommended Books and Readings		55
	3.3 Question to Ponder		56
	3.4 Introduction		56
	3.5 Risk		56
	3.6 The Risk Society		77
	3.7 Measuring Risk		78
	3.8 Sea Level Rise and Subsidence		87
	Further Reading		91
	3.9 Summary		91
	3.10 Case Study: 1998 Ice Storm in Eastern Canada and Northeastern United States		91
	End Notes		96
4.	**Hazard, Vulnerability, and Resilience**		**103**
	4.1 Why This Topic Matters		105
	4.2 Recommended Readings		105
	4.3 Hazard		106
	4.4 Introduction to Vulnerability and Resilience		111
	4.5 Vulnerability		113
	4.6 Resilience		122
	4.7 Grassy Narrows		138
	Further Reading		142
	4.8 Responsibility and Response Ability—Comments on Vulnerability and Community by John (Jack) Lindsay		142
	End Notes		145
5.	**Disasters and Complexity**		**151**
	5.1 Why This Topic Matters		153
	5.2 Recommended Readings		153

	5.3 Introduction	154
	5.4 Characteristics of Complex Systems	155
	5.5 Normal Accident Theory	166
	5.6 Discussion	169
	5.7 Close Calls or Near Misses	171
	Further Reading	174
	5.8 Conclusion	174
	5.9 Case Study: Flooding along the Mississippi and Missouri Rivers, Hurricane Katrina and the New Orleans Catastrophe	174
	Further Reading	187
	End Notes	188
6.	**Disaster Models**	**193**
	6.1 Why This Topic Matters	194
	6.2 Recommended Readings	195
	6.3 What Is a Model?	196
	6.4 Philosophical Approaches	198
	6.5 Disaster Models	202
	6.6 Conclusion	222
	6.7 Case Study: Sarno Landslides	222
	Further Reading	223
	6.8 A Comment by Joe Scanlon	224
	End Notes	226
7.	**Myths and Fallacies**	**229**
	7.1 Why This Topic Matters	230
	7.2 Recommended Readings	231
	7.3 Myths of Fact	231
	7.4 Myths of Human Behavior	237
	7.5 Fundamental Myths of Our Relationship to the World	240

	7.6	Conclusion	244
	7.7	Fables: of Little Pigs and Ants	244
	7.8	Case Study: The Great Flood	247
	7.9	A Comment by Joe Scanlon	248
		End Notes	250
8.	**The Poetry of Disaster**		**253**
	8.1	Why This Topic is Important	254
	8.2	An Essay by Nicole Cooley	254
	8.3	Some Thoughts	258
	8.4	Case Study: Burning of the Library at Alexandria	262
		Further Reading	266
	8.5	A Comment by Joe Scanlon	266
	8.6	A Comment by Robin Cox	269
		End Notes	270
9.	**Ethics and Disaster** *with Peter Timmerman*		**273**
	9.1	Why This Topic Matters	274
	9.2	Recommended Readings	275
	9.3	Introduction	277
	9.4	Ethics	279
	9.5	Ethics and the Construction of Risk—a Reflection	298
	9.6	Conclusion	299
	9.7	Example of an Ethical Dilemma: Temporary Settlement versus Permanent Housing	300
	9.8	Jean Slick on Ethical Dilemmas	303
	9.9	Commentary by Naomi Zack	304
		End Notes	306

10.	Workshop on Principles of Disaster Management *with Ian Davis*	**311**
	10.1 Why This Topic Matters	312
	10.2 Recommended Readings	313
	10.3 Why are Principles Needed for Disaster Management?	314
	10.4 The Complexity of Current Principles	318
	10.5 Two Models: Clarifying Principles	319
	10.6 Tasks for Breakout Groups	323
	End Notes	326
11.	Final Reflections	**329**
	End Notes	333

Appendix 1: Selected Disaster Data	335
Appendix 2: Statistics Canada: Factors and Measures Related to Community Resilience	337
Appendix 3: Interviews with Ian Burton and Ken Hewitt	341
Index	355

Digital Assets

Thank you for selecting Butterworth Heinemann's *Disaster Theory: An Interdisciplinary Approach to Concepts and Causes*. To complement the learning experience, we have provided a number of online tools to accompany this edition. Please consult your local sales representative with any additional questions.

For the Instructor

Qualified adopters and instructors need to register at the following link for access: http://textbooks.elsevier.com/web/manuals.aspx?isbn=9780128002278.

- *Test bank* composes, customizes, and delivers exams using an online assessment package in a free Windows-based authoring tool that makes it easy to build tests using the unique multiple-choice and true or false questions created for *Policing in America*. What is more, this authoring tool allows you to export customized exams directly to Blackboard, WebCT, eCollege, Angel, and other leading systems. All test bank files are also conveniently offered in Word format.
- *PowerPoint lecture slides* reinforce key topics with focused PowerPoints, which provide a perfect visual outline with which to augment your lecture. Each individual chapter has its own dedicated slideshow.
- *Instructor's guides* design your course around customized learning objectives, discussion questions, and other instructor tools.

Foreword, by Ian Burton

It is a pleasure for me to respond to the author's request for a foreword to this most unusual and readable book. Although based in large part on David Etkin's lectures at York University in Toronto, this is not a standard textbook. It is rather a very broad, sweeping analysis, and eclectic tour of the field of disaster studies, both local and global, ancient and modern, theoretical and practical. David is by profession a physicist and a meteorologist and he brings some of that background expertise and insight into his book. He also casts a much wider net and draws upon an impressive range of fields and literatures to raise questions about disasters. Most pedagogic books provide or attempt to provide answers. This book is more Socratic in approach. Questions are raised and explored. Relevant information is provided but the author makes few definitive statements. This book is not designed to provide answers but to provoke thought and analysis.

Hence, what is learned from this book depends very much on the reader and what the reader brings to the study of disasters – be it a fascination with definitions and epistemology; a love of data and statistics; the imagination of a modeler; or an orientation to the literary, historical, poetic, ethical dimensions of life and its risks, or some personal experiences. I know of no other book quite like this.

Reading it stimulated me to think of the evolution of the social construction of disasters. Disasters were once quite rare and infrequent, or occurred in faraway places and were not heard about or reported. They were also mostly considered to be local or place-based events. Of course, some really major disasters did occur or were believed to have occurred – like Noah's flood, for example, or the Lisbon earthquake, or the epidemics of the plague. Thus they were either overwhelming Acts of God or problems that could be largely managed and contained within the regions where they occurred.

In the modern era, the idea that disasters are Acts of God persists despite overwhelming evidence to the contrary. The growth of science and technology from the seventeenth century onward led to an expectation that science and technology could be advanced and deployed to help control the natural forces themselves (flood control, for example), or to reduce losses by improving capacity for forecasting and warning and understanding where and when extreme events would be more likely to occur. This science-based approach was coupled with a humanitarian concern in which external help was made available in situations where the affected places and people could not cope. More recently, over the last five decades or more, such approaches have gradually come to be seen as too limited. It has been argued that disasters result more from the day-to-day choices made by people over long periods of time. Disasters are made "by design." Two big elements in the design of disasters are exposure and vulnerability. The rapid growth of human populations and the associated wealth, infrastructure, and other assets has put much more at risk. A currently prevailing view in the disaster risk management and reduction communities is that such steps as improved land-use planning (keeping development away from high risk areas) and improved building designs and standards could play a major role. There has been much expansion in an apparent capacity to apply and implement such actions but experience suggests that they have not been well or effectively applied. After a disaster, new plans and regulations are frequently adopted amid resolutions and promises that things will be "built back better." Such resolutions can soon be forgotten in the face of other development priorities that may themselves be unsustainable. While efforts to correct past mistakes are sometimes successful, there often remains a lack of attention to the creation of new and greater risks.

This book points to the need for new thinking about disasters triggered by extreme events but which are largely created by human choices, and it challenges the reader to contribute to the new thinking. Why, despite much greater scientific understanding of the frequency, timing, and location of extreme events, does exposure and vulnerability continue to grow? Why have decades of experience and policy innovations not led to an effective adoption of planning and vulnerability reduction? To what extent might it be true to say that disasters are not simply unique, place-based events but share some common underlying root causes that could help to explain the growing disaster epidemic?

A careful reading of this book can lead to creative ideas in response to this and the many other questions raised.

Ian Burton
March 19, 2014

Preface

Life moves very fast. It rushes from Heaven to Hell in a matter of seconds.
Paulo Coelho

Disasters engage us on many levels; they create a sense of awe in terms of the damage and chaos they create, they saturate and numb our emotions, and they arouse our curiosity. My interest in disasters began when I was a meteorologist forecasting weather in the provinces of Nova Scotia and Ontario, Canada. As part of a (now defunct) government research team called the Adaptation and Impacts Research Group in Environment Canada, I was fortunate to have Dr Ian Burton as a director. As a global expert in the area of hazard management, he connected me with others in the field and enabled my education about disasters from a social science perspective, a challenging transition from the physics background I entered the field from. Thank you Ian…

Extreme events like tornadoes or severe storms were always interesting to me from several perspectives, particularly in terms of their impact on people and communities. As I read and observed more about them and studied case studies of disasters, it became clear to me that their occurrence generally had much more to do with how people live and construct their physical and social spaces than about the actual hazard itself. I have continued to study disasters both as a government scientist located at the University of Toronto and later as a professor in the field of disaster management at York University, and to be fascinated by their power and complexity, saddened by their impacts, and frustrated by the actions people continue to take that make themselves vulnerable.

Over time our knowledge has increased enormously, along with our technical competence, and yet disaster losses continue to escalate. People create most of the disasters that affect us in the presence of the knowledge it takes to avoid them. Clearly, we are doing something wrong, and understanding why this is so is fundamental to disaster risk reduction.

I wish to dedicate this book to my family, who have supported my journey into academia, to my children Jonathan, Stephanie and Heidi and my granddaughter Alexandra, who represent our future, and especially to my wife Deborah, who edited this volume. As well, I want to acknowledge the many dedicated people who devote themselves to reducing the horrendous impact disasters have on so many of us. First responders are high profile, often thought of as heroes, and always appreciated. But there are many emergency managers and other people in the field of hazard management who work in the background to mitigate or prevent disasters. They are rarely recognized or thanked, and they deserve to be.

Introduction

> *The more things change, the more they remain the same.*
> *(Plus ça change, plus c'est la même chose.)*
> **Alphonse Karr ("Les Guêpes")**

This book is about understanding disaster. There are many ways that this could be approached, depending on the discipline, profession, and the author's purpose. Although my education was originally in physics and mathematics, and I come from a professional background of meteorology and environmental issues, I am taking a broad and holistic approach that emphasizes the social sciences. I do this because I see human behavior as the primary reason we create vulnerable communities that experience disaster—this is the area where we can exert the most influence on reducing disaster risk. I do not mean to diminish the importance of the physical sciences in the study of this field; they are essential. But they are also insufficient, and one of the main purposes of this book is to provide an understanding of the various ways people relate to and cope with disaster risk.

In the field of disaster studies there is a noteworthy gap between theory and the actual practice of disaster risk reduction. This gap is diminishing, but the newness of the academic side of the field and the emergence of the practitioner community from civil defense has created a situation where much that is known is not used. In the medical field a transition to evidence-based practice was formally introduced in 1992—an approach that has spread to other disciplines. More emphasis needs to be placed on this approach in the practice of disaster and emergency management. Doing this means incorporating the knowledge gained from good theoretical and empirical research into how we reduce disaster risk. It is this research that is the driving force behind the organization of this book, linking the theoretical underpinnings of this field of study with empiricism and practice.

Since the mid-20th Century there have been an enormous number of disasters that have taken a terrible toll on people and their communities. Because of this, government, nongovernmental organizations, and universities have all begun to pay increasing attention to the study of disaster and disaster management. One of the results of this growing awareness is programs such as the one I work in at York University, Toronto, Canada, and the graduate course on disaster theory that I teach. In Canada, such programs have been around for about a decade, but over time they will transform the profession of disaster and emergency management into one with a much more holistic approach than has historically been used.

Predicting the future is not an easy task, but it seems to me that the world in which we are living is going through a transformation. The massive trends of globalization, urbanization, climate change, terrorism, population growth, species extinction, the rise of sea level, and the growth of novel technologies combine to alter old hazards, create new ones, and make societies vulnerable in ways that they have not been before. Coping with these changes is more than challenging and requires a constant questioning of the theories, models, and assumptions upon which our understanding of disaster risk construction is based.

This book was written with students in mind, but it is also appropriate for professionals working within the field who want to gain a better understanding of theory and how it affects the practice of disaster risk reduction. It is also appropriate for the educated layperson who simply has an interest in this field of study. As much as possible, I have illustrated theory with examples and case studies to avoid the "ivory tower syndrome." The most fundamental question to be asked in the study of disasters is, What are they? That is the subject of the first chapter. The second chapter is a report on global disaster data with a commentary on the uncertainties inherent in the data sets and how biases can be created, depending on how they are constructed and used. Disaster data suffer from the irony that as they become more robust, they suffer from problems of representativeness. For example, the insurance industry has good data on what disasters cost them, but that alone is a rather poor indicator of most disasters. In the third chapter I address various aspects of risk theory that are particularly relevant to disaster risk; in particular, I focus on social constructionism, risk homeostasis, risk perception, and issues related to the measurement of risk. Many fields of study address the issue of risk, and each tends to have its own terminology and set of definitions. A discussion of this area must begin with simply accepting a terminology and then moving forward. Chapter 4 deals with the most important aspects of vulnerability and resilience theory. *Resilience* is the new catchword within the disaster field, and it is certainly a very useful approach to risk reduction. One should not jump too quickly on the bandwagon, though—approaches that address vulnerability and robustness are still critical. Chapter 5 addresses the fascinating topic of complexity and how our understanding of complex systems may be relevant to understanding the evolution of disasters, as well as which management strategies might be most effective. This area is not yet well developed and is ripe for future research. Chapter 6 overviews a number of different disaster models that are used by various groups and organizations as tools to assist with managing disaster risk. The choice of which model to use can be important in terms of outcomes, and professionals in this field should have a good sense of the strengths and weaknesses of each. A model is not a theory and is not right or wrong in itself; the usefulness of models is contingent on their utility with respect to a person's or institution's particular goal. Chapter 7 deals with the topic of disaster myths and fallacies. These misunderstandings sometimes play out in significant ways in terms of planning processes and response, and theoreticians and practitioners should be familiar with the research that has been done on them. The idea for Chapter 8 came from a Leonard Cohen concert I attended with my wife. Enthralled by his musical poetry, I wondered to what extent people use this and other forms of literature to understand disaster. The answer, I learned, is that it is used a great deal! It seems natural for people to use stories, poetry, and art as tools to help them come to terms with loss. Chapter 9 deals with the issue of ethics and disaster, a topic addressed too infrequently within the emergency management community, although it is commonly addressed in other fields such as humanitarian aid or health. Ethics must be the basis for management principles but often are not addressed. The final chapter provides an opportunity for students to engage in a workshop, to develop sets of principles for various scenarios.

As I often do, in 2012 I attended the annual Federal Emergency Management Agency conference on higher education in emergency management, held during June on the beautiful Federal Emergency Management Agency campus in Emmitsburg, Maryland, where I sat in on a session addressing emergency management theory. One of the questions asked was which theories and scholars each of us base our teaching on. There was little agreement! Coming at the field from a variety of disciplines and focusing on various aspects of the disaster problem results in a multitude of perspectives. Disaster theory has not yet coagulated, and given the nature of the problem, it may always be a fuzzy beast. Not everybody will agree with my choice of topics for this book, but I consider them to be the most critical in terms of understanding why and how disasters happen and how we need to deal with them.

As an academic in this field I have read about many historical and recent events, while at the same time watching new ones unfold. Some disasters such as the Haiti earthquake are sad replays of old themes, whereas others such as the Fukushima nuclear meltdown offer new twists as a result of the changing world in which we live. Yet all of them revolve around human behavior and decision making, which in many ways are not very different from ancient times. To illustrate this, I offer the following two quotes.

> *As it was, their judgment was based more on wishful thinking than on a sound calculation of probabilities: for the usual thing among men is that when they want something they will, without any reflection, leave that to hope, while they will employ the full force of reason in rejecting what they find unpalatable.*
>
> *Thucydides, 425 BC, The Peloponnesian Wars*

> *Thereupon one of the priests, who was of a very great age, said: O Solon, Solon, you Hellenes are never anything but children, and there is not an old man among you. Solon in return asked him what he meant. I mean to say, he replied, that in mind you are all young; there is no old opinion handed down among you by ancient tradition, nor any science which is hoary with age. And I will tell you why. There have been, and will be again, many destructions of mankind arising out of many causes; the greatest have been brought about by the agencies of fire and water, and other lesser ones by innumerable other causes.... In the first place you remember a single deluge only, but there were many previous ones.*
>
> *Egyptian Priest to Solon, the Athenian lawgiver, from Plato, Timaeus, 360 BC*

Disasters are holistic, unbounded by disciplinary or political boundaries. They have historical and cultural roots and need to be understood in context. They affect people differently, depending on culture, class, race, socioeconomic status, worldview, and psychology. There are winners and there are losers, not only in the aftermath of disaster but particularly as a result of the construction of risk, which benefits some while diminishing others. Our most fundamental concerns about disasters are rooted in ethics and empathy, and it is through those windows that efforts to manage them must be viewed.

It made sense to me to begin this book with a description of a disaster, and I chose the Lisbon disaster of 1755 (Figure 1). This is a fascinating event not only because it was so terribly destructive but also because it engaged society in a religious and philosophical debate that challenged their most fundamental assumptions about God and the world in which they lived. As well, it led to a restructuring of political power in Portugal and of the built environment. How those changes happened and the long-term outcomes from them demonstrate important lessons.

FIGURE 1 Lisbon, Portugal, during the great earthquake of November 1, 1755. This copper engraving, made that year, shows the city in ruins and in flames. Tsunamis rush upon the shore, destroying the wharfs. The engraving is also noteworthy because it shows highly disturbed water in the harbor, which sank many ships. Passengers in the left foreground show signs of panic. Original in Museu da Cidade, Lisbon. Reproduced in O Terramoto de 1755, Testamunhos Britanicos = The Lisbon Earthquake of 1755, British Accounts. Lisbon: British Historical Society of Portugal, 1990. *Source: The Earthquake Engineering Online Archive - Jan Kozak Collection KZ128.*

Lisbon 1755: The First Modern Disaster

…this extensive and opulent city is now nothing but a vast heap of ruins.

Rev. Charles Davy

Before November 1, 1755, Lisbon was one of the major cities in Europe, with a population of 200,000–250,000 and a thriving port trade that was the center of the Portuguese colonial empire. It is described as being opulent; much of the wealth came from its colonies, particularly Brazil, in the form of gold and precious stones. Within the city were 40 churches, 75 convents, 33 palaces, many large public buildings, and about 20,000 4–5 story houses built mostly of stone and wood. The city plan was typically medieval, with narrow winding streets strewn with refuse. About 10% of the population was homeless and unemployed; it was a city of contrasts.

The region had a history of earthquakes, and Lisbon had experienced them before, in 1356, 1531, 1551, 1597, 1598, 1699, 1724, and 1750. The epicenter of the 1755 earthquake was about 200 km west-southwest of Lisbon; with a magnitude of between 8.5 and 9.0 MM, it became one of the most destructive natural disasters recorded in European history (hundreds of aftershocks were felt during the following year). There were actually 3 quakes over a period of 3–10 min that resulted in the massive collapse of buildings within the city, which was exacerbated by the liquefaction of soils under much of the area. Fires quickly broke out throughout the city because of the many candles on the altars of the churches and from fires in hearths; in addition, the city was largely constructed from flammable wood. Much of the population fled to open areas and to the coast for safety, but about 40 min after the earthquake a tsunami up to 10–15 m swept through the harbor and downtown area; tsunami heights of up to 20–30 m occurred elsewhere, particularly in the Algarve region in southern Portugal. The first tsunami was followed by 2 more waves and the fires, fanned by strong winds, raged for 5–6 days in areas not submerged by the sea water. By the end of the week Lisbon was a smoking ruin.

It went on consuming everything the earthquake had spared, and the people were so dejected and terrified that few or none had courage enough to venture down to save any part of their substance; everyone had his eyes turned towards the flames, and stood looking on with silent grief, which was only interrupted by the cries and shrieks of women and children calling on the saints and angels for succor, whenever the earth began to tremble, which was so often this night.

Rev. Charles Davy, 1755[1]

The particular day on which the Lisbon earthquake happened was one of the most important aspects of the disaster. It occurred on November 1, 1755—the holiday of All Saint's Day—at 9:40 a.m. local time, when most of the population of Lisbon was celebrating mass. Collapsing churches crushed many of the worshippers. Historical estimates of fatalities in Lisbon range from 60,000 to 100,000, although more recent estimates have been downgraded to 20,000–30,000 deaths. Most of the buildings in the city were destroyed or badly damaged, including all the palaces, the opera house, 65 convents, all prisons, hospitals, and libraries. Included in the loss was the 70,000-volume royal library, hundreds of works of art, and the royal archives. A recent estimate of the direct damage done by the earthquake comes to between 32% and 48% of the Portuguese gross domestic product.

The person in command of response and reconstruction, Senhor de Carvalho, later the Marquis de Pombal, Portugal's secretary of state, was an ambitious, competent, and ruthless politician. When asked by King José I what they should do, he is reported to have answered "bury the dead and feed the living." Impressed with his calmness and common sense, the king put him in command of response and recovery. Pombal enforced a rapid and effective response, including the mass burial of corpses at sea, using troops to conscript able-bodied men to clear rubble, distribute food, create rent and price controls, establish temporary hospitals, and erect tents and huts for the homeless. A number of gallows were erected in plain sight, and looters and arsonists were hanged (Figure 2).

Following the disaster, Pombal commissioned an inventory of property and public works, prohibited any construction before official approval, created a 4% tax on manufacturing and merchandising, and put together a team of designers who created three options for rebuilding the city: (1) a option with no changes, (2) a option widening the streets, and (3) a "clean slate" option including a new road pattern, lower density, and new construction standards. The clean slate option—which included shorter building heights (because higher buildings tended to collapse more easily than lower ones during the earthquake), the elimination of arches, and the use of masonry and framed construction methods that were more earthquake resistant—was selected. It created the first building codes in Lisbon. The final city architecture was based on a grid street layout, a reduction in the visibility of the Church (philosophically in line with the controls Pombal put on Jesuits and the Inquisition) and a city structure suited to a mercantile lifestyle. It symbolized a shift from a royal and religious center to a center of trade. It was not a change embraced by the nobility or Church, but it did represent the shift in power toward the merchant, the common man, and government bureaucracy.

FIGURE 2 1755 German copperplate image, "The Ruins of Lisbon." Survivors camp in a (rather fanciful) tent city outside the city of Lisbon following the November 1, 1755, earthquake. The image shows criminal activity and general mayhem, as well as the hanging of quake survivors under constabulary supervision. Priests are present, one holding a crucifix, one possibly a prayer book, and appear to be giving last rites to persons being hanged.
Source: The Earthquake Engineering Online Archive - Jan Kozak Collection.

The Lisbon disaster happened at a point of particular interest in European philosophy. There was a tension between rationalists such as Gottfried Wilhelm Leibniz[2] and Alexander Pope, who argued that we lived in the best of all possible worlds that is part of God's unknowable plan, and contrary views such as expressed by Voltaire in *Candide* and his "Poem on the Disaster of Lisbon," which took a much more critical approach to the human condition as it relates to God's justice. Jean Jacques Rousseau put forth a different argument, commenting that the damage would have been much less had people not lived in six- to seven- story buildings. That the disaster happened on All Saints Day and killed so many worshippers was a serious blow to traditional Christian understanding about the relationship between man and God. Why would God allow such an event to happen?

Much of the debate in the 18th century dealt with whether God was benign and the degree to which man was powerless in the presence of an all-powerful God. Common interpretations of disasters revolved around divine retribution for man's sins and a warning for mankind to be more virtuous and faithful. Religious perspectives included the views that God punished Lisbon because of the Catholic Inquisition (as suggested by John Wesley from England), that the Inquisition was not rigorous enough, that the king was not sufficiently pious, and so forth. Another interesting interpretation was put forth by Rondet, who viewed the disaster as part of a symbolic discourse dealing with prophecies: Disasters "are nothing but language"[3] that can be used to unveil the truth of prophecies.

Interestingly, this event was also the trigger for some of the first scientific investigations into the causes of earthquakes; previous studies had been done after a minor earthquake in London in 1750. For example, Pombal created a survey sent to all parishes asking the following questions.

- How long did the earthquake last?
- How many aftershocks were felt?
- What kind of damage was caused?

- Did animals behave strangely?
- What happened in wells and water holes?

Some pursued a more scientific approach, such as Kant (who suggested natural causes) and Reverend John Michell of Cambridge (who argued that earthquakes were caused by waves originating within the earth, in what can be called the birth of seismology).[4] The notion of divine determinism gradually was replaced by social constructionism[5]; such existential questions were commonly discussed in the 18th century but are largely absent from the 21st century discussion of disasters, an illustration of the importance of understanding the culture in which events occur to gain insight into the meaning that is assigned to them.

> *As a metaphor for a new world order, the disaster encouraged an understanding of nature, and led naturalists and philosophers to channel their reasoning towards making scientific use of the laws of matter when dealing with terrestrial physics, as Kant did. But the event also became memorable in Europe's intellectual tradition for raising the scandal of evil, for reviving the issue of God's justice, questioning the relationship between nature and morals, for bringing new and plausible arguments to the public debate about time no longer being sacred, and about the course of history. As the century progressed, writers began to insist on the idea of a 'before' and 'after' the 1755 earthquake, using the event as a dividing line and a moment of common awareness associated with its cultural, philosophical and aesthetic legacy, which divided the century without compromising the otherwise pervasive Enlightenment.*
>
> *Ana Cristina Araujo*[6]

Lisbon often is referred to as the first modern disaster,[7] not only because the state response was so well organized and coordinated but also because of the search for scientific explanations. It also was used as an opportunity for the shift and consolidation of political power and the restructuring of an economy, much akin to what Naomi Klein[8] discusses in her book on disaster capitalism. According to some analyses, the disaster ended up benefiting the Portuguese economy because of the restructuring that happened after 1755.[9] The role that power plays in disaster is critical, and its application and distribution, both in terms of how risk is constructed in the first place and how power can be reconstructed following these kinds of events, does not receive enough attention in disaster studies.

Like the Hurricane Hazel disaster discussed in Chapter 2, the Lisbon disaster was a focusing event that allowed for a restructuring of society. There are many examples of communities or cities rebuilt "as they were." In this case, society embraced a different vision.

Further Reading about the Lisbon Disaster

Alexander D: Nature's impartiality, man's inhumanity: Reflections on terrorism and world crisis in a context of historical disaster, *Disasters* 26(1):1–9, 2002.

Araujo AC: Focus: Lisbon earthquake: part 2, *European Review* 14(3):313–319, 2006.

Araújo AC: The Lisbon earthquake of 1755: public distress and political propaganda, *e-Journal of Portuguese History* 4(1), 2006.

Chester S: Faith, doubt, aid and prayer: the Lisbon earthquake of 1755 revisited, *European Review* 14(03):321–328, 2006.

Bressan D: *The Earthquake of Lisbon: Wrath of God or Natural Disaster?* Scientific American Blog, 2011. http://blogs.scientificamerican.com/history-of-geology/2011/11/01/november-1-1755-the-earthquake-of-lisbon-wraith-of-god-or-natural-disaster/.

Chester DK: The 1755 Lisbon earthquake, *Progress in Physical Geography* 23(3):363–383, 2001.

Dynes Russell R: The Lisbon earthquake in 1755: the first modern disaster. In Theodore ED, Radner John B, editors: *The Lisbon earthquake of 1755. Representations and reactions*, Oxford: SVEC, 2005, pp 34–49.

Dynes Russell R: *The Lisbon Earthquake in 1755: Contested meanings in the first modern disaster*, University of Delaware Disaster Research Center, 1997. http://www.udel.edu/DRC/.

Dynes RR: The dialogue between Voltaire and Rousseau on the Lisbon earthquake: The emergence of a social science view, *International Journal of Mass Emergencies and Disasters* 18(1):97–115, 2000.

Holmes OW: *The Lisbon Earthquake, 1755*. The Gallery of Natural Phenomena: The earth, the sea, the sky–and beyond,(n.d.) http://www.phenomena.org.uk/earthquakes/earthquakes/lisbon.html.

Lukes S: Questions about Power: Lessons from the Louisiana Hurricane. In *Understanding Katrina: Perspectives from the Social Sciences*, 2006. http://understandingkatrina.ssrc.org/Lukes/.

Mullin JR: The Reconstruction of Lisbon following the earthquake of 1755: a study in despotic planning, *Planning Perspective* 7(2):157–179, 1992.

Murteira H: The Lisbon earthquake of 1755: the catastrophe and its European repercussions, *The Economia Global e Gestão (Global Economics and Management Review), Lisboa* 10:79–99, 2004.

Pereira AS: The opportunity of a disaster: the economic impact of the 1755 Lisbon earthquake, *Journal of Economic History* 69(2):466, 2009.

End Notes

1. *Modern History Sourcebook: Rev. Charles Davy: The Earthquake at Lisbon, 1755. Fordham University*, http://www.fordham.edu/halsall/mod/1755lisbonquake.asp.
2. Leibniz created the word theodicy in 1710, referring to attempts to reconcile God's love, justice and omnipotence on the one hand and human suffering on the other.
3. Huet, M.J., *The Culture of Disaster*, (Chicago: University of Chicago Press, 2012).
4. Chester, D. K., "The 1755 Lisbon earthquake," *Progress in Physical Geography* 25, no. 3 (2001): 363-383.
5. Huet, M.J., *The Culture of Disaster*, (Chicago: University of Chicago, 2012).
6. Araujo, A. C., "Focus: Lisbon earthquake: Part 2," *European Review* 14, no. 3 (2006): 313–319.
7. Dynes, Russell R., "The Lisbon earthquake in 1755: the first modern disaster," in *The Lisbon earthquake of 1755. Representations and reactions*, eds. Braun, Theodore E. D. and Radner, John B. (Oxford: SVEC, 2005), 34–49.
8. Klein, N., *The shock doctrine: The rise of disaster capitalism*, (Metropolitan Books, 2007).
9. Pereira, A. S., "The opportunity of a disaster: the economic impact of the 1755 Lisbon earthquake". *Journal of Economic History*, 69, no. 2 (2009): 466.

1

What Is a Disaster?

To name the catastrophic demon won't slay it. But it can help chase our fears out of the shadows and into the sunlight.

Sam Tanenhaus

FIGURE 1.1 The disaster of homelessness. An aged German woman in 1945, overcome by the worry of trying to find a home, breaks down and cries, head in hand. In her other hand she holds a small handbarrow containing her few belongings. *Source: German Federal Archives.*

FIGURE 1.2 A cooking disaster.

CHAPTER OUTLINE

1.1 Why This Topic Matters	2
1.2 Recommended Readings	3
1.3 The Meaning of Disaster	3
1.4 Summary	10
1.5 Case Study: The 2003 Heat Wave in Europe	10
Further Reading about the 2003 Heat Wave	18
End Notes	19

CHAPTER OVERVIEW

The most logical place to begin a book on disaster theory is by defining the term disaster. However, this turns out to be a complicated exercise, since it has been used and understood in a variety of ways. There are specific operational definitions that are very important because they trigger political decisions and the flow of resources. Definitions that are more theoretical focus on the scale of impact and the degree to which outside resources are required in order to cope with these events. It has become clear that disasters are more than just large emergencies, and that catastrophes are more than just large disasters; they exhibit different qualitative characteristics that require different management strategies. For this reason, the development of better disaster taxonomies is important for both research and operations.

KEYWORDS

- 2003 European heat wave.
- Catastrophe.
- Disaster definition.
- Disaster taxonomy.
- Emergency.

1.1 Why This Topic Matters

The importance of defining a disaster is a matter of contention. Susan Cutter[1] makes this point when she says, "I submit that disaster studies… are spending too much time and intellectual capital in defining the phenomena under study, rather than in researching more important and fundamental concerns of the field."

The alternate perspective, that such theoretical questions are important, is argued by Quarantelli[2] when he says that "there are going to continue to be serious problems in our data gathering and analysis" unless we achieve a better conceptual grasp of the issue. How we understand the term disaster determines many other things: what kind of data is gathered, how it is analyzed and ranked, what models are used to manage disaster risk, and how some policies are created and implemented.

Certainly the question "What is a disaster?" is interesting from an academic perspective, and I tend to agree with Professor Quarantelli of its importance, though duly noting the critique of Professor Cutter. I suspect that the energy and attention given to this topic does not seriously detract from operations or other important empirical studies. How the word disaster is defined (which says much about how we understand these events (Figures 1.1 and 1.2)) has significant implications for what research is undertaken and what strategies are used to manage them. It is one of those intellectually and emotionally potent words that should be used with care.

There is sometimes a perspective, particularly amongst some practitioners, that theory is not important, but I believe the opposite is true. It is theory that provides new insights, and the theory one uses plays out in practice in important ways. Good theory underlies good practice. A number of years ago I was involved, along with a number of other very good scientists, in a project to create a hazards map for Canada. I argued for the inclusion of drought, but because it is a slow onset diffuse event many of the other scientists did not consider it a disaster. Their concept of a disaster was of a rapid onset well-defined event like a volcanic eruption or a tornado. I thought then, and I think now, that a broader vision of disaster is needed; it is, however, a good illustration of how theory influences practice.

■ ■ 1.2 Recommended Readings ■

- Perry, R.W. and E. L. Quarantelli (2005). *What Is A Disaster? New Answers to Old Questions*, Xlibris Corporation.
- Perry, R.W. (2006). What is a Disaster? In *Handbook of Disaster Research*, Rodriguez, Quarantelli and Dynes (eds.), Springer, New York, pp. 1–15.
- Quarantelli, E. (1998). *What Is a Disaster? Perspectives on the Question*. London: Routledge.

■ ■ Questions to Ponder ■

- What kind of phenomenon is a disaster? (e.g., object, spectacle, feeling, physical event, etc.).
  - Can it be more than one kind of thing?
  - If it is, what are the implications of that, in terms of trying to define it?
- Why do various academics and institutions disagree about the precise definition of a disaster?
- Do we need a single universally accepted definition, or is there value in diversity?

STUDENT EXERCISES

- Name the greatest disaster, in your opinion, to affect humanity in recorded history.[3]
  - What criteria did you use to select this event?
  - Why did you use those criteria?
  - What are the implications of your choice?
  - What other criteria might also be important?
- Close your eyes and imagine yourself in a landscape where a rising tide of destruction has surrounded you with devastation, misery, and ruin.[4]
  - Immediately and without thought, how do you interpret this experience?
  - How is this thought experiment relevant to defining disaster?

1.3 The Meaning of Disaster

Every now and then, something happens that really makes me think about the meaning of the word disaster, a word that can be highly subjective. I was fascinated by the images of destruction in 2005 after Hurricane Katrina devastated New Orleans and other southern states in the United States. In particular, I remember seeing two similarly impacted middle class families, both apparently insured, being interviewed after they returned to see their destroyed homes. One father exclaimed that their loss was a disaster—everything they owned was gone. The other family gave thanks that nobody had been killed and exclaimed that it "could have been a disaster!"

A few years ago, my then-90-year-old father was in a bumper pool tournament in his retirement home. He made it to the finals but was ultimately defeated (good naturedly) by his archrival, Franklin.[a] Later when he was talking about it, he said, "the game was a disaster." Of course, it was a great victory for Franklin! Similarly, the attack on the World Trade Center on September 9, 2001, was a disaster for New York City but was exalted by al-Qaeda. People interpret events differently according to their purpose, worldview, and values. David Alexander[5] identified seven different schools of thought and expertise on disasters including geography, anthropology, sociology, development studies, health sciences, geophysical sciences with engineering, and social psychology. Is it possible to create a universal objective definition for disaster?

The word "disaster" is a combination of the prefix "dis," which means bad or ill-favored, and suffix "aster," which means star. Its literal meaning, therefore, has an astrological context where calamity results from the unfavorable position of a planet or star. This meaning is interesting not only for historical reasons but also because of the fatalistic philosophy that underlies it.

Two important books, both edited volumes, have explored the meaning of the word disaster: *What Is a Disaster? Perspectives on the Question* by Enrico Quarantelli (1998) and *What Is a Disaster? New Answers to Old Questions* by R. W. Perry and E. Quarantelli (2005). Quarantelli was the founding director of the Disaster Research Center at the University of Delaware and is one of the fathers of disaster sociology. I strongly recommend that you remember his name and seek out his writings.

These two volumes are interesting for a number of reasons. Though there are some common threads about how we understand disaster, there is also vast disagreement on the specific meaning of the word, and even of the importance of trying to define it. Some definitions are operational while others are much more academic and theoretical. Both are useful and have their place in the field. Traditional definitions typically "revolve around four key ingredients: agent description, physical damage, social disruption and negative evaluation,"[6] while more recent approaches emphasize social constructionism and the disruption created during and following disasters.

The Centre for Research on the Epidemiology of Disasters (CRED) has what is probably the best international disaster database and is frequently used by organizations such as the Red Cross and United Nations. If you go to their Web site you can browse through their database and plot a variety of interesting charts. Their formal definition of disaster is "a situation or event which overwhelms local capacity, necessitating a request to the national or international level for external assistance, or is recognised as such by a multilateral agency

---

[a] Not his real name.

or by at least two sources, such as national, regional or international assistance groups and the media." The criteria[7] they use for data to be entered into their database (see Chapter 2) requires one of the following to be met:

- 10 or more people reported killed.
- 100 or more people reported affected.
- Declaration of a state of emergency.
- Call for international assistance.

There are some subtleties to disaster criteria that the CRED definition avoids. For example, it does not discuss the issue of geographical scale. Gross rather than normalized data is considered, yet the difference between total damage and per capita damage can be significant. As well, time scale is not included, which suggests that there is no difference between 10 deaths in 1 min or in 1 month.

■ ■ Question to Ponder ■

Is scale (geographical and/or temporal) important to the notion of disaster? If so, why?

The phrase "overwhelms local capacity" reflects a notion common to most definitions of disaster. For example, one used by the UN International Strategy for Disaster Reduction[8] is "A serious disruption of the functioning of a community or a society causing widespread human, material, economic or environmental losses which exceed the ability of the affected community or society to cope using its own resources." Definitions like this view disaster as a social phenomenon and are fundamental to most currently accepted definitions.

Within the U.S. federal government, the definition of disaster has a strong political flavor. The Stafford Act[9] defines a major disaster as "Any natural catastrophe... or, regardless of cause, any fire, flood, or explosion, in any part of the United States, which, in the determination of the President, causes damage of sufficient severity and magnitude to warrant major disaster assistance under this Act to supplement the efforts and available resources of States, local governments, and disaster relief organizations in alleviating the damage, loss, hardship, or suffering caused thereby." Though such a definition reflects the amount of damage that occurs, it creates opportunities for disaster declarations that are, in part, a function of political, economic, or personal relationships. The declaration of a state of emergency or disaster is important since it triggers a variety of political, bureaucratic, and practical responses, including mutual aid agreements and disaster financial assistance.

Public Safety Canada uses the following definition: A disaster is "essentially a social phenomenon that results when a hazard intersects with a vulnerable community in a way that exceeds or overwhelms the community's ability to cope and may cause serious harm to the safety, health, welfare, property or environment of people; may be triggered by a naturally occurring phenomenon which has its origins within the geophysical or biological environment or by human action or error, whether malicious or unintentional, including technological failures, accidents and terrorist acts." The Canadian Disaster Database (available to the public on their Web site) is similar to the CRED site and includes events if:

- 10 or more people are killed.
- 100 or more people are affected/injured/infected/evacuated or homeless.
- There is an appeal for national/international assistance.
- The event has historical significance.
- There is significant damage/interruption of normal processes such that the community affected cannot recover on its own.

Another approach is the extent to which there is a failure of cultural norms. This definition is very appealing from a sociological perspective but is rather difficult to operationalize; its usefulness is probably greater for research.

Some approaches differ substantially from those described above. For example, Jigyasu,[10] using an Eastern perspective, views disaster as part of the endless cycle of creation and destruction.[11] He describes the cause as being internal chaos, resulting from a lack of understanding that the world is an illusion, and that truth must be experienced and can only be accessed by a pure heart. Erikson[12] in his classic set of case studies on disasters in his book *A New Species of Trouble* (I highly recommend reading it) suggests that "instead of classifying a condition as trauma because it was induced by disaster, we would classify an event as disaster if it had the property of bringing about traumatic reactions." This approach puts the notion of disaster squarely within the social constructionist paradigm. The earlier example of two families' different reaction to the losses of their home in Hurricane Katrina is a good example of this.

Gunderson and Holling[13] present an ecological perspective on disaster in the book *Panarchy: Understanding Transformations in Human and Natural Systems*. From this perspective, disaster represents a natural change of state (Figure 1.3) that a system undergoes as part of its life cycle during the release (Ω) stage. An example of this would be a forest fire, when an old growth forest burns due to the accumulation of flammable ground clutter and makes way for the regrowth of young trees. From this perspective, a disaster is not necessarily something to be avoided but rather a natural part of system evolution. A more detailed description of the panarchy hypothesis is beyond the scope of this chapter, though it has important implications for disaster research.

I would like to add one perspective to the above, which resonates with the comment by Quarantelli[14] that "we often are trying to use only one concept, that is, the label of disaster to attempt to capture too much" and that "we should stop trying to squeeze relatively heterogeneous phenomena under one label."

The issue of how one should understand the word disaster is important but has no single correct answer. Some words can be defined with mathematical precision; the definition of a circle (the locus of points equidistant from a defined point) is one example. Other words have meanings that are fuzzy in nature and perhaps (as discussed by the

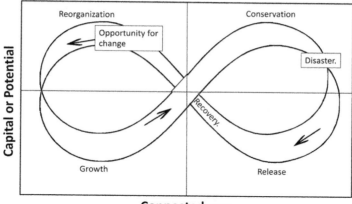

FIGURE 1.3 Panarchy: The adaptive cycle. Panarchy identifies four basic stages of complex systems: exploitation, conservation, release, and reorganization. Exploitation is associated with rapid expansion. During conservation the slow accumulation and storage of energy and material is emphasized. Release occurs rapidly, as during a disaster. Reorganization can also, but not always, occur rapidly, like disaster recovery. Potential sets the limits to what is possible. Connectedness determines the degree to which a system can control its own destiny through internal controls, as distinct from being influenced by external variables. *Source: Modified from: Gunderson, L. H. (2001). Panarchy: Understanding Transformations in Human and Natural Systems. Island Press.*

philosopher Ludwig Wittgenstein[15]) do not require a precise definition to be used successfully. I believe that the word disaster fits into the latter category.

Meaning may also be understood through shared sets of characteristics. For example, we know a cat is a cat when we see it because it has enough cat-like characteristics. Not all cats have fur (Figure 1.4) or can meow (and not all animals that have fur or meow are cats, though if they have both traits they are likely cats), but they are still cats because they resemble the family of cats strongly enough for us to label them that way. Similarly, one might analyze an event to see if it exhibits enough disaster-like characteristics.

FIGURE 1.4 It's still a cat, even though it has no fur. (Not a pretty sight though…). *Source: The Featured Creature.* http://www.thefeaturedcreature.com/2011/12/6-strange-breeds-of-hairless-cats.html.

What are disaster characteristics? One might (and people have) created disaster taxonomies,[16] similar to ones used in biology. Burton, Kates, and White[17] created a typology for natural hazards based upon frequency, duration, area, speed of onset, spatial dispersion, and temporal spacing that might form a base for a disaster taxonomy. The most common way of differentiating disaster types is by labeling them (1) natural,[18] technological, or human caused, and (2) rapid or slow onset; however, this is a rather coarse categorization.

What is the difference between a disaster and an emergency, or a disaster and a catastrophe? Are the differences just one of size, or are there qualitative or functional differences? These questions are interesting, important, and useful to keep in mind as one continues to study this field. Quarantelli[19] addressed some of these questions in his paper *Emergencies, Disasters and Catastrophes Are Different Phenomena*. He concludes that there are several important factors that differentiate a catastrophe from a disaster. In a catastrophe:

- Most or all of the community built structure is heavily impacted.
- Local officials are unable to undertake their usual work roles, and this often extends into the recovery period.
- Most, if not all, of the everyday community functions are sharply and simultaneously interrupted.
- Help from nearby communities cannot be provided.

These comments "mean that more innovative and creative actions and measure will be required in a catastrophe than even a major disaster." This notion will be explored further in Chapter 5. For now, I will simply make the following points:

- Scale is important. What might be a catastrophe for a household may be disaster for a community, and an emergency for a state or province.
- In an emergency, help beyond the affected community is not required to cope with response and recovery.
- In a catastrophe, the people and institutions involved in response and recovery have themselves become victims in a significant way.
- Larger events, like catastrophes, increasingly show the characteristics of chaotic systems, such as emergent properties and surprising outcomes.
- One size does not fit all. Different risk reduction and management strategies are required for emergencies, disasters, and catastrophes.

> **STUDENT EXERCISE**
> - List two events that are emergencies, two that are disasters, and two that are catastrophes.
>   - Explain your choices.

How important is this labeling? The answer to this question depends upon the purpose of the classifier. From a legal perspective and the assignment of blame, labeling can be very important for determining liability. For example, Huffman[20] says, "To prevail in an action for damages under the common law of negligence, a plaintiff has to prove that the defendant breached a duty owed to the plaintiff and that the defendant's act caused injury to the plaintiff... The duty and causation requirements reflect that the law of negligence is rooted in the concept of fault." The difference between a human-caused and natural disaster will remain important as long as fault is central to liability. From a legislative and policy perspective, the naming of an emergency or a disaster can trigger special powers for government and make financial and other resources available to affected communities. Another example comes from the field of psychology. Different types of disasters affect people differently. "Quarantelli... makes a most convincing case for investing effort in taxonomy to create meaningful classification systems. He points out that the many empirical studies of 'disasters' have begun to produce anomalous findings; using only one example, we know that serious mental health consequences, rare in most studies based on floods, tornadoes, hurricanes, and earthquakes, appear to be greater in cases associated with conflict situations."[21]

The perspective and purpose of the user of a classification system is central, and Kreps[22] points out that because disaster research is so interdisciplinary, it may be that "no single paradigm can encompass all of this complexity." If this is true then a number of taxonomies may be required. Underlying this issue is the question of whether this is simply an empirical exercise to create useful tools for some purpose, or whether it is theoretically based. Given that disaster theory is, to a large extent, imported from a variety of disparate fields, it may remain mostly an empirical exercise until the theory has become well developed—and McEntire[23] has suggested that a robust theory of disasters may not be attainable. Thus, empiricism may continue to dominate this field of study.

Several important factors must be acknowledged in disaster taxonomy. One is the tension between universal core characteristics that are invariant in space, time, and culture as opposed to those that are contextual.[24] A second relates to cause and effect. A third relates to physical versus social, economic, or environmental attributes. Complexity and scale may also be important, particularly with respect to management strategies and emergent properties.

STUDENT EXERCISE

Make a list of words that you would use to characterize disaster. Organize those words into a taxonomic structure using a tree diagram.

- Apply your list to the following events:
  - Hurricane Katrina, United States, 2005
  - Chernobyl nuclear meltdown, Ukraine, 1986
  - The burning of the Ancient Library of Alexandria, Egypt in BC 48 (Figure 1.5).

FIGURE 1.5 The burning of the library at Alexandria in 391 AD, illustration from "Hutchinson's History of the Nations," c.1910 (litho), Dudley, Ambrose (fl. 1920s). *Source: The Stapleton Collection, Bridgeman Art Library.*

1.4 Summary

How disaster is defined is important for a number of reasons. Operationally, it establishes when declarations occur, but it also affects how we conduct research and create policy. Defining disaster is a challenging task for which there may never be a single universally accepted answer, though there are common elements that appear in almost all definitions. Particularly, the notion that disasters are social events that harm people and communities is prevalent. It may be that by defining disaster we are trying to force too much meaning into a single word, and that phrases based upon some taxonomic structure are needed. This chapter has introduced you to some of the complexities of this topic and given you an opportunity to consider many of the possibilities that might be used in a classification scheme.

1.5 Case Study: The 2003 Heat Wave in Europe

Silent and invisible killers of silenced and invisible people.

Eric Klinenberg

The year 2003 was exceptional from a climate perspective. Globally, it was one of the three warmest years as recorded up to that time. In Europe an extreme heat wave during the summer resulted in unprecedented impacts. Figure 1.6 shows global summer temperature anomalies, with the highest temperatures occurring in central Europe. Temperature anomalies exceeded three standard deviations in several places, and in some regions European summer temperatures were the warmest since 1500;[25] Britain recorded its first-ever temperature over 38 °C on August 10. Accompanying the heat wave was a serious

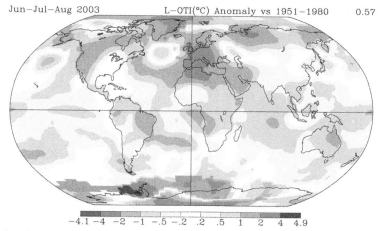

FIGURE 1.6 Global surface air temperature anomalies (in °C, with respect to the 1951–1980 reference period), June–August 2003. Note the very high anomalies over Western Europe. *Source: NASA Goddard Institute for Space Studies, Earth Sciences Division.* http://data.giss.nasa.gov/gistemp/maps/.

drought with significant crop damage and massive forest fires. Many rivers ran dry by the end of August. One important aspect of this heat wave was that minimum temperatures in many locations stayed above 25 °C, which did not allow people a cooling off period at night. Temperatures tended to be hotter in cities due to the urban heat island effect; cities tend to warm up more during the day, cool off less at night, and are less windy with more air pollution. Most of France was affected by this event, with temperatures exceeding 35 °C for at least 9 days in 61 of the 96 French *departements*.[26]

The dry conditions prior to and during the heat wave were significant. Without the preceding drought, temperature anomalies would have been up to 50% lower, since the presence of moisture cools the surface as a result of evapotranspiration and latent cooling. Large portions of Europe received about half their normal precipitation from February to May. As well, the effects of the heat wave were exacerbated by low wind speeds that reduced ventilation and high levels of ground level ozone.

Historically, there have been a number of killer heat waves: 11,000 died in Beijing in 1743, 2000 were killed in Athens in 1987, over 55,000 in Russia in 2010,[27] and over 700 in Chicago in 1995.[28] From 1999 to 2003, an average of 688 deaths per year were attributed to heat waves in the United States. The heat wave in London, England, in 1858 resulted in thousands of deaths from cholera, typhoid, and other infections because of the growth of bacteria in the River Thames; subsequently, Parliament rebuilt their sewer system. During the New York heat wave of 1896, Theodore Roosevelt, the police commissioner at that time, gained public favor by distributing free ice in tenement neighborhoods and providing ambulance services. No heat wave event, though, reached the massive death toll of the 2003 event, with estimates ranging from 30,000 to 80,000.[29] Table 1.1 below highlights historical European heat wave events.[30]

Table 1.1 Heat Wave Events and Mortality in Europe

Heat Wave Event	Attributable Mortality (% Increase)
1976—London, UK	9.7% increase England and Wales and 15.4% Greater London
1981–Portugal	1906 excess deaths (all cause, all ages) In Portugal, 406 in Lisbon (month of July)
1983—Rome, Italy	35% increase in deaths in July; 83 in 65+ age group
1987—Athens, Greece July 21–31	Estimated excess mortality >2000
1991—Portugal July 12–21	997 excess deaths
1995—London, UK July 30–August 3	11.2% (768) in England and Wales, 23% (184) Greater London
1994—Netherlands July 19–31	24.4% increase, 1057 (95% CI 913, 1201)
2003—Italy, June 1–August 15	3134 (15%) in all Italian capitals
2003—France August 1–20	14802 (60%)
2003—Portugal August 1–31	1854 (40%)
2003—Spain August 1–31	3166 (8%)
2003—Switzerland, June 1–August 31 (3 months)	975 deaths (6.9%)
2003—Netherlands June 1–August 23	1400 deaths
2003—Baden-Wuerttemberg, Germany August 1–24	1410 deaths
2003—Belgium	1297 deaths for age group 65+
August 4–13, 2003—England and Wales	2091 (17%). Mortality in London Region: 616 deaths (42% excess)

One calculation puts the return period of the 2003 heat wave in France, based upon current climate, at more than 1000 years,[31] which translates into a 0.1% annual probability. Climate change is very likely to increase the occurrence of heat waves in the future, especially of extreme heat waves due to nonlinearity in the climate system. According to an article published in the magazine *Nature*, "it seems likely that past human influence has more than doubled the risk of European mean summer temperatures as hot as 2003,… with the likelihood of such events projected to increase 100-fold over the next four decades."[32]

From a climatological perspective, it is difficult to unravel cause and effect. Since physical processes are so connected with each other, it becomes a chicken-and-egg problem. The contributing factors, however, were:

- "an extremely persistent blocking.
- the northward displacement of the Azores anticyclone and the African Inter-Tropical Convergence Zone.
- a strong positive phase of the East Atlantic teleconnection pattern.
- the southward shift of the extratropical storm tracks.
- the strong amplitude of the summer Northern Annular Mode.
- tropical Atlantic anomalies in the form of wetter-than-average conditions in the Caribbean basin and the Sahel.
- the anomalously clear skies and the downward net radiative fluxes.
- the prolonged high sea surface temperatures (SST) in the northern North Atlantic and the Mediterranean.
- the intense negative soil moisture anomaly in central Europe and the resulting feedback mechanism."[33]

Human response to heat is negligible up to a critical threshold after which people begin to suffer. That threshold and the sensitivity of people to temperatures beyond it vary by geography. For example, in Europe, Andalusians have a critical threshold of 41°C, while people living in Belgium are less adapted to heat and begin to feel ill at 27.5°C.[34] Figures 1.7a and 1.7b shows several relationships for heat related mortality.

Hajat,[36] in a global study, found temperature-mortality slopes above heat thresholds ranged from zero (for Dublin, Ireland) to 19% (for Monterrey, Mexico). Statistically, slopes were significantly related to the following variables: percentage of population over 65 years old, level of health care, and city gross domestic product (GDP). Older people tend to be more sensitive to extreme heat and suffer dehydration and other complications; as well, they are more likely to be on medications that may interfere with systems that regulate body temperature, such as sweating. The presence of pulmonary, cardiovascular, or psychiatric conditions were found to be significant factors for increased risk. Although many deaths occurred during and shortly after the heat wave, health damage from heatstroke can contribute to deaths up to 1 year after illness.[37]

Studies indicate a lack of a harvesting effect (the death of people who are about to die from other causes[38]). Fatalities spiked toward the latter part of the heat wave (lagging the wave by 1–3 days on average), as people passed their coping ability (Figure 1.8); high levels of ground level ozone may have been a contributory factor. In particular, the length of the heat wave and the high nighttime temperatures made it very difficult for people to cope unless they had air conditioning. Most of the deaths occurred in France, though Italy, Spain, Portugal, and the United Kingdom were also affected.

Climatology influences adaptation and therefore impacts of the heat wave. In Italy, more deaths occurred in northwestern cities that are usually cooler.[40] The pattern of

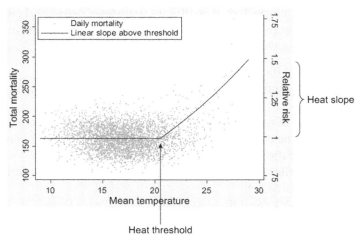

FIGURE 1.7a Relationship between daily mortality and summertime daily mean temperature in London, 1976–2003.[35] *Source: Hajat, S., and Kosatky, T. (2010). Heat-related mortality: a review and exploration of heterogeneity. Journal of Epidemiology and Community Health, 64(9), 753–760.*

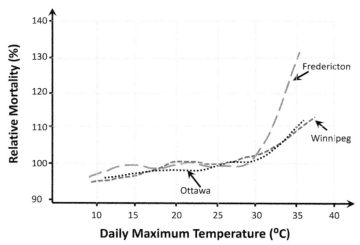

FIGURE 1.7b Relative mortality as a function of daily maximum temperature for three cities in Canada; (1) Fredericton, New Brunswick, (2) Ottawa, Ontario and (3) Winnipeg, Manitoba during the summertime (June–August) of 1986–2005. For all three cities 30C represents a critical threshold, beyond which heat contributes to mortality. The slope of the curve is much steeper in Fredericton, suggesting a less adapted population. *Source: Adapted from: Casati, B., Yagouti, A., and Chaumount, D. (2013). Regional Climate Projections of Extreme Heat Events in Nine Pilot Canadian Communities for Public Health Planning. Journal of Applied Meteorology and Climatology, 52(12), 2669–2697.*

deaths was also very age dependent. In France, excess mortality began at age 35 and increased with age; 91% of victims were over age 65. In this heat wave, women were more at risk than men, having a 15% higher death rate at 45 years; other events such as the 1995 Chicago heat wave have shown men to be more at risk,[41] and so this must be assumed to be

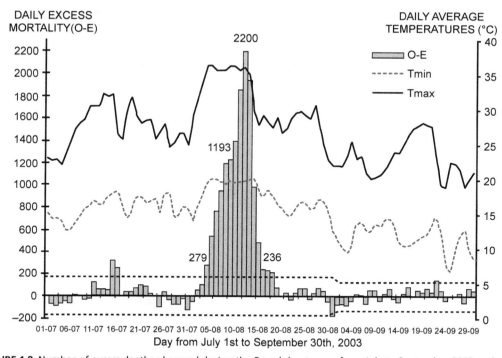

FIGURE 1.8 Number of excess deaths observed during the French heat wave from July to September 2003 and average daily maximum (T_{max}) and minimum (T_{min}) temperatures recorded during the period. The X-axis shows the 92 days from July 1 to September 30, 2003. The histogram shows the numerical values of the daily excess mortality for August 4, 8, 12, and 16. The dotted horizontal lines show the limits of the 95% fluctuation intervals of the daily number of deaths.[39] *Source: Fouillet, A., Rey, G., Laurent, F., Pavillon, G., Bellec, S., Guihenneuc-Jouyaux, Clavel, J. Jougla, E. and Hémon, D. (2006). Excess mortality related to the August 2003 heat wave in France. International Archives of Occupational and Environmental health, 80(1), 16–24.*

culturally dependent. More deaths were observed at people's homes (+74%) and in retirement homes (+91%) than at hospitals (+45%) and private clinics (+22%). Being married reduced mortality rates. Living in the top floor of a house usually, but not always, increased risk since they tend to be hotter (where there were no elevators, people living on top floors tended to be more physically fit and therefore more resilient). Physical and social isolation were critical. People tend to die alone in dense urban cities (due to the urban heat island effect and lower wind speeds) behind locked doors and closed windows. This issue was exacerbated by the timing of the heat wave. In August, much of the French population takes their holidays, with the result that many elderly people were left on their own.

Mortality varied depending upon where you lived; in France, excess mortality ranged from +4% in Lille to +142% in Paris. Risk factors in the 1995 Chicago heat wave were social isolation related to poverty, geographical segregation, neglect of physical infrastructure, reduced at-home services, and poor coordination of health services. Borrell[42] comments that some of these factors likely played a role in deaths in Barcelona, Spain, particularly due to weak welfare services. Managing risk includes drinking water frequently, taking frequent showers, opening windows at night but closing them during the day, and wearing light clothes. Living in a place surrounded by high vegetation also reduces risk. Air conditioning is important, but may be less accessible during heat waves due to power shortages. The huge electricity demand in France coincided with a 19% reduction in hydroelectric production and a 4% drop in nuclear energy production due to low water levels (Figure 1.9) and warm water that made meeting demand challenging. Global trends toward urbanization combined with climate change seem likely to make heat waves an increasingly serious hazard in the future.

FIGURE 1.9 Streamflow drought in River Gardon, France, August 2003. *Source: Watch: Water and Global Change, WorkBlock 4 – Extremes: Frequency, Severity and Scale. Photo by H. A. J. van Lanen.* http://www.eu-watch.org/Research-Work/workblock-4.

In France, the political response[43] was initially not helpful, to put it mildly. Eric Klinenberg refers to this kind of response as "Deny, Deflect, Defend", a pattern evident not only during this disaster but also during the Chicago event. Initial communications from hospitals were "dismissed as one more instance of biased recriminations against hospital budgetary cutbacks," media sensationalism, or politically motivated criticisms. On August 11, the French Director General of Health Lucien Abenhaim said that "everything's under control" in an attempt to reassure the public. On August 19, he resigned in disgrace.[44] No doubt there are times when people cry wolf, but there is a strong tendency for alarms that are not politically palatable to be dismissed. "Those were difficult times indeed for whistle blowers. Damned if you do issue a warning, damned if you don't, damned if you failed to convince those who were intent on dismissing your warnings anyway. Disputing that 'everything is under control,' that 'there is no cause for alarm, nothing is happening' was unacceptable. Institutions are bound to dismiss and reject these alerts, which they see as something akin of a treason."[45]

There were several systemic flaws in the response system. One was that, during August, most senior civil servants were on holiday. The bureaucratic process of converting medical data to memos to be read by politicians and senior civil servants takes days to accomplish, and ministerial cultures in France (and many other countries,

I am sure) are based upon building files using "clear and fixed partitions of areas of competence and levels of responsibility, top-down dynamics, and a programmed time frame. This culture is so prevalent that it permeates monitoring, even when obviously incompatible with the task at hand and its time frame." This problem was exacerbated by the nature of the event; with no clear onset or boundaries such as happens with tornadoes or terrorist attacks, there was no precise trigger for government to respond to. The onset was diffuse and slow, and evidence was slow to move up the chain of command.

Patrick Legadec, Director of Research at the Ecole Polytechnique in Paris, provides a valuable summary of the bureaucratic inadequacies.[46] His main points are bulleted below, and I highly recommend reading his thoughtful critique on this disaster.

Inadequacies in meeting the challenge were significant and unavoidable. The basic managerial culture of most organizations is characterized by the following visions and operational frameworks:

- A stable and surprise-free world
- A response culture
- A step-by-step, top-down, centralized approach
- Public information, as a high risk move.

These deep-rooted cultural frameworks create the following social traps during crisis:

- Inability to detect, or search for, signals
- Laborious mobilization
- Divisions, partitions, demarcation lines
- Vertical isolations
- Dramatic errors in communication
- Scapegoat searching.

What is required in crisis situations is:

- Sharp and wide open surveillance abilities
- Swift reports
- Upgraded monitoring capacity, crisis teamwork and data sharing
- Ability to mobilize expertise in crisis
- Sharing of leadership, network-based decision making, far from the command-and-control approach
- High-quality communication from start to finish
- Management of the crisis to the very end
- Strategic intelligence
- After the crisis, a careful healing process.

He notes the following traps prevented a good response:

- Inadequate mindsets
- Noisy context

- Unusual geographical pattern
- Monitoring difficulties
- An unusual killer
- Inadequate data monitoring
- Unusual data
- Inadequate focus of attention
- Stealth problems
- Scientific gaps.

"Heat waves receive little public attention because they fail to generate the massive property damage and fantastic images produced by other weather related disasters, but also because their victims are primarily social outcasts—the elderly, the poor, the isolated—from whom we customarily turn away."[47] Effectively responding to these sorts of events requires a social vision that is more inclusive than that required for other types of disasters, with an emphasis on social networks.

After the 2003 heat wave, France implemented an alerting system and preventative measures. In one study of the 2006 heat wave, the number of deaths dropped by about two-thirds, compared to what would have been expected without those interventions.[48] Another study of a 2-week heat wave in June 2005 found no extra mortality due to collective risk awareness of the public and actions taken to help the most vulnerable people.[49] Aside from having heat response plans, cities with pronounced urban heat island effects can do much to reduce heat wave risk by creating more green spaces and green roofs, maintaining local water bodies, and creating more reflective and light-colored surfaces.

This disaster is particularly interesting from several perspectives. It reveals a social vulnerability that was previously unknown in its magnitude, particularly in France. In particular, it points toward the importance of informal social networks to reduce the vulnerability of the elderly. The inability of government to adequately respond in a timely manner highlights the dysfunction of too political a response and also the mismatch between normal bureaucratic processes and those needed during crisis. It should also be a wake-up call in terms of future risks, given the trends toward urbanization, climate warming, and a more elderly population.

Further Reading about the 2003 Heat Wave

Barriopedro D, Fischer EM, Luterbacher J, Trigo RM, Garcia-Herrera R: The hot summer of 2010: redrawing the temperature record map of Europe, *Science* 332(6026):220–224, 2011.

Beniston M, Diaz HF: The 2003 heat wave as an example of summers in a greenhouse climate? Observations and climate model simulations for Basel, Switzerland, *Global and Planetary Change* 44(1):73–81, 2004.

Brücker G: Vulnerable populations: lessons learnt from the summer 2003 heat waves in Europe, *Euro Surveillance: Bulletin Europeen Sur Les Maladies Transmissibles = European Communicable Disease Bulletin* 10(7):147, 2005.

Foroni M, Salvioli G, Rielli R, Goldoni CA, Orlandi G, Sajani SZ, et al.: A retrospective study on heat-related mortality in an elderly population during the 2003 heat wave in Modena, Italy: the Argento Project, *Journals of Gerontology Series A: Biological Sciences and Medical Sciences* 62(6):647–651, 2007.

Fouillet A, Rey G, Laurent F, Pavillon G, Bellec S, Guihenneuc-Jouyaux C, et al.: Excess mortality related to the August 2003 heat wave in France, *International Archives of Occupational and Environmental Health* 80(1):16–24, 2006.

Garc'Ia-Herrera R, D'Iaz J, Trigo RM, Luterbacher J, Fischer EM: A review of the European summer heat wave of 2003, *Critical Reviews in Environmental Science and Technology* 40:267–306, 2010.

Hajat S, Kosatky T: Heat-related mortality: a review and exploration of heterogeneity, *Journal of Epidemiology and Community Health* 64(9):753–760, 2010.

Klinenberg E: *Heat Wave: A Social Autopsy of Disaster in Chicago*, University of Chicago Press, 2003.

Lagadec P: Understanding the French 2003 heat wave experience: beyond the heat, a multi-layered challenge, *Journal of Contingencies and Crisis Management* 12(4):160–169, 2004.

Poumad'ere M, Mays C, Le Mer S, Blong R: The 2003 heat wave in France: dangerous climate change here and now, *Risk Analysis* 25(6), 2005.

Salagnac JL: Lessons from the 2003 heat wave: a French perspective, *Building Research and Information* 35(4):450–457, 2007.

Sardon JP: The 2003 heat wave, *Euro Surveillance: Bulletin Europeen Sur Les Maladies Transmissibles = European Communicable Disease Bulletin* 12(3):226, 2007.

Stott PA, Stone DA, Allen MR: Human contribution to the European heat wave of 2003, *Nature* 432(7017):610–614, 2004.

End Notes

1. Cutter S., "Are We Asking the Right Questions?" in *What is a Disaster: New Answers to Old Questions*, ed. Perry and Quarantelli (USA: Xlibris Corporation, 2005), 39–48.

2. Quarantelli E.L., "A Social Science Research Agenda for the Disasters of the 21st Century: Theoretical, Methodological and Empirical Issues and Their Professional Implementation," in *What is a Disaster: New Answers to Old Questions*, ed. Perry and Quarantelli (USA: Xlibris Corporation, 2005), 325–96.

3. As a class exercise, this is interesting due to the variety of choices made by students, reflecting different values, and sometimes an astonishing ignorance of historical disasters. It leads into a very interesting discussion of the relative importance of life, economic damage, and culture, including the difference between gross impacts and those that are normalized by population or GDP, for example. Interesting answers given by students in the past include the biblical flood, the burning of the Royal Library of Alexandria, Egypt, in 48 BCE by Julius Caesar, the latest Ice Age, and the Lisbon earthquake and fire of 1755. In particular, one question I have found useful for instigating a discussion is, "Which is a worse disaster: (a) 1000 deaths from a population of 1,000,000, or (b) 500 deaths from a population of 100,000?

4. One might think of a scenario such as a swath of tornado destruction, earthquake, or tsunami when answering this question. But suppose it was the September 11 attack on New York City's World Trade Center? Contrast the perspective of a native New Yorker to a member of al-Qaeda.

5. Alexander D, *Confronting Catastrophe*, (Harpenden, England: Terra Publishing, 2000).

6. Kreps G.A., "Description, Taxonomy and Explanation in Disaster Research," *International Journal of Mass Emergencies and Disasters* 7, no. 3, (1989): 277–280.

7. EM-DAT: The International Disaster Database, Centre for Research on the Epidemiology of Disasters – CRED, http://www.emdat.be/criteria-and-definition.

8. International Strategy for Disaster Reduction. *Defining a few key terms. ISDR.* (2004), Available online at: http://www.unisdr.org/eng/media-room/facts-sheets/fs-Defining-a-few-key-terms.htm.

9. The Stafford Act, Robert T. *Stafford Disaster Relief and Emergency Assistance Act* (Public Law 93-288) as amended. FEMA 592, (2007) https://www.fema.gov/library/viewRecord.do?fromSearch=fromsearch&id=3564.

10. Jigyasu R., "Disaster: A Reality or Construct? Perspective from the East," in *What is a Disaster: New Answers to Old Questions*, ed. Perry and Quarantelli, (USA: Xlibris Corporation, 2005), 49–59.

11. This approach is reminiscent of an ecological understanding of disaster. The book Panarchy by Gunderson and Holling has an extensive discussion of this perspective, with a particular focus on the notion of resilience.

12. Erikson K. T., *A New Species of Trouble: Explorations in Disaster, Trauma, and Community* (New York: W. W. Norton & Company, 1994).

13. Gunderson and Holling, *Panarchy: Understanding Transformations in Human and Natural Systems*, (Island Press, 2002).

14. See note 2 above.

15. Wittgenstein L., *Philosophical investigations* (Wiley-Blackwell, 2010).

16. See note 6 above.

17. Burton I., *The Environment as Hazard*, (The Guilford Press, 1993).

18. It is important to mention that the term natural disaster can be critiqued as being an oxymoron, since human factors that cause vulnerability inevitably intrude. As well, many technological or human-caused disasters often have strong natural components, such as wave height or wind direction. Cause and effect can be complicated.

19. E. L. Quarantelli, *Emergencies, disasters and catastrophes are different phenomena*, (University of Delaware DRC Preliminary Paper #304, 6, 2000). E. L. Quarantelli, "Catastrophes are Different from Disasters: Some Implications for Crisis Planning and Managing Drawn from Katrina," *Understanding Katrina; Perspectives from the Social Sciences*, (2006). Retrieved from http://understandingkatrina.ssrc.org/Quarantelli/.

20. Huffman J.L., "Law, Comparative Legal Study, and Disaster Taxonomy," *International Journal of Mass Emergencies and Disasters*, 7, no. 3, 1989: 329–348.

21. R.W. Perry, "What is a Disaster?" In *Handbook of Disaster Research*, ed. Rodríguez H. (New York: Springer, 2007), 1–15.

22. See note 6 above.

23. D.A. McEntire, "Emergency Management Theory: Issues, barriers, and recommendations for improvement," *Journal of Emergency Management*," 3, no. 3 (2005): 44–54.

24. See note 6 above.

25. Garc'Ia-Herrera, R., D'Iaz, J., R.M. Trigo, Luterbacher J., and E.M. Fischer, "A Review of the European Summer Heat Wave of 2003," *Critical Reviews in Environmental Science and Technology*, 40, (2010): 267–306.

26. Fouillet A., Rey G., Laurent F., Pavillon G., Bellec S., Guihenneuc-Jouyaux C. and Hémon D. "Excess mortality related to the August 2003 heat wave in France," *International Archives of Occupational and Environmental Health*, 80, no. 1, (2006): 16–24.

27. Barriopedro, D., Fischer, E.M., Luterbacher, J., Trigo, R.M., and García-Herrera, R. (2011). The hot summer of 2010: *redrawing the temperature record map of Europe. Science*, 332(6026), 220–224.

28. Bouchama A., "The 2003 European heat wave," *Intensive Care Medicine*, 30, no. 1, (2004): 1–3.

29. J. M. Robine, S.L Cheung, Le Roy S., Van Oyen H., and F. R. Herrmann, "Report on excess mortality in Europe during summer 2003," *EU Community Action Programme for Public Health, Grant Agreement 2005114, 28*, (2007). Brücker G., "Vulnerable populations: lessons learnt from the summer 2003 heat waves in Europe," *Euro surveillance: bulletin europeen sur les maladies transmissibles=European communicable disease bulletin*, 10, no. 7, (2005):147.

30. R. S. Kovats, and Hajat S., "Heat stress and public health: a critical review," *Annu. Rev. Public Health*, 29, (2008): 41–55.

31. R. S. Kovats, and Kristie L.E, "Heatwaves and public health in Europe," *The European Journal of Public Health*, 16, no. 6, (2006): 592–599.

32. Peter A. Stott, D.A. Stone and M.R. Allen. "Human contribution to the European heatwave of 2003," *Nature*, 432, no. 7017, (2004): 610–613.

33. See note 25 above.

34. J.P. Sardon, "The 2003 heat wave," *Euro Surveillance: bulletin europeen sur les maladies transmissibles=European communicable disease bulletin*, 12, no. 3, (2007): 694. Retrieved from http://www.eurosurveillance.org/ViewArticle.aspx?ArticleId=694.

35. Hajat S., and Kosatky T., "Heat-related mortality: a review and exploration of heterogeneity," *Journal of Epidemiology and Community Health*, 64, no. 9, (2010): 753–760.

36. Ibid.

37. See note 27 above.

38. See note 26 above.

39. Ibid.

40. Conti S., Masocco M., Meli P., Minelli G., Palummeri E., Solimini, R., and Vichi M., "General and specific mortality among the elderly during the 2003 heat wave in Genoa (Italy)" *Environmental Research*, 103, no. 2, (2007): 267–274.

41. Klinenberg E., *Heat wave: A social autopsy of disaster in Chicago* (University of Chicago Press, 2003).

42. Borrell C., Marí-Dell'Olmo M., Rodríguez-Sanz M., Garcia-Olalla P., J.A. Caylà, Benach J., and Muntaner C., "Socioeconomic position and excess mortality during the heat wave of 2003 in Barcelona" *European Journal of Epidemiology*, 21, no. 9, (2006): 633–640.

43. Lagadec P. "Understanding the French 2003 Heat Wave Experience: Beyond the heat, a Multi-Layered Challenge," *Journal of Contingencies and Crisis Management*, 12, no. 4, (2004): 160–169.

44. *Heat Crisis: French official quits. CNN* (August 18, 2003). Retrieved from http://edition.cnn.com/2003/WORLD/europe/08/18/paris.heatwave/.

45. See note 42 above.

46. Ibid.

47. See note 40 above.

48. See note 26 above.

49. J.L. Salagnac, "Lessons from the 2003 heat wave: a French perspective," *Building Research & Inforjmation*, 35, no. 4, (2007): 450–457.

2

Disaster Data: A Global View of Economic and Life Loss

> *You can use all the quantitative data you can get, but you still have to distrust it and use your own intelligence and judgment.*
>
> **Alvin Toffler**

CHAPTER OUTLINE

2.1 Why This Topic Matters	24
2.2 Recommended Readings	25
2.3 Introduction	26
2.4 Measuring Loss	26
2.4.1 Measuring Loss	29
2.5 Data Quality	30
2.5.1 Sources and Methodology	30
2.5.2 Biases in Disaster Data	33
2.5.3 Availability	35
2.6 Databases	35
2.6.1 Center for Research on the Epidemiology of Disasters	36
2.6.2 World Bank	39
2.6.3 Munich Reinsurance Database	40
2.7 Conclusions	41
2.8 Case Study: Hurricane Hazel and Toronto	46
Further Readings	50
End Notes	50

CHAPTER OVERVIEW

Disaster data are needed to test theories, to develop empirical studies, and for policy development. There are, however, numerous difficulties in the data bases that exist, and there are no universally

Susan McGregor is a co-author for this chapter.

accepted methodologies to measure disaster impacts. The choice of indices, for example, can strongly bias analyses in different directions; one example is total disaster cost as compared to per capita cost. These choices are underlain by institutional mandates, data availability, and values. Thus, users of disaster data must be careful in terms of how they interpret such information, in order to avoid, or be transparent about, biases and errors.

KEYWORDS

- Affected
- CRED
- Dead
- Disaster data
- Economic impact
- EM-DAT
- Hurricane Hazel
- Statistics

2.1 Why This Topic Matters

Data is not information, information is not knowledge, knowledge is not understanding, understanding is not wisdom.

Clifford Stoll

Everybody observes the world around them. We do so in order to arrive at an understanding of it, and to make judgments about how to change it for the better, or to make better decisions about how to react. For these reasons many disciplines emphasize the gathering of data, to create mental models, to develop new theories, or to validate existing ones. But the process of observing, recording, and analyzing data is far from simple, and must be done with care if evaluations are to be unbiased and useful.

The study of disasters is to a very large extent an empirical field, which means that an understanding of data, the way in which they are measured, and how they are interpreted are critical. There are some serious difficulties around the gathering and analysis of disaster data, and the purpose of this chapter is to clarify them. There is an old quote mostly attributed to Benjamin Disraeli (Figure 2.1) and popularized by Mark Twain that says, "There are three kinds of lies: lies, damned lies, and statistics." There is much truth in this statement, and the different ways that disaster data can be accumulated, sorted, and analyzed can lead to varying interpretations. Data are not truth.

FIGURE 2.1 Benjamin Disraeli.

■ ■ 2.2 Recommended Readings ■

- Arnold M., Chen R. S., Deichmann U., Dilley M., Lerner-Lam A. L., Pullen R. E., and Trohanis Z., Natural Disaster Hotspots Case Studies, *World Bank Disaster Risk Management Series* No. 6, (Washington D.C., 2006), 204.
- Asgary A., (2005), *Technological Disasters' Cost/loss Data: Current Issues and Future Challenges.* 2nd Toulouse–Montreal Conference: The Law, Economics and Management of Large-Scale Risks, September 30 to October 1, 2005, Montreal, Canada.
- Below R., and Guha-Sapir D., (2006), *An Analytical Review of Selected Data Sets on Natural Disasters and Impacts.* UNDP/CRED Workshop on Improving Compilation of Reliable Data on Disaster Occurrence and Impact, April 2–4, 2006, Bangkok, Thailand.
- CRED (2004), *Expert Consultation on Collection and Validation of Economic Data Related to Disasters.* Organized by Centre for Research on the Epidemiology of Disasters (CRED), hosted by World Bank, Hazard Management Unit, Washington DC, December 9–10.
- ISDR (2002), *Comparative Analysis of Disaster Databases.* Final Report. Working Group 3 of the Inter-Agency Task Force of the International Strategy for Disaster Reduction on Risk, Vulnerability and Impact Assessment, LARED, November 30, 37.

■ ■ Question to Ponder ■

What is the relationship between disaster data and disaster theory?

> **STUDENT EXERCISE**
>
> Using the table of disaster data in Appendix 1:
>
> - Order the disasters in the following ways:
>   - Total deaths
>   - Deaths per million of population
>   - Total cost
>   - Cost as a % of GDP.
> - Do you see these differences as important? Why? Which ranking do you think is best?

2.3 Introduction

In the following discussion, natural disaster data are used to illustrate a number of concepts. Disaster data are becoming increasingly important for policy and decision makers in an era when the frequency and intensity of natural disasters appear to be increasing. There is an expectation that loss of life and the economic costs from natural disasters will increase, as a result of climate change, environmental degradation, and growing vulnerability due to such factors as increased gaps in wealth, urbanization, loss of resilience, and population growth in risky areas. (As an aside, please note that the phrase "natural disaster" is something of an oxymoron. Although natural events are the trigger of many disasters, the damage that ensues is largely the result of poor development decisions that place people and property at risk.[1])

As part of their contract with citizens, governments, particularly national governments, are expected to provide a reasonable level of protection against disasters. When they do not provide this protection, censure and disenchantment follow and can potentially lead to varying levels of unrest, particularly in the presence of already stressed political climates. There are many international institutions devoted to the cause of disaster risk reduction, including the World Bank's Disaster Management Facility, the United Nations International Strategy for Disaster Reduction (ISDR), the International Federation of Red Cross and Red Crescent Societies (IFRC), and, from a data collection and research perspective, the Centre for Research on the Epidemiology of Disasters (CRED).

This chapter provides a global overview of human and economic losses that result from natural disasters. First, we consider issues around measuring loss, and which values and assumptions underlie different metrics. Second, we review data quality and methodologies. Finally, we present a summary of two global data sets and discuss what reasonable inferences can be drawn from the data.

2.4 Measuring Loss

The choice of metric used to measure loss is critically important. Choices are driven by both practical and ideological considerations; some are practical because certain types of data are easier to gather than others, and some are ideological because values underlie which data are gathered and how they are analyzed. As an example, consider how the metric "number of deaths" can be used as a measure of disaster (Table 2.1). Similar lists could be made for economic impacts.

Table 2.1 Different ways of expressing disaster fatalities[2]

- Deaths per million people in a political or geographical boundary
- Total number of deaths per event
- Average number of deaths per decade
- Total number of years of life expectancy lost[a]
- Deaths per million people within an affected area
- Total number of deaths within an affected area
- Deaths per facility
- Deaths per unit of concentration of a toxin
- Deaths per ton of toxin released
- Deaths per ton of toxin absorbed by people
- Deaths per ton of chemical produced
- Deaths per million dollars of product produced

Although each metric is based on human deaths, disaster rankings vary depending on the choice of metric.
[a]Not commonly used in disaster loss research.

Different values underlie which definition of damage is used. Consider the statistic of "deaths per event." First, it emphasizes human life over economics. Second, it equates all lives as being equal, for example, the life of an elderly person to the life of a child, or the life of an altruist to that of a criminal. Those who adhere to a value that all lives are of equal value would not object to such valuations. For those who value children over the elderly because of the longer potential life span of the former or because of a special duty owed to children, "loss of life expectancy" would be a better choice of metric. Deaths per event emphasize rare but larger disasters over more frequent smaller ones that may cumulatively kill a greater number of people.

Ethical issues can affect how disasters are perceived, but are not explicitly accounted for in many statistics.[3] For example, how should one assess society's duty to people who have knowingly chosen to place themselves at risk, as compared to victims who lacked this knowledge?

Lists that use total economic damage are biased toward events that occur in wealthier countries. Lists that use economic damage as a fraction of gross domestic product (GDP) would place greater emphasis on developing nations. To the extent that disaster rankings reflect how harm is measured, the indices chosen reflect the values of the analyst. Consider two disasters—Hurricane Katrina in the United States, in 2005 (Figure 2.2), and the Bam, Iran (Figure 2.3), earthquake in 2003. Hurricane Katrina killed approximately 1300 people[4] and had an economic impact of around $200 billion.[5] The Bam earthquake killed as many as 32,000 people[6] and cost about $1–2 billion; the former ranks high in terms of total economic costs, but low in terms of deaths. In the latter case, the reverse is true. Even

FIGURE 2.2 Hurricane Katrina, August 28, 2005. *source: NOAA.*

FIGURE 2.3 Bam earthquake, (a) before and (b) after. *Before sources: Arad Mojtahedi, Wikipedia,* http://en.wikipedia.org/wiki/File:Arge_Bam_Arad_edit.jpg. *After source: FEMA.*

when events are analyzed relative to geographical or political boundaries, arbitrary choices can result in very different statistics. For example, the importance of geographical boundaries is illustrated by noting that a loss of $200 billion is not large relative to the size of the United States economy, about $13 trillion, but it is much more significant as a fraction of the economy of Louisiana (where most of those losses occurred), which is about $193 billion.

2.4.1 Measuring Loss

Economic loss can be divided into two general categories: direct (stock losses) and indirect (flow costs). Both categories include tangible and intangible assets.[7] Most economic impacts that are tabulated in disaster lists are direct impacts, because indirect impacts are complex and much more difficult to measure, requiring a detailed study. Care is required to avoid double counting.[8]

Clay and Benson[9] have even argued that from a macro-economic perspective, disasters may have a positive long-term effect, for example, by bringing about a construction boom and opportunities for upgrading critical infrastructure or technology. However, what to include as a cost and what to include as a benefit are not always clear. Potential benefits depend greatly on the capacity of the affected region and can be very sector dependent. Some developing nations experience the opposite of a boom following catastrophe. This will occur due to an increase in national debt resulting from loans taken to enable recovery, and has the effect of reducing their capacity to recover.[10] The effect of Hurricane Mitch on Honduras and Nicaragua is an example of this; the recognition of this effect resulted in debt cancellations by France, The World Bank, and the IMF.[11] Similarly, Haiti received debt relief after their devastating earthquake of January 2010.

Many aspects of disasters can be and are measured, particularly direct economic costs, fatalities, and number of people affected (though this last variable is only vaguely defined). Other metrics could be measured but generally are not because of the difficulty or cost of doing so. Additionally, there are aspects of disasters that are particularly subjective and cannot be measured (at least in a way that would reflect a societal consensus), because diverse groups in society value them very differently. Examples are degree of stress or suffering, loss of confidence, loss of personal memorabilia, loss of opportunity, increased marginalization of disadvantaged groups, damage to natural ecosystems and habitats, and heritage or cultural impacts. A disaster victim is likely to give great importance to many of these factors, yet they are not counted in disaster databases.

Aggregate numbers of either lives lost or economic impact that forms the basis of a global perspective on loss generally say little about who was affected. Disaster losses and victims are not equally stratified along class lines but tend to accumulate among the disadvantaged and vulnerable. Averaging statistics at large scales obscures this important distinction. There is literature on vulnerability which demonstrates that disaster victims tend to be disproportionately poor, elderly, racial minorities, children, and women.[12] In a way, disasters are like a social autopsy, exposing inherent inequalities of risk in society.[13]

A more humanistic perspective is that disasters can be more completely understood by going beyond data and incorporating the stories and narratives of victims.[14] An example of this is the story of the father who struggled to save his five children from a flood, only to be forced to let his four daughters go, one by one, in order to save his son.[15] This story is far more compelling than any dry statistic, but such narratives are not incorporated in global data sets.

Disasters are highly complex events that are not well represented by single or even many numbers. This paper, which has a global perspective, is based on aggregated data; the authors therefore suggest that users of such information do so with care and consideration of the points raised above.

2.5 Data Quality

2.5.1 Sources and Methodology

In order to understand the quality of disaster data it is important to look from where they came. The following table (Table 2.2) includes a number of types of agencies that may serve as a source of disaster data and the types of data they are likely to collect.

Table 2.2 Examples of agencies that collect disaster data, and the type of metric collected

Type of Agency	Examples	Type of Data Gathered
• National governments	Ministry of Health, Economic Departments, Census Bureau, Public Safety	Mortality and morbidity, population baseline data, GNP, GDP, accidents requiring public services responses
• Insurance companies	Insurance and Reinsurance Agencies	Costs of infrastructure damage, recovery/replacement costs
• UN agencies	World Food Program (WFP), World Health Organization (WHO), United Nations Office for the Coordination of Humanitarian Affairs (UNOCHA), and United Nations International Strategy for Disaster Reduction (UNISDR)	Affected populations, mortality and morbidity, displaced, affected infrastructure, existence and effectiveness of mitigation efforts
• World Bank	Global Risk Identification Unit	Wide range of data
• International organizations (with UN mandates)	International Committee of the Red Cross (ICRC), Federations of the Red Cross and Red Crescent Societies (IFRC), International Organization for Migration (IOM)	Number of affected individuals, number of displaced/homeless, mortality rates, morbidity rates
• International donors	United States Office for Foreign Disaster Assistance (OFDA), European Community Humanitarian Office (ECHO)	Number of affected individuals, number of displaced/homeless, mortality rates, morbidity rates
• Nongovernmental organizations (NGOs)	CARE, World Vision, Save the Children, Mercy Corps, Doctors without Borders	Number of affected individuals, number of displaced/homeless, mortality rates, morbidity rates

Table 2.2 Examples of agencies that collect disaster data, and the type of metric collected—cont'd

Type of Agency	Examples	Type of Data Gathered
• Academic researchers	CRED	Deaths, affected people, economic costs
• Media representatives	Local outlets, international outlets such as CNN, BBC, New York Times, and the Associated Press	Number of international responders, number of deaths, number of homeless, number of international individuals affected
• Affected population	Local community leaders, social clubs, service groups such as Lion's Club, Rotary Club	Homeless, displaced, loss of businesses

Since there is no current global standard for data collection and the above agencies do not necessarily collect data for inclusion in global databases as their motivation, there are resultant incompatibilities. Different agencies collect data for their own purposes, and as such each has their own methodologies and motivations. Some of the motivations and methods are intuitive, such as a hospital collecting information on injuries and deaths as part of their normal routine; thus, a disaster researcher should have an easy time gathering morbidity and mortality information based on hospital records. Attribution can be difficult though, where cause and effect are unclear. For example, should a deceased person whose surgery was delayed due to a disaster situation be included as a disaster victim?

Nongovernmental organizations involved in providing emergency shelter gather data relating to the number of families left homeless as a result of a disaster as well as the number of newly constructed buildings. Organizations working in a temporary shelter camp (Figure 2.4) may collect similar data, but might use "number of individuals" rather than "number of families" as their base for counting. When including the number of homeless in an internationally recognized database one should decide which agency has the more credible or accurate data.

FIGURE 2.4 Refugee Camp in east Zaire, 1994. *Source: Center for Disease Control.*

Insurance companies collect and record data based on filed claims and thus are most likely to have consistent data over events and countries. However, these data are only for insured individuals and companies, and are only gathered as they relate to insured losses. This means that their data would be particularly lacking for developing countries where personal property and business insurance are rare to nonexistent. The above examples show that a weakness in the collection of disaster data is determined by institutional mandate.

Another issue related to the collection of disaster data is that the terminology used by many agencies is not clearly or commonly defined. For example, most organizations collect data on what they determine to be affected people, though there does not appear to be a standard definition for this term. To an agency that deals with health issues, affected could mean a person who has an injury or disease as a result of the disaster situation. To an agency that builds shelters, affected can mean those individuals who have been made homeless due to a disaster. Similarly, agencies focused on education, economic development, or water and sanitation will identify their affected populations by their own specific criteria. Given this broadly undefined standard it can be unclear just how many people are actually affected by any given disaster situation and it explains why some agencies report different statistics for the same disaster response.

> **STUDENT EXERCISE**
> - Write what you consider to be a good working definition of an affected person.
>   - Compare your definition with your fellow students' definitions.

The international community has recognized a need for more consistency in disaster data, and through a series of World Bank meetings in the 1980s attempted to evaluate the quality, accuracy, and completeness of three existing global disaster datasets (Table 2.3).

Table 2.3 Comparison of source materials of major disaster databases[16]

EM-DAT	NatCat	Sigma
• UN agencies • US government • Governmental sources • IFRC • Research centers • Lloyd's • Reinsurance sources • Press • Private	• Insurance-related media and publications • Online • Databases and information systems from news agencies • Governmental and nongovernmental organizations (REUTERS, IFRC, OCHA, USGS, etc.) • Media reports • Worldwide network of scientific and insurance contacts • Technical literature • Munich Re clients • Branch offices	• Daily newspapers • Lloyd's list • Primary insurance and reinsurance periodicals • Internal reports • Online databases

These are NatCat (maintained by Munich Reinsurance Company), Sigma (maintained by Swiss Reinsurance Company), and EM-DAT (maintained by CRED). Each of these databases measures a number of similar statistics such as number of deaths, number of those affected, and total damage in dollars. Each has also established its own criteria based on the source data that it deems to be most credible. These criteria undoubtedly introduced bias.

2.5.2 Biases in Disaster Data

Each organization, government agency, and individual collecting disaster data does so with a focus on their specific needs. They also collect data over different time periods and have differing interests as to what data are the most important to collect and preserve. In addition they will be biased in the types of disasters from which they consider collecting data. For example, the National Oceanic and Atmospheric Administration (NOAA) will have little interest in earthquakes and will therefore not collect any data about them. They will, however, be interested in data such as deaths from cyclones. Along the same vein, insurance companies are only concerned with those individuals and companies who hold insurance, and have a vested interest in determining the least costly method of reimbursing these claims. Conversely, reporters and journalists may wish to collect and report data only on those most severely impacted as this makes for more sensational stories.

> **STUDENT EXERCISE**
> - Find a news article quoting disaster statistics that reveals bias.
>   - Explain the reason behind the bias.

In addition to who is doing the collecting, when disaster data are collected also may play a role in their quality. First-response agencies such as fire fighters, search and rescue personnel, and the International Federation of Red Cross and Red Crescent Societies are often on site first and thus have the potential to gather some of the most credible data in the initial stages of disaster response. In reality though, these agencies have the priority of assisting victims and do not prioritize data collection; in the days immediately following a disaster these agencies are likely to keep records of the number of assisted, injured, and dead, but less likely to collect data on other victims. Sometimes even the numbers of dead and injured are not easy to access or estimate. In these cases reliable numbers may take months or years to obtain, if they are even attainable. In addition, agencies such as those noted above are generally concerned with loss of life and injury, and collect little or no data on economic losses.

Postresponse collection of disaster data is often more systematic, and will access records kept by governments, NGOs, the IFRC, social service agencies, hospitals, and financial institutions. There is a danger at this point of double counting of data from the initial response figures. For example, the IFRC may initially report that 1500 people were injured in a given disaster and 300 were killed. Once these people have been transported to hospitals and morgues some of those who were injured may die and those already dead will have death certificates issued. If 200 of the 1500 injured people die of those injuries en route to or at the hospital, there is a danger of the hospital then reporting 300 dead as a result of the disaster and the morgue reporting 200 deaths. The media then, with careful research, may dutifully report 500 deaths and 1500 injured when in fact there were only 500 dead and 1300 injured (300 dead at the site and 200 who later died in hospital).

Attribution is important; for example, when do the effects of a disaster end? Depending on severity, effects can be measured in days, months, years, or even decades. These seemingly arbitrary timelines are set by agencies depending on a variety of factors. One author's experience in Banda Ache was that the government of Indonesia, in response to the tsunami in 2004, declared the disaster over in 90 days as a way to control visas issued to foreign aid workers. The US Office for Foreign Disaster Assistance (OFDA) often sets timelines on disaster funding of 90 days in order to conserve their funds. This process requires agencies to rollover their funding requests to agencies such as the Office for Transition Initiatives or the Agency for International Development. In these cases, the gathering of data stop, or the methods for gathering data and the type of data collected may change with the funding mechanism.

The amount and quality of data collected may also be affected when the disaster itself is the impetus for its collection. For example, some areas of the world have limited medical facilities and thus many patients and diseases go untreated and unreported. The influx of relief workers following some disasters results in clinics and field hospitals being set up and staffed in areas that might not have had much predisaster medical care. For example, during the South East Asian Tsunami in 2005, the town of Meulaboh had 14 doctors. In the period following the disaster, the number of doctors grew to 41; this for a population that was 80,000 prior to the Tsunami and only 45,000 after the event. This represents a change from one doctor for every 5700 people to one doctor for every 1100 people.[17] For this reason, any data collected from these new facilities will skew regional and national trend statistics with relation to morbidity and mortality rates.[18]

Statistics often show that deaths from diseases such as malaria spike following floods. For example, in the Southern Africa flood in 2000 a study by a Japanese Disaster Relief Medical Team that assisted in the town of Chokwe where the damage from the flood was extensive found that infectious diseases were found in 85% of patients and that the "incidence of malaria increased by four to five times over non disaster periods." This was based on a 9 day study where 2611 patients were seen in the clinic. What is not noted in the study is that malaria is so prevalent in Mozambique that the average Mozambican would not visit a clinic for malaria under normal circumstances. This means that a clinic or hospital would not normally be able to record these cases. Given the flood situation and the abundance of doctors and clinics, people were more likely to visit clinics and thus have their illnesses recorded.

Economic data may also be incomplete given the fact that many residences are informal and may not be on record or insured. In these cases the financial losses associated with lost homes and belongings will not show up on official records, even though the costs of providing shelter for them in the short term and relocation in the long term will have major financial impacts on social services and response agencies.

Small business enterprises may make up a large percentage of local employment and income, yet may not be insured or registered and thus might also be missed on reports of economic losses that rely on insurance claims. Communities such as these will experience lost capital investment that may take years or even generations to recoup. In addition, government data for income and businesses usually rely on tax returns as source documents. In countries where income is very low or the government is unable to police small enterprise, baseline data for a community's economic position may be poor, resulting in underreported losses. In addition, the above discussion does not take into account the informal or criminal sector of a nation's economy which, for obvious reasons, does not have public records.

In all the above cases the amount and quality of disaster data were reduced by collection issues, quality of predisaster information, or circumstances of disaster response or recovery. However, there are sometimes political or economic motivations for communities, governments, or relief agencies to purposefully bias reports of losses for organizational, public relations, fundraising, or national security purposes. For example, agencies that receive public donations to fund their efforts may overreport losses for fund raising purposes, whereas governments that rely on foreign investment or tourism may underreport losses so as to not scare away future foreign investment. The government of China, for example, was heavily criticized for not sufficiently reporting SARS infections in 2002–2003.

2.5.3 Availability

Disaster data have issues related to quality and consistency, but also with respect to availability. Some data and reports by agencies are internal documents, not generally available for public use because of legal, financial, or ethical reasons for keeping data confidential. For example, hospitals may not be able to release details about those who died under their care prior to notifying families, and some businesses may regard their data as proprietorial. Similarly, academics may hesitate to release their data immediately following its gathering as they intend to publish papers on the topic. All of these factors can make data collection challenging, even when information does exist.

2.6 Databases

Many national and regional disaster databases exist, but only two will be analyzed in this chapter due to its global focus. One is the disaster database from the Center for Research on the Epidemiology of Disasters, and the second is the database from the reinsurance company Munich Re. For a comprehensive comparison of these datasets, the reader is referred to reports by ISDR,[19] Guha-Sapir and Below,[20] and Tschoegl.[21] Data by the World Bank are also presented; additionally, the PreventionWeb[1] provides an online listing of disasters from several sources, by country, region, or year since 2007.

2.6.1 Center for Research on the Epidemiology of Disasters

The Center for Research on the Epidemiology of Disasters International Disasters Data Base (EM-DAT) can be accessed at no charge from the CRED Website http://www.emdat.be/, which is housed at the Catholic University of Louvain, Belgium. As of April 2013 the database[22] contained over 18,000 entries and was growing at an average of 700 entries per year. It is updated daily and is made available to the public each month after validation of the figures. Events are included in the database if they meet at least one of the following criteria: 10 or more people killed, 100 or more people reported affected; declaration of a state of emergency; call for international assistance. Data are entered according to country affected and include location, date, number of people killed/injured/affected, number of people homeless, and estimated damage costs. Sources include governments, UN agencies, NGOs, research institutions, insurance institutions, and press agencies. From an international perspective, EM-DAT is the most complete database available to the public.

Tschoegl[23] identified several problems with EM-DAT:

- "Over the past three decades, economic losses were reported for less than 30% of all natural disasters, and since 1995, the percentage of entries with economic losses has been declining.
- The percentage of large disasters with reported economic losses has remained relatively constant, but the percentage of small and medium disasters with reported economic losses has decreased.
- The percentage of entries with economic losses is higher for developed countries such as the United States and Japan than for developing countries.
- There is no systematic collection of economic data.
- An evaluation of economic damage is not systematically done.
- There are no standardized methodologies for the reporting of economic losses.
- Only direct losses are generally reported for major disasters. Little mention is made of indirect losses and even less of secondary costs.
- There is no information on the breakdown of losses by and no further indication on the content of the costs."

The data suggest that the number of natural disasters has increased over time (Figure 2.5). However, inferences must be drawn from these data with caution due to the many problems outlined above. Much of the increase may reflect reporting biases, specifically better reporting over time due to improved telecommunications and globalization. Increases in wealth and population are also significant factors. Some regional studies[24]

---

[1] http://www.preventionweb.net/english/professional/statistics/.

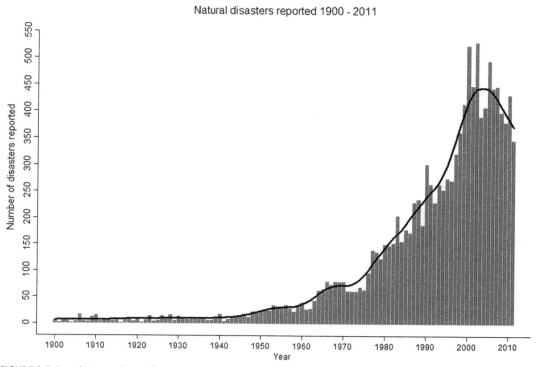

FIGURE 2.5 Trends in numbers of natural disasters. Note the dramatic increase in the number of events since the mid-twentieth century. Care must be taken when making inferences from these data, due to better reporting over time, as well as change. *Source: EM-DAT, The OFDA/CRED International Disaster Database – www.emdat.be, Université Catholique de Louvain, Brussels (Belgium).*

have documented increases in vulnerability because more people are living in riskier locations; thus it may well be that there has been a real per-capita increase in global disaster vulnerability,[25] though it is unproven. This increase is not evenly distributed by hazard. An analysis by Neumayer and Barthel[26] found that in recent decades the increase was due to weather-related events (increasing from about 300/year in 1980 to about 680/year in 2010), with geophysical disasters having no trend, hovering around 100/year. Okuyama and Sahin,[27] using EMDAT and Munich Re NatCat data, show that this diverging trend between meteorological and geological disasters began in the mid-twentieth century. Though climate change may well exacerbate this trend in the future, Changnon et al.[28] suggest that it is driven by demographic shifts, settlement patterns, and social trends as opposed to changes in physical hazards.

Figure 2.6 summarizes reported costs of disasters over time, not corrected for inflation. Like Figure 2.5, it shows a large increase in costs over the past few decades. Some events, such as the Kobe earthquake and Hurricane Katrina, are spikes that dominate the graph.

Trends in the numbers of deaths from natural disasters are shown in Figure 2.7. This graph shows the stochastic nature of life loss from natural disasters and a trend different

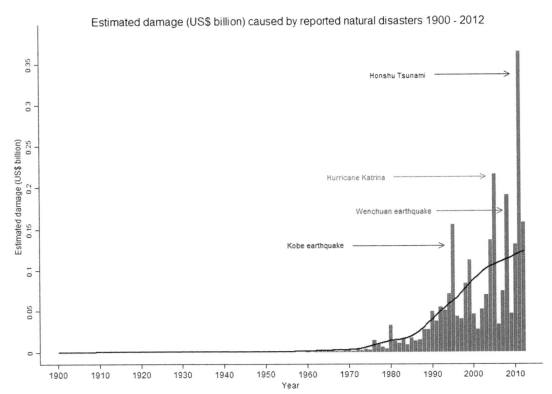

FIGURE 2.6 Trends in economic impact of natural disasters. Data prior to 1980 are very incomplete, and the large upward trend must be considered with caution. Note the dominance of a few extreme catastrophes. *Source: EM-DAT, The OFDA/CRED International Disaster Database – www.emdat.be, Université Catholique de Louvain, Brussels (Belgium).*

from that of the previous two graphs. Generally, the number of fatalities has been decreasing due to better warning systems and improved construction practices. These improvements, however, are not the same everywhere; wealthy countries have benefited from them to a much greater extent than lesser developed countries.[29]

Figure 2.8 shows trends in the number of people affected by natural disasters. This statistic is particularly problematic because it is difficult to determine exactly what the word affected means; thus the number of people that should be placed in this category is problematical. Nevertheless, the number of people reported to be affected by disasters has increased over time.

The number and impact of disasters differ by type and location as illustrated in Figure 2.9, which shows the number of people reported killed from different causes. Globally, windstorms, drought, and volcanic eruptions have been particularly important during the period of record. The highly variable spatial distribution of disasters is illustrated in Figure 2.10, which shows windstorm disasters, aggregated according to country.

Regional disaster mortality data show that Asia is the most affected continent (Figure 2.11). It is likely, however, that some areas such as Africa suffer from significant

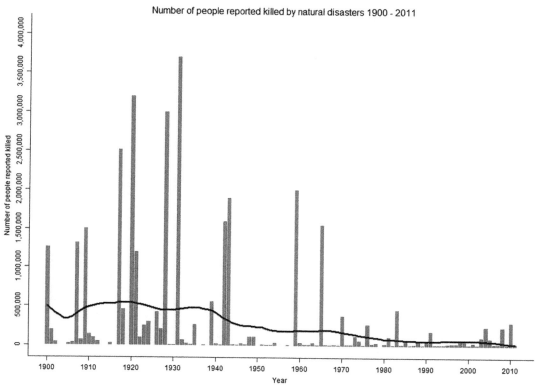

FIGURE 2.7 Trends in deaths from natural disasters. This graph, unlike the previous figures, shows a general decrease in deaths over the past century. This is generally ascribed to better warning systems and health care. *Source: EM-DAT, The OFDA/CRED International Disaster Database – www.emdat.be, Université Catholique de Louvain, Brussels (Belgium).*

underreporting. It is worth noting that Figure 2.11 is based on total numbers of people killed; if the data were presented as a percentage of population, regions with large populations, such as Asia, would be deemphasized relative to regions such as Oceania, which have much smaller populations.

2.6.2 World Bank

Though the World Bank does not maintain a disaster database, they have funded analyses of risk on a global scale. Columbia University's Earth Institute analyzed disaster hotspots to provide information useful to the World Bank and policymakers.[30] Figures 2.12–2.14 show estimates of economic and mortality risk. Note how the inclusion of GDP alters economic risk patterns (Figures 2.12 and 2.13). The hotspots tend to show that certain regions are more at risk than others, either as a result of the prevalence of one or more hazards (for example, the southeast coast of the United States to hurricanes) and/or due to the presence of large vulnerable populations (such as in China, Africa, and India).

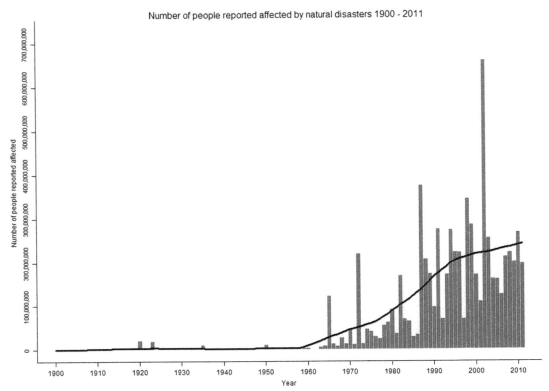

FIGURE 2.8 Trends in number of people affected by natural disasters. As in previous figures, the large upward trend must be considered with caution, due to biases in the data set and better reporting over time. Nevertheless, in a globalizing world, it seems reasonable that increasing numbers of people are affected by disasters. *Source: EM-DAT, The OFDA/CRED International Disaster Database – www.emdat.be, Université Catholique de Louvain, Brussels (Belgium).*

STUDENT EXERCISE

- Compare Figures 2.12–2.14 and explain the different patterns.

2.6.3 Munich Reinsurance Database

Munich Re[2] maintains a database for natural catastrophes (MRNatCatSERVICE or NatCat) on human and material losses and publishes annual maps of natural disasters, as well as a global map of natural hazards. Data on losses include information on the type of disaster and its location, as well as insured and total economic impacts that are direct costs and losses. As of February 2008 there were more than 25,000 events listed, which increase by 800–1000 events per year.[31] Analyses conducted by Munich Reinsurance[32] show increasing

---

[2] www.munichre.com/.

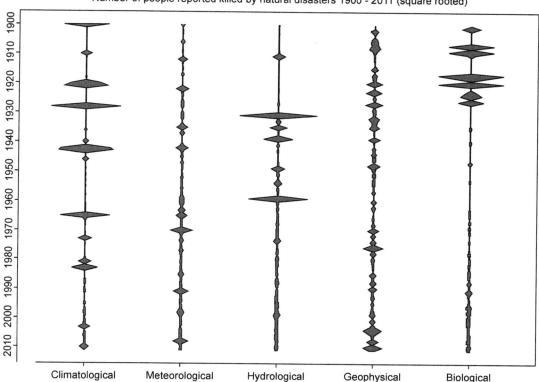

FIGURE 2.9 Deaths from natural disasters, by type. As in previous figures, trends over time must be considered with caution. This figure suggests that windstorms, droughts, and volcanoes kill the most people, of all natural hazards. *Source: EM-DAT, The OFDA/CRED International Disaster Database – www.emdat.be, Université Catholique de Louvain, Brussels (Belgium).*

trends in the number and cost of great natural catastrophes (Figures 2.15 and 2.16). Data on insured losses are accurate since insurance companies are very good at tracking claims. Estimates of total economic loss, however, suffer from aforementioned difficulties. Using Catastrophe Loss models, Munich Re has also estimated the cost to the insurance industry of a number of great natural disasters in the United States (Table 2.4).

2.7 Conclusions

Most natural disaster data are fraught with problems, in terms of both the methodologies of disaster analysis and the collection of the data. Data for the same disaster are often collected by agencies with different or even competing objectives. There are no agreed-on criteria for defining specific disaster data measures, nor is there a specified time period that is used to define the beginning and end of a data collection period for any given disaster. In fact, there is no globally agreed on threshold for what is considered a disaster. To add

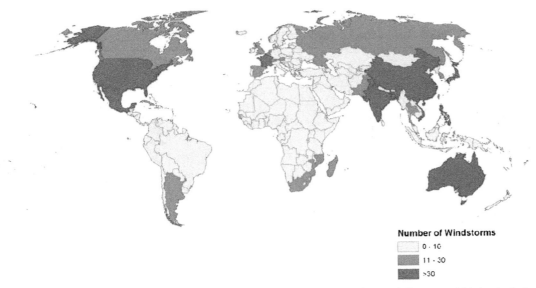

FIGURE 2.10 Windstorm disasters. This map illustrates the national disparity of natural disasters, which is a typical feature. Hazards are not uniformly distributed, as is vulnerability, and large variations in the frequency of natural disasters result. *Source: EM-DAT, The OFDA/CRED International Disaster Database—www.emdat.be, Université Catholique de Louvain, Brussels (Belgium).*

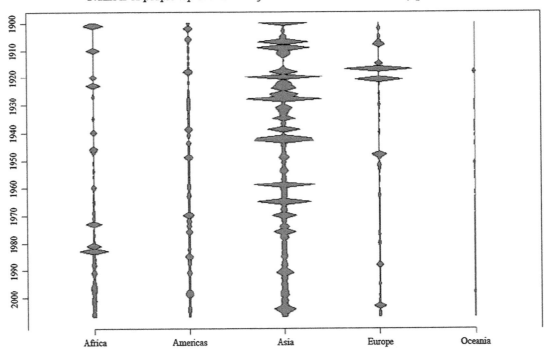

FIGURE 2.11 Deaths from natural disasters by region. This figure illustrates regional variations in disaster fatalities, with Asia showing the greatest number. Care should be taken when making inferences from it, since some regions such as Africa likely experience underreporting relative to others such as the Americas. *Source: EM-DAT, The OFDA/CRED International Disaster Database – www.emdat.be, Université Catholique de Louvain, Brussels (Belgium).*

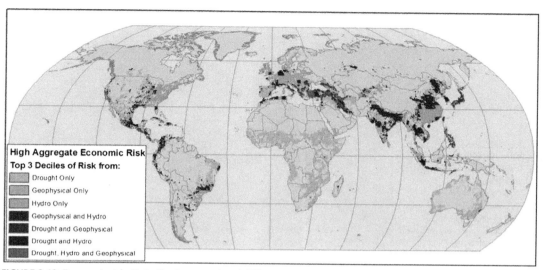

FIGURE 2.12 Economic risk. Note the large regional differentiation of disaster impact as a function of hazard type. Since this map considers only total economic impact, it is biased toward richer regions. *Source: The Earth Institute, Columbia University. Arnold M., Chen R. S., Deichmann U., Dilley M., Lerner-Lam A. L., Pullen R. E., and Trohanis Z., Natural Disaster Hotspots Case Studies, World Bank Disaster Risk Management Series No. 6, (Washington D.C., 2006), 204.*

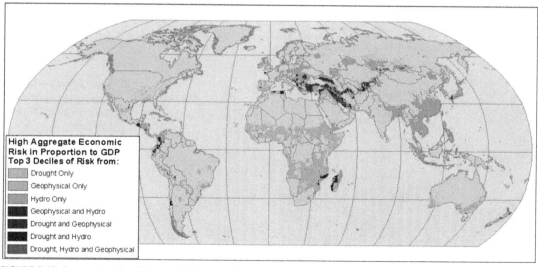

FIGURE 2.13 Economic risk relative to GDP. Note the large regional differentiation of disaster impact as a function of hazard type. Regional impacts, though, are very different from those shown in Figure 2.12, which did not consider the size of local economies. *Source: The Earth Institute, Columbia University. Arnold M., Chen R. S., Deichmann U., Dilley M., Lerner-Lam A. L., Pullen R.E., and Trohanis Z., Natural Disaster Hotspots Case Studies, World Bank Disaster Risk Management Series No. 6, (Washington D.C., 2006), 204.*

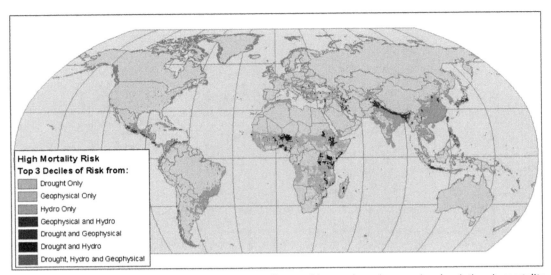

FIGURE 2.14 Mortality risk by hazard. Similar to previous figures, this map shows large regional variations in mortality risk. Note the dominance of drought (in Africa) and flood (especially in Asia) as the hazards involved in mortality risk. Note the dominance of drought (in Africa) and flood (especially in Asia) as the hazards involved in mortality risk. *Source: The Earth Institute, Columbia University. Arnold M., Chen R. S., Deichmann U., Dilley M., Lerner-Lam A. L., Pullen R. E., and Trohanis Z., Natural Disaster Hotspots Case Studies, World Bank Disaster Risk Management Series No. 6, (Washington D.C., 2006), 204.*

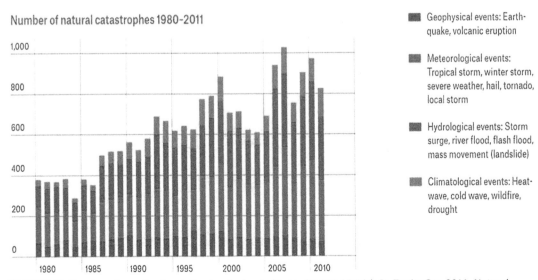

FIGURE 2.15 Number of great natural disasters by hazard type. *Source: Munich Re Topics Geo 2011, Natural Catastrophes 2011, Analyses, assessments, positions.*

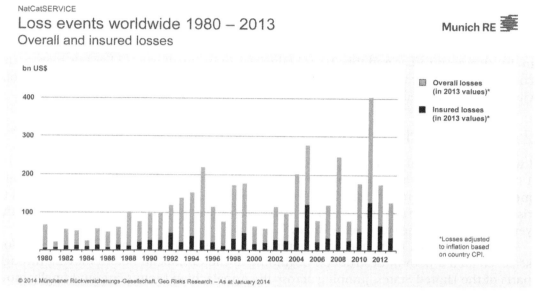

FIGURE 2.16 Economic impact of great natural disasters. Similar to the EMDAT data shown previously; note the increase over time, as well as the large annual variation. *Source: Munich Re Topics Geo 2011, Natural Catastrophes 2011, Analyses, assessments, positions.*

Table 2.4 The top 10 natural catastrophes between 1900 and 2006, as measured by insured losses standardized to 2007 U.S. dollars[33]

Event	Estimated loss ($2007)
1. Great Miami hurricane (1926)	$80 billion
2. San Francisco earthquake (1906)	$78 billion
3. Great Galveston hurricane (1900)	$53 billion
4. Great Okeechobee hurricane (1928)	$49 billion
5. Hurricane Andrew (1992)	$42 billion
6. Hurricane Katrina (2005)	$41 billion
7. Ft. Lauderdale hurricane (1947)	$36 billion
8. Great New England hurricane (1938)	$31 billion
9. Southwest Florida (1944)	$28 billion
10. Galveston hurricane (1915)	$27 billion

complexity to the already difficult to understand disaster data analysis, much of the data related to disasters is collected by private agencies that have no obligation to share this with the greater research community.

Due to the myriad of issues outlined in this chapter, trend analysis and regional comparisons of disaster occurrences are difficult, and it is important to take into consideration the difficulties, biases and constraints on raw data before making broader generalizations. Even so, the data suggests that the number and economic costs of natural disasters have been increasing over time, and this trend seems likely to continue.

Analysis of global datasets shows that some regions of the world seem to be more disaster prone than others. Regional differences are due to variations in patterns of hazard, exposure, and vulnerability. As global populations increase, people are more likely to live in disaster-prone areas and thus be at higher risk of being affected.[34] For this reason, the continued and improved collection of disaster data is important for the creation of policies related to risk reduction.

2.8 Case Study: Hurricane Hazel and Toronto

I was 5 years old and living in Toronto when Hurricane Hazel hit the city on October 15, 1954. My father was away on a business trip, and my mother was alone in the house with me and my older sister, trying to decide how she could evacuate if the water continued to rise. Fortunately, she did not have to. I do not have a clear memory of it myself, but over the years I have heard the story many times.[35]

Hurricane Hazel, at its maximum strength a category 4 storm on the Saffir-Simpson scale, created a path of destruction beginning in the Caribbean before moving through parts of the United States, jumping across the Appalachian Mountains, and finally moving northward into Canada. For Canadians this was a very rare event. It is very unusual for hurricanes, as they move north, to track into southern Ontario and to be reenergized by a mid-latitude system; however, this is what happened to Hurricane Hazel. Such storms are called extratropical storms and are especially dangerous because they retain some of the characteristics of tropical storms (especially heavy rainfall), even though they have transitioned. When Hazel hit Toronto it was equivalent to a Category 1 hurricane in terms of its strength. Winds reached speeds of 110 km/h and a maximum of 225 mm of rain fell during a 24-h period. Because the soil was already saturated by above average rainfall during the previous month (similar to what happened at Sarno), it is estimated that 90% of the rainfall could not infiltrate into the ground and, as a result became runoff. The maximum of rainfall occurred just northwest of Toronto (Figures 2.17–2.19), which added to local flooding because of the effect of local topography. "It was like dumping a lake the size of Lake Simcoe on the Humber River drainage area and having it all trying to get out by way of the river at once," Turnbull told the Toronto Star (October 23, 1954).[36] It took three events to come together for the hazard to be so severe in the Toronto area—a very wet two weeks before Hazel, a strong mid-latitude cold front,[37] and a tropical storm. A search on YouTube will allow you to see its path and some of its impacts.[38]

As a result of Hurricane Hazel about 1000 people died in Haiti (also, the storm destroyed 40% of its coffee trees and 50% of its cacao crop), 95 in the United States, and over 100 in Canada, 81 of which were in the Toronto area. In 1954 many people lived in the floodplains of the city. The worst of the flooding occurred after midnight (the time when the center of the storm passed over Toronto), when flash floods swept through the area. The death toll would have been much higher had not Jim Crawford, a 23 year old off-duty policeman, and Herb Jones, a contractor, worked heroically through the night in a small boat to rescue 50 people from their porches, second floor windows, or roofs. In the City of Toronto thousands

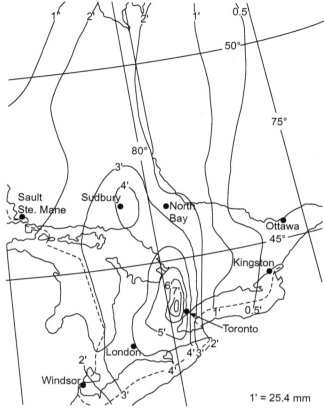

FIGURE 2.17 Rainfall from Hurricane Hazel over Ontario. Source: Hurricane Hazel—Storm Information. *Source:* © *Environment Canada, 2008.*

were left homeless as the flood waters washed out homes, trailers, bridges, roads, railways, and dams. Though the Dominion Weather Office had provided warnings the day before the storm hit,[39] they were not taken seriously by people who had no experience with such an event. There were no evacuations called for by authorities, although residents of Long Branch, with experience in annual spring floods, began to evacuate the evening before the storm. As a result, fast moving flood waters that increased water levels by up to 6–8 m were devastating to many communities.

The process of recovery was difficult and protracted, and involved the use of 800 troops. A Hurricane Relief Fund was set up, which distributed about $5.1 million dollars. Government funds were also made available for reconstruction subject to various restrictions, such as owners having to pay 20% of the damage costs. The total cost of the storm is estimated to be about $100 million in 1954. From a policy perspective, however, the most interesting aspect of this disaster is how it triggered major changes in flood management processes in the Province of Ontario.

FIGURE 2.18 Memorial Plaque for Hurricane Hazel. *Source:* http://www.creditplaques.com.

FIGURE 2.19 High water mark from Hurricane Hazel. *Source:* Rebecca Hanson.

Policy changes and organizational learning are most likely during what has been identified as "focusing events" or "windows of opportunity." "A focusing event is a sudden, exceptional experience that, because of how it leads to harm or exposes the prospect for great devastation, is perceived as the impetus for policy change.[40]" Policy windows are created as a result of the convergence of three streams (pardon the pun); the political stream, the policy stream, and the problem stream. Hurricane Hazel turned out to be such an event, because it was rare, sudden, harmful, revealed the potential for greater harm, addressed an issue with a recent history of attention at political and professional levels, and came to the attention of the public and policymakers simultaneously. Interviews with professionals working in flood management in the region clearly associate Hurricane Hazel with changes in floodplain protection and policy, even 50 years after the storm. In Ontario, the standard for flood planning is the 1/100 year event or a storm of record, and Hurricane Hazel is used in Southern Ontario.

In 1945 the Government adopted the Conservation Authorities Act, which assisted with the creation of Conservation Authorities in Ontario. These Authorities were based on hydrological basin scales, not political boundaries, and were a very progressive holistic approach to water management issues; by 1950 there were 14 authorities that had been created by 250 municipalities. Flood management was, therefore, a current policy issue at both the provincial and the federal levels, and the Canada Water Conservation Assistance Act of 1953 included cofunding at both levels of government for approved public works.

Prior to Hazel, Conservation Authorities had been advocating for the abandonment of development in flood plains in addition to engineering solutions. The Hazel flood created an opportunity for the existing conservation agenda to be advanced. There were political forces at work as well. An area at the mouth of the Etobicoke River was viewed as an undesirable semirural slum by the Long Branch Council, who wanted to clear it out and create a waterfront park as part of a larger vision of developing a greenbelt for the region. In Long Branch 192 properties were expropriated and the area became a park, as per the council's vision. The storm created an opportunity for them to take action. In addition, the notion of expropriating properties and preventing flood plain development was supported by the press. All the pieces needed to make Hazel into a focusing event were in place. The provincial government amended the Conservation Authorities Act, giving power to the authorities to acquire and regulate flood prone lands by prohibiting development and zoning them for recreation and conservation. Local Conservation Authorities were amalgamated to create the Toronto and Regional Conservation Authorities. After the storm some, but not all, of the proposed dams were built.

There is an interesting postscript to this case study. In the summer of 1986 a series of extreme rainfall events moved from Michigan into Ontario. Flood damage in Michigan was about $500 million, while in Ontario it was under $500,000, even though flood yields were larger in Ontario.[41] The researchers studying this determined that the reason for the difference was that Conservation Authorities in Ontario had mostly prohibited development within flood plains, whereas the U.S. National Flood Insurance Program, which is largely voluntary, had a net effect of allowing or even encouraging it.

...flood plain land use regulation may be the single adjustment most likely to reduce flood losses. Structural measures, flood warning systems and flood proofing will be of little value if the reduction in damages is more than offset by new damage potential in the flood plains[42]

Dan Huber, in a 2012 analysis of the National Flood Insurance Program[43] in the United States noted that it was in debt by more than $18 billion. An insurance approach to managing flood risk has the potential to be useful, but can create significant moral hazard and is not a substitute for avoiding development in hazardous regions, which inevitably increases exposure and vulnerability.

Further Readings

Brown DW, Moin SMA, Nicolson M: A comparison of flooding in Michigan and Ontario: "Soft" data to support "soft" water management approaches, *Canadian Water Resources Journal* 22(2), 1997.

Hurricane Hazel, Environment Canada Website, http://www.ec.gc.ca/ouragans-hurricanes/default.asp?lang=En&n=4343267B-1.

Hurricane Hazel, 50 Years Later, http://www.hurricanehazel.ca/ssi/evolution_flood_control.shtml.

Kennedy B: *Hurricane Hazel*, Macmillan of Canada, 1979.

Michaels S, Goucher NP, McCarthy D: Policy windows, policy change, and organizational learning: watersheds in the evolution of watershed management, *Environmental Management* 38:983–992, 2006.

Kreutzwiser R, Woodley I, Shrubsole D: Perceptions of flood hazard and floodplain development regulations in Glen Williams, *Ontario Canadian Water Resources Journal* 19(2), 1994.

Robinson D, Cruikshank K: Hurricane Hazel: disaster relief, politics, and society in Canada, 1954–1955, *Journal of Canadian Studies* 40(1), 2006.

STUDENT EXERCISE

Some communities can be very effective at mitigating risk, in terms of both reducing hazard and minimizing vulnerability. Other communities, such as Sarno, are not.

- Make a list of community characteristics that lead to both situations.
- Pick a community that you know well, and rate it according to your list.

End Notes

1. Mileti D., *Disasters by design: A reassessment of natural hazards in the United States* (Joseph Henry Press, 1999); ISDR (2002). Comparative Analysis of Disaster Databases. Final Report. Working Group 3 of the Inter-Agency Task Force of the International Strategy for Disaster Reduction on Risk, Vulnerability and Impact Assessment, LARED, November 30, 37; Wisner B., Blaikie P., Cannon T., and Davis I., *At Risk: Natural Hazards, People's vulnerability and Disasters (2nd ed.)* (New York: Routledge, 2004).

2. Slovic P., and Weber E. U., Perception of Risk Posed by Extreme Events. Decision Research and University of Oregon, *Conference on Risk Management Strategies in an Uncertain World*, Palisades, New York, April 12–13 (2002).

3. Stefanovic I. L., "The Contribution of Philosophy to Hazards Assessment and Decision Making," *Natural Hazards* 28, no. 2–3 (2003): 229–47.

4. Cooper, C., and Block R., *Disaster: Hurricane Katrina and the Failure of Homeland Security* (New York: Times Books, 2006).

5. Canton L. G., *Emergency Management: Concepts and Strategies for Effective Programs* (Hoboken NJ: Wiley Inter-Science, 2007).

6. Ghafory-Ashtiany M., and Hosseini M., "Post-Bam Earthquake: Recovery and Reconstruction," *Natural Hazards* 44, no. 2 (2008): 229–41.

7. National Research Council. *The Impacts of Natural Disasters: A Framework for Loss Estimation*. Committee on Assessing the Costs of Natural Disasters, National Research Council, National Academy of Sciences, Washington (1999).

8. Asgary A., *Technological Disasters' Cost/loss Data: Current Issues and Future Challenges*. 2nd Toulouse-Montreal Conference: The Law, Economics and Management of Large-Scale Risks, September 30 to October 1, Montreal, Canada (2005).

9. Clay E., and Benson C., (2005). *Aftershocks: Natural Disaster Risk and Economic Development Policy*, Overseas Development Institute Briefing Paper, November, London, U.K, http://www.odi.org.uk/resources/odi-publications/briefing-papers/2005/natural-disaster-risk-economic-development-policy.pdf, accessed 21.01.11.

10. Metoyer C. C., Hurricane Mitch, Alemán, and Other Disasters for Women in Nicaragua, *International Studies Perspectives* 2, no. 4 (2001): 401–15.

11. Harding G., (1998) Hurricane Mitch Forces Debt Relief to Top of Agenda. EuropeanVoice.com. Retrieved from http://www.europeanvoice.com/article/imported/hurricane-mitch-forces-debt-relief-to-top-of-agenda/37481.aspx, accessed 21.01.11; IMF (2000). IMF and World Bank Support Debt Relief for Honduras. Press Release No. 00/41, July 10, *International Monetary Fund*, 00(41), Washington, D.C.

12. Enarson E., and Morrow B., eds. (1998). *The Gendered Terrain of Disaster: Through Women's Eyes*. International Hurricane Center, Laboratory for Social and Behavioral Research, Westport, Praeger (1998).

13. Wisner B., Blaikie P., Cannon T., and Davis, I., *At Risk: Natural Hazards, People's vulnerability and Disasters (2nd ed.)* (New York: Routledge, 2004).

14. Heijmans A., "From vulnerability to empowerment," in *Mapping vulnerability: disasters, development & people* (115–126), eds. G. Bankoff, G. Frerks, and D. Hilhorst, (London: Earthscan, 2007).

15. See note 12 above.

16. Guha-Sapir B., and Below R., *The Quality and Accuracy of Disaster Data: A Comparative Analysis of Three Global Data Sets*. World Bank, Disaster Management Facility, ProVention Consortium (2005).

17. Lee V., and Low E., "Coordination and Resource Maximization during Relief Efforts," *Prehospital and Disaster Medicine* 21, no. 1 (2006), S8.

18. de Ville de Goyet C., and Zeballos C., "Communicable Diseases and Epidemiological Surveillance after Sudden Natural Disasters," *Medicine for Disasters*, (1988) 252–69.

19. International Strategy for Disaster Reduction, *Comparative Analysis of Disaster Databases*. Final Report, Working Group 3 of the Inter-Agency Task Force of the International Strategy for Disaster Reduction on Risk, Vulnerability and Impact Assessment, LARED, November 30 (2002).

20. Below R., and Guha-Sapir D., *An Analytical Review of Selected Data Sets on Natural Disasters and Impacts*. UNDP/CRED Workshop on Improving Compilation of Reliable Data on Disaster Occurrence and Impact. April 2–4, Bangkok, Thailand (2006).

21. Tschoegly L., *An Analytical Review of Selected Data Sets on Natural Disasters and Impacts*. UNDP/CRED Workshop on Improving Compilation of Reliable Data on Disaster Occurrence and Impact. April 2–4. Bangkok, Thailand. Centre for Research on the Epidemiology of Disasters, Université catholique de Louvain School of Public Health (2006).

22. See note 20 above.
23. See note 21 above.
24. See note 13 above.
25. Schipper L., and Pelling M., "Disaster risk, climate change and international development: scope for, and challenges to, integration," *Disasters* 30, no. 1 (2006), 19–38.
26. Neumayer E., and Barthel F., Normalizing Economic Loss From Natural Disasters: A Global Analysis," *Global Environmental Change* 21, no. 1 (2011), 13–24.
27. Okuyama Y., and Sahin S., Impact Estimation of Disasters: A Global Aggregate for 1960 to 2007. *World Bank Policy Research Working Paper* #4963, 40 (2009).
28. Changnon S. A., Pielke R. A., Changnon D., Sylves R. T., and Pulwarty R., "Human Factors Explain the Increased Losses from Weather and Climate Extremes," *Bulletin of the American Meteorological Society* 81, no. 3 (2000), 437–42.
29. International Strategy for Disaster Reduction, *Living with Risk: A Global Review of Disaster Reduction Initiatives* 2 (United Nations Publications, 2004): 429.
30. Dilley M., Chen R. S., Deichmann U., Lerner-Lam A. L., and Arnold M., *Natural Disaster Hotspots: A Global Risk Analysis*, no. 5 (World Bank Publications, 2005), 145;
    Arnold M., Chen R. S., Deichmann U., Dilley M., Lerner-Lam A. L., Pullen R. E., and Trohanis Z., *Natural Disaster Hotspots: Case Studies*, no. 6 (World Bank Publications, 2006): 204.
31. Hoeppe P., *Geohazards: Minimizing Risk, Maximizing Awareness: The Role of the Insurance Industry*. Munich Re Group, International Year of the Planet Earth, Paris, February 13 (2008).
32. Munich Reinsurance (2007). *Knowledge Series: Topics Geo Natural Catastrophes 2007, Analyses, Assessments, Positions*. Munich Re. http://www.extremeweatherheroes.org/media/32331/302-05699_en.pdf.
33. Ibid.
34. Intergovernmental Panel on Climate Change (2007). *Climate Change 2007 – The Physical Science Basis*. Agenda 6(07) (Cambridge: University Press, 2007).
35. Environment Canada, *Hurricane Hazel – Storm Information*. Retrieved from, http://www.ec.gc.ca/ouragans-hurricanes/default.asp?lang=En&n=5C4829A9-1.
36. Ibid.
37. The strength of the front can be illustrated by noting that the temperature just west of City of Toronto was 8 °C while it was 16 °C downtown.
38. Videos of Hurricane Hazel can be seen at http://www.youtube.com/watch?v=piBDs9eupQc and http://www.youtube.com/watch?v=-gGW7SzFJZQ.
39. Fred Turnbull, head of the weather office at Malton, issued a statement to the Telegram newspaper in which he stated that the rain they were expecting "could be the heaviest ever recorded in the city's history," exceeding the rainfall in 1887 that dropped close to four inches [101.6 mm] of rain, http://www.ec.gc.ca/ouragans-hurricanes/default.asp?lang=En&n=5C4829A9-1.
40. Michaels S., Goucher N. P., and McCarthy D., "Policy Windows, Policy Change, and Organizational Learning: Watersheds in the Evolution of Watershed Management," *Environmental Management* 38, no. 6 (2006): 983–92.
41. Brown D. W., Moin S. M. A., and Nicolson M., "A Comparison of Flooding In Michigan and Ontario: 'Soft' Data To Support 'Soft' Water Management Approaches." *Canadian Water Resources Journal* 22, no. 2 (1966).
42. White G. F., and Haas J. E., *Assessment of research on natural hazards*, (MIT Press, 1975).
43. Huber D., Fixing A Broken National Flood Insurance Program: Risks and Potential Reforms. *Center for Climate and Energy Solutions*, (2012), http://www.c2es.org/docUploads/flood-insurance-brief.pdf.

3

Disaster Risk

It is probable that the improbable will happen.
Aristotle (Figure 3.1)

FIGURE 3.1 Aristotle.

CHAPTER OUTLINE

3.1 Why This Topic Matters 55
3.2 Recommended Books and Readings 55
3.3 Question to Ponder 56
3.4 Introduction 56
3.5 Risk 56
    3.5.1 Risk as a Social Construct 59
    3.5.2 Risk Homeostasis 62
    3.5.3 Risk Perception 66
    3.5.4 Risk as a Feeling 73
3.6 The Risk Society 77

- **3.7 Measuring Risk** **78**
  - 3.7.1 Methodology 1: An Indices Approach 79
    - *3.7.1.1 Examples of Indices Studies* 80
    - *3.7.1.2 Issues in Risk Indices* 83
  - 3.7.2 Methodology 2: An Engineering Approach 83
  - 3.7.3 Methodology 3: A Case Study Approach 85
- **3.8 Sea Level Rise and Subsidence** **87**
- **Further Reading** **91**
- **3.9 Summary** **91**
- **3.10 Case Study: 1998 Ice Storm in Eastern Canada and Northeastern United States** **91**
- **End Notes** **96**

CHAPTER OVERVIEW

Risk is a concept used by many disciplines as an overarching analytical theme, and it is extraordinarily useful for that purpose. The definition used in this book is Risk = Hazard × Vulnerability. In this pseudo-equation, exposure is implicit. Although having the appearance of objectivity, risk is best viewed as being largely socially constructed. There is no optimum risk assessment or management strategy; context determines which approach is most suitable. Also, there are many issues or traps that can result in poor or biased risk estimations. These include adherence to specific worldviews, heuristics, bounded rationality, emotions, and values.

KEYWORDS

- Affect
- Heuristics
- HIRA
- Homeostasis
- Insurance
- Moral hazard
- Myths
- Psychometric paradigm
- Risk indices
- Risk management
- Risk measurement
- Risk perception
- Risk society
- Sea level rise
- Social construct

3.1 Why This Topic Matters

The word risk is so commonly used in everyday discourse, professionally, and within the academic community that it is important to have an understanding of its meaning. Risk is more than simply a variable. It is a paradigm that can be used to create a generalized approach to understanding and managing disaster; how that paradigm is understood and applied has major repercussions in terms of what strategies are selected to manage emergencies and disasters. Just a quick look at the bookshelves in my office reveal nine books with the word risk in the title—and that does not include the many texts that include it in chapters. There are journals, institutions, and professions that are all devoted to the study and practice of risk management, and as a student of disaster, it is essential that you be exposed to a portion of it. This is especially true given the international focus on disaster risk reduction as a management model.

You are entering an area of risk

 ■ ■ 3.2 Recommended Books and Readings ■

- Adams J., *Risk* (London: UCL Press, 1995).
- Beck U., *World Risk Society* (Wiley-Blackwell, 1999), 495–499.
- Haque C. E., and Etkin D., eds., *Disaster Risk and Vulnerability: Mitigation through Mobilizing Communities and Partnerships* (MQUP, 2012).

- Slovic P. E., *The Perception of Risk* (Earthscan Publications, 2000).
- Slovic P., *The Feeling of Risk: New Perspectives on Risk Perception* (Routledge, 2013).
- Wisner B., *At Risk: Natural Hazards, People's Vulnerability and Disasters* (Psychology Press, 2004).

■ ■ 3.3 Question to Ponder ■

Why has the concept of risk become so prevalent?

> **STUDENT EXERCISE**
>
> - What is the first thought or image that comes into your mind when you hear the word risk?
> - Compare your answer to those of your fellow students, and discuss what convergence or divergence there is in your various responses.

> **RISK: FROM THE ONLINE ETYMOLOGY DICTIONARY**
>
> risk (n.) 1660s, risque, from French risque, from Italian risco, riscio (modern rischio), from riscare "run into danger," of uncertain origin. The anglicized spelling first recorded 1728. Spanish riesgo and German risiko are Italian loan-words. Risk aversion is recorded from 1942; risk factor from 1906; risk management from 1963; risk taker from 1892.

3.4 Introduction

The literature on risk is so diverse and complex that it was difficult to decide what to include in this chapter. In the end, I decided to focus on definition, a few key themes, major debates in the field, some issues around measurement, and a discussion of how different approaches to risk emphasize different risk management strategies. In particular, the importance of risk being socially constructed and the two theories of risk homeostasis and the risk society are explored, in terms of their relevance to disaster.

As an overarching theme, I have found the notion of risk to be extremely useful. When separated into its component parts, it is an effective tool for analysis, not only quantitative analysis, but also as a method to identify different stakeholder views and understand much of the social discourse around various modern and complex hazards. In particular, it is a useful approach for one area that I have been involved with for many years: climate change.[1]

3.5 Risk

> *Whoever controls the definition of risk, controls the rational solution to the problem at hand.*
>
> *Paul Slovic (Figure 3.2)*

FIGURE 3.2 Paul Slovic. *Source:* http://www.decisionresearch.org/people/slovic/.

The first thing to know about the term risk is that it means different things to different people; there is no universal agreement on its meaning. Any serious conversation about risk must therefore begin with a common understanding of its definition. It matters less which definition is used than that we agree to a particular usage and move forward.

Wayne Blanchard (now retired), who for many years ran the Federal Emergency Management Agency (FEMA) higher education program put together a useful document[2] summarizing various definitions of terms used in emergency management. The different definitions of the term risk covers eight pages! Interpretations include risk as various combinations of probability, threat, consequence, exposure, hazard, and vulnerability. In this book I will use a definition of risk based on the Pressure and Release (PAR) disaster model from the book *At Risk*[3] (see Section 6.5.2). In brief, this model considers risk as a multiplicative function of hazard and vulnerability.[4] Exposure is implicit in the equation.

$$\text{Risk} = \text{Hazard} \times \text{Vulnerability}$$

This equation is the basis for many risk assessments done by emergency managers. The Hazard Identification and Risk Analysis (HIRA) approach is favored by many and is used by Emergency Management Ontario.[5] Figure 3.3 is an illustration of the HIRA approach, which ranks the frequency of a hazard versus the consequences of its occurrence. It is an intuitive and useful approach to estimating risk, but also is impoverished in terms of the factors that it incorporates.

A more sophisticated methodology based on HIRA was developed by the Provincial Emergency Program in British Columbia; it is called the Hazard, Risk and Vulnerability Analysis Tool Kit[6] and is available online. In the fall of 2012 while attending the Canadian Risk and Hazards Conference in Vancouver, BC, I asked several community-level emergency managers if they used this kit. Their response was that it was too involved, difficult, and resource intensive to use. This highlights a fundamental problem in operationalizing risk assessment: doing it well takes time, expertise, and resources, which are simply not

	Consequences				
Frequency	1	2	3	4	5
5	Minor	Moderate	Major	Severe	Severe
4	Minor	Moderate	Major	Severe	Severe
3	Minimal	Minor	Moderate	Major	Severe
2	Minimal	Minimal	Minor	Moderate	Major
1	Minimal	Minimal	Minimal	Minor	Moderate

FIGURE 3.3 A typical Hazard Identification and Risk Analysis chart. Risk increases with frequency and consequences, and the highest risk is shown in red in the upper right hand part of the chart.

available in smaller communities. Without the support of higher levels of government, it will always be a challenge for communities with scarce resources to achieve an in-depth understanding of their risks.

Another more sophisticated approach incorporates notions of uncertainty, catastrophic potential, and causal connections[7] and classifies risks into several types (Figure 3.4). The categories are defined as follows:

- Damocles: high catastrophic potential, probabilities (widely) known (e.g., hurricanes)
- Cyclops: no reliable estimate on probabilities, high catastrophic potential at stake (e.g., terrorism)
- Pythia: causal connection confirmed, damage potential and probabilities unknown or indeterminable (e.g., H1N1 influenza A virus)
- Pandora: causal connection unclear or challenged, high persistency and ubiquity (e.g., bioaccumulation)
- Cassandra: intolerable risk of high probability and great damage, but long delay between causal stimulus and negative effect (e.g., climate change)
- Medusa: perception of high risk among individuals and large potential for social mobilization without clear scientific evidence for serious harm (e.g., genetically modified foods)

Optimal risk management strategies depend on the characteristics of the risk being considered; three different but not exclusionary strategies can be used:

- Risk-based or risk-informed management strategies (Damocles and Cyclops)
  - Sufficient knowledge of key parameters
- Precautionary or resilience-based strategies (Pythia and Pandora)
  - High uncertainty or ignorance

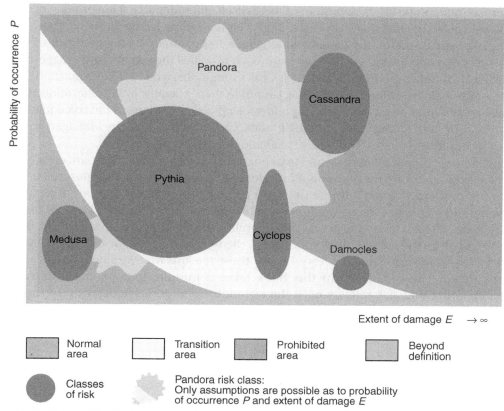

FIGURE 3.4 Risk types. The Y-axis represents frequency, and the X-axis represents extent of damage. The specific types of risks include consideration of other factors, such as uncertainty and catastrophic potential.[8]

- Discursive (social discourse) management strategies (Cassandra and Medusa)
  - High ambiguity

It is important to have an understanding of the nature of a specific risk to develop appropriate management strategies. *A "one size fits all" approach to risk management does not work well.* Taking a conventional risk-based approach to the problem of climate change,[9] for example, would only capture a subset of appropriate strategies because of the complexity, catastrophic potential, varying vested interests, and uncertainties associated with it.

3.5.1 Risk as a Social Construct

Historically, the major debate in the field of risk is the degree to which it is objective and measureable,[10] versus the degree to which it is subjective and sometimes immeasurable. (You can measure subjective information, but the meaning of the numbers is always debatable.) The former perspective is called the rationalist approach and emphasizes statistics, decision theory, and scientific management to control risk. The latter is called a constructionist approach, and proposes that nothing is a risk in itself, but rather that it is a product of cultural, political, social, and historical ways of seeing.[11] This is a typical debate that occurs between traditional physical and social scientists, although it is less common than it used to be. There are other less extreme positions; the realist approach modifies the rationalist by suggesting that social and cultural processes distort or bias risk estimations, while a middle position suggests that while objective criteria do exist, risk cannot in theory be absolutely determined because social and cultural processes inevitably add subjective components to the risk equation. The book *Risk* by John Adams[12] and his article "Cars, Cholera and Cows: The Management of Risk and Uncertainty"[13] are both useful and readable discussions on this topic.

There are, of course, many aspects to hazard and vulnerability that are clearly objective, quantifiable, and measurable. Examples are the number of tornadoes that occur per year in Southern Ontario[14] (more than you probably think), the number of people who live in flood plains in Quebec or along the Mississippi River (too many), or building code standards. There are also aspects of vulnerability that depend on social and cultural norms, such as the value of life or suffering, the worth of a park, or the relative vulnerability of age as compared to poverty. It is for these reasons that we say that risk is socially constructed. The remainder of this book takes a middle position on understanding risk, meaning that risk reflects both the rationalist and constructionist interpretations (Figure 3.5).

FIGURE 3.5 The rationalist-constructionist spectrum.

Saying that risk is socially constructed has another interpretation: that the decisions made by society determine what and who is at risk. For example, allowing housing construction near hazardous chemical plants is a social/political decision that puts people who live there in harm's way. The proximity of residential areas to hazardous industrial ones has become increasingly important due to urban growth, and in Toronto, the city I live in, this was highlighted by the propane explosion at the Sunrise plant on August 10, 2008. A more extreme example of this is the Bhopal gas tragedy of 1984.

The notion that risk is socially constructed has important consequences in terms of the manner in which risk is assessed, communicated, and managed. It means that the traditional top-down expert-driven approach is insufficient, although experts are clearly needed to understand the many technical aspects of hazard and vulnerability. Risk assessments need to include the views of stakeholders and reflect what they value. A few years ago, I was on the examining committee of a PhD student who researched nuclear contamination in Port Hope, Ontario.[15] Her description of meetings of the local community with subject matter experts was fascinating and highlighted the problems that occur when experts tell community members that they do not need to worry about radioactive-related risks because they are at more risk from smoking or driving. It just does not work well.

> **STUDENT EXERCISE**
>
> - As part of a government team tasked with doing a risk analysis of a community exposed to radioactive waste, list three objective measures and three subjective factors that would contribute to your risk estimation.

A typical risk management framework is shown in Figure 3.6. Although it is often used in a top-down manner, one could also use this framework with an emphasis on stakeholder engagement and social discourse. The latter approach is far more robust, but also much more expensive and time consuming, allowing for a shift in power that is not always comfortable for vested interests.

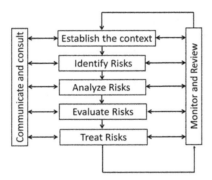

FIGURE 3.6 A traditional risk management framework. This model works best when hazards, vulnerabilities, and the methods used to treat them are well understood.

This particular model is applicable when risks are subject to robust analysis and have a solution, but such is not always the case. It is almost impossible to get consensus on some risks because they are wicked problems or there is insufficient information (e.g., genetically modified foods, climate change, and nuclear power). Other risks are not easily subject to being treated. The disposal of nuclear waste is a good example of this, since many solutions are not socially acceptable (often due to the not-in-my-backyard or NIMBY effect). And some risks are viewed by many stakeholders as solutions, not problems (e.g., war, if you are part of the military industrial complex). The complex relationship between the construction of risk as a benefit or solution versus its construction as a cost or problem greatly complicates risk management strategies.

3.5.2 Risk Homeostasis

> **Answer the following questions (be honest; say what you really think):**
> 1. Are you safer in a car crash if you are wearing a seat belt?[16]
>   ☐ Yes ☐ No ☐ Not sure
> 2. Does seat belt legislation save lives?
>   ☐ Yes ☐ No ☐ Not sure
> 3. Do dams and dykes control flood waters?
>   ☐ Yes ☐ No ☐ Not sure
> 4. Does the building of dams and dykes reduce flood damage?[17]
>   ☐ Yes ☐ No ☐ Not sure

Consider the effect of seat belts on driving safety. Many studies have shown that drivers are safer wearing seat belts if they are in an accident.[18] For that reason, many countries created legislation to compel drivers to wear them (for example, in Ontario, Canada, fines for not wearing seat belts range from $200 to $1000, and the guilty driver loses two demerit points[19]). Data following such legislation showed a decrease in fatalities, apparently supporting the policy change.

Not all countries created laws at the same time though, and it is interesting to look at driving-related deaths in countries that had a law as compared to those that did not. What would you expect the comparison to show? Consider Figure 3.7, which separates driving deaths during the period 1970–1978 between countries that created laws forcing drivers to wear seat belts, as compared to those that did not. It shows that those without laws experienced a greater decline in driving fatalities than those that had them. This is highly counterintuitive (to me at least, when I first saw it)! How can this be, and what are the implications of this relationship?

More insight into this phenomenon can be gained by examining the well-known

> Wearing seat belts saves lives. Does seat belt legislation save lives? Maybe not!

Munich Taxi Experiment[20]. In this experiment, it was found that when drivers were not aware that antilock braking systems had been installed on their cars, accident rates were reduced, but when they were aware they were driving with the antilock braking system, accident rates stayed about the same because they drove in a more unsafe manner.

Risk is reflexive, because people are reflexive. When people perceive that their risk environment has changed, they alter their behavior to suit the new environment (particularly if there are incentives or disincentives); hence, the taxi drivers change their driving behavior. The drivers had a set rate of acceptable risk (this is the homeostasis part). They would drive as fast and as safely (or unsafely) as their acceptable level of risk would allow. To drive slower would mean less income, but to drive faster would mean too much risk, hence the tendency to remain at set levels of risk, which is called risk homeostasis. If you are interested in reading

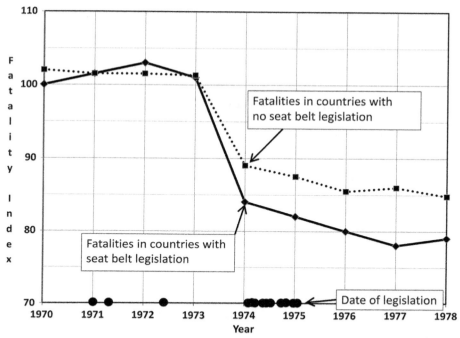

FIGURE 3.7 The effect of seat belt legislation. Bars at the bottom of the chart indicate years in which law came into effect in different countries. Note how countries without seat belt legislation had a greater reduction in fatalities than those with the legislation. *Source: Adapted from Adams J., Risk (London: Routledge, 1995).*

more on this topic, I recommend the book *Target Risk* by Professor Gerald Wilde from Queen's University (it is available at no cost on the web). The definitions below are quoted from him.

Target risk: The level of risk a person chooses to accept to maximize the overall expected benefit from an activity.

Homeostasis: A regulating process that keeps the outcome close to the target by compensating for disturbing external influences. For example, the human body core temperature is homeostatically maintained within relatively narrow limits despite major variations in the temperature of the surrounding air.

Risk homeostasis: The degree of risk-taking behavior and the magnitude of loss due to accidents and lifestyle-dependent disease are maintained over time, unless there is a change in the target level of risk.[21]

The basic principle behind risk homeostasis is that people transform safety measures into performance measures, particularly if there are incentives such as financial gain (Figure 3.8). This principle is behind the well-known levee effect (Figure 3.9), in which the construction of flood control works results in increased development in the now-perceived-to-be-safer areas (except that they often turn out not to be as safe as was thought or hoped). Along a similar vein, John Handmer in a study of flood warnings in

FIGURE 3.8 Principle of risk homeostasis.

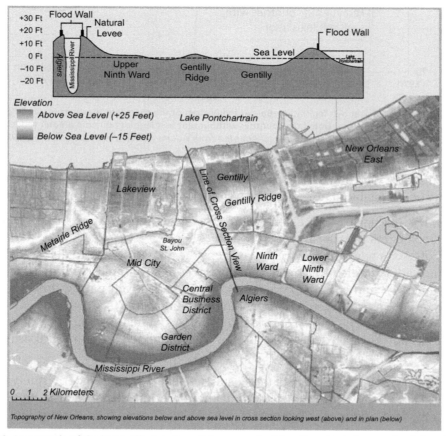

FIGURE 3.9 An example of the levee effect. In New Orleans, much of the city exists below sea level. People believed that the flood walls and levees protected them, although that protection only existed up to a threshold that nature was able to exceed. *Source: Federal Emergency Management Agency.*

Australia noted that, "Flood warnings provide a margin of safety that is readily consumed by or traded for behaviour which benefits someone economically."[22] This process tends to confound the original intent of many engineering-based policies. A psychological strategy, instead of or combined with an engineering approach, would include the addition of incentives for desirable behaviors and disincentives for undesirable ones.

The principle of risk homeostasis is of fundamental importance in the management of disaster risk. One application is in the area of disaster financial assistance, which has been critiqued for creating moral hazard,[23] because of the tendency of people to be more likely to accept risk when the cost is borne by others.

> **STUDENT RESEARCH PROJECT**
>
> Discuss how government policy (where you live) for disaster assistance should balance the needs and rights of citizens with the obligations of the state, while minimizing the effect of moral hazard. Then compare this with those of your own state, province, or country.

Moral hazard is also a large issue for the insurance industry, and is one of the main problems with providing insurance for flood prone areas; only those at risk would tend to buy the insurance, while those who live outside of flood zones would tend to avoid it (this is called "adverse selection"). While writing this section, I read a news article[24] titled "Floods Devastate Britain," which commented that "Britain has refused to bankroll a fund to subsidize insurance for households in flood-prone areas, derailing talks over the scheme and potentially leaving 200,000 homes without protection." Apparently they are very concerned with costs associated with flooding and may have been looking at the US experience. The US has a federal flood insurance program that began in 1968. It is both public and optional, but not considered to be actuarially sound because of subsidized insurance rates authorized by Congress. About 75% of the flood claims go to policy holders located in designated flood prone areas, and as of January 31, 2011, the outstanding debt and accrued interest cost was almost $27,000,000,000.[25] Moral hazard can be very expensive! I have heard presentations by the insurance industry in Canada proposing the development of residential flood insurance policies within a public-private partnership. A discussion paper, "Making Flood Insurable for Canadian Homeowners,"[26] supporting this approach presents an interesting set of arguments for this proposal. In order for an insurance scheme to be viable, six conditions need to be met[27]:

- Mutuality: A large number of people must combine to form a risk community;
- Need: There must be a need for insurance cover when the anticipated event occurs;
- Assessability: The peril must be assessable in terms of possible losses;
- Randomness: The event must be independent of the will of the insured, and the time at which the insured event occurs must not be predictable;
- Economic viability: The risk community must be able to cover flood-loss financial needs;

- Similarity of threat: The risk community must be exposed to the same threat, and the occurrence of anticipated damages must result in the need for funds in the same way for each member of the community.

The discussion paper argues that through a combination of bundled insurance policies (this means that flood risk would be "bundled" with other risks, as opposed to being separate) and having risk-based premiums (those exposed to greater flood risk would pay higher premiums), all six conditions can be met, moral hazard can be avoided, and flood insurance can be made viable. Of the various country models considered, the UK flood insurance one comes closest to this proposal. As a means to enhance disaster recovery, there are arguments for it (recall that insurance is not a risk reduction strategy, but a risk transfer strategy), which is why a number of countries have adopted this model. However, I believe that it should be viewed with great caution because of the potential for: (1) undermining land use planning as a way to reduce flood risk, (2) encouraging unsafe development, and (3) inappropriate transfer of risk from individuals to the state. In the UK, the "rate of new development in areas of high flood risk has remained fairly constant over the last 10 years. The cumulative impact of these new developments has potentially increased vulnerability to flood risk. In the last decade, between 12,000–16,000 new homes have been built every year in areas of high flood risk. This has remained a fairly constant proportion (around 10%) of all new residential development. This compares with a stock of approximately 1.3 million homes currently located in areas of high flood risk (equivalent to 4.5% of the total housing stock."[28] The agenda of the insurance industry may be that the public sector absorbs most of the risk (this is called the socialization of risk) while the private sector makes profit (after all, insurance companies are part of the private sector). From a governmental perspective, the issue of importance may be less whether the scheme is financially viable, but rather its greater social implications in terms of the construction of flood-vulnerable communities.

■ ■ Questions to Ponder ■

- What criteria would you use to decide whether or not to implement a flood insurance program for your country?
- On balance, would you be inclined toward creating such a program, or not?

3.5.3 Risk Perception

People do not respond directly to the risks they are exposed to; rather they respond to their perceptions of those risks. There has been a good deal of research on risk perception and the various biases and distortions that people are subject to. Understanding this area of study is essential to understand how to have successful disaster management and risk communication. Historically the literature focuses on cognitive aspects of risk perception, but in recent years there has been an increase in studies emphasizing the importance of the affective side.[29] The rest of this section is devoted to a brief description of the main issues important to risk perception.

The Gap between Experts and the Public—There is a large gap between how experts tend to judge risk and how the lay public does. Expert judgment tends to rely on narrow technical metrics such as annual fatalities. There are, however, many ways of expressing fatality statistics, including deaths/population, normalized deaths, deaths/facility, deaths/political boundary, or loss of life expectancy. Which measure is used requires a judgment call on the part of the risk assessor and reflects social values and the purpose of the risk assessment; different metrics can result in very different analyses, as discussed in Chapter 2. The public tends to have a much broader perspective on how to assess risk and include factors such as the benefit the hazards may provide to society (energy is one example), dread, voluntariness, controllability, catastrophic potential, uncertainty, and equity.

The Psychometric Paradigm—Paul Slovic's research on risk perception identified a number of factors that contribute to varying perceptions of risk,[30] which differ by hazard. For example, X-rays and nuclear power have very different risk profiles, as shown in Figure 3.10. This is part of the psychometric paradigm, which scales risk according to the quantitative judgments people make about the riskiness of various hazards.

STUDENT EXERCISE

On Figure 3.10, draw what you imagine the risk profiles would be for the following hazards:

- Tornadoes
- HIV
- Tsunamis
- Genetically modified foods

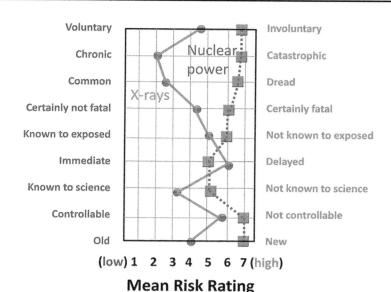

FIGURE 3.10 Risk factors for X-rays and nuclear power.[31] *Source: Adapted from Fischhoff B., Slovic P., Lichtenstein S., Read S., Combs B., "How Safe Is Safe Enough? A Psychometric Study of Attitudes toward Technological Risks and Benefits,"* Policy Sciences 9, (1978): 127–152.

The sets of factors identified by Slovic can be broadly combined into two categories, because many of the factors are strongly correlated with each other. Factor one (called dread risk) and factor two (called unknown risk) can then be used to differentiate hazards (Figure 3.11). Overlain on the four quadrants are the preferred management strategies, as per Hovden.[32] Where dread and uncertainty is high, the precautionary principle combined with social discourse become important approaches. Only in the bottom left quadrant, where dread and uncertainty are low, is the traditional risk-based approach suitable by itself.

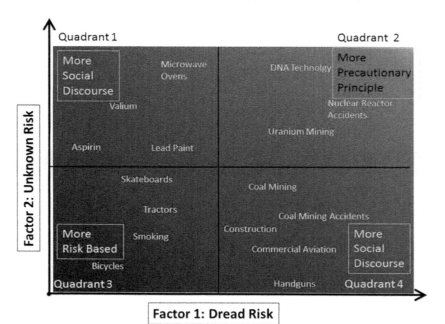

FIGURE 3.11 The psychometric paradigm of risks. Factor one is composed of the variables controllability, dread, globally catastrophic, level of fatality, equitability, scale, risk to future generations, reducibility, trend, and voluntariness. Factor two is composed of observability, degree known to exposed, delayed versus immediate, new versus old, and degree known to science. Superimposed on the chart are risk management strategies appropriate to the quadrant. *Source: Modified from Slovic P. E., The Perception of Risk (Earthscan Publications, 2000).*

THE TRAP OF THE EXPERT

I see no good reasons why the views given in this volume should shock the religious sensibilities of anyone.

Charles Darwin, The Origin of Species, 1869.

Experts tend to be far better at solving particular problems, but are more likely to frame a problem within a narrow perspective. The different ways experts and the public bind problems mean that disagreements often do not disappear in the presence of evidence, particularly for wicked problems.

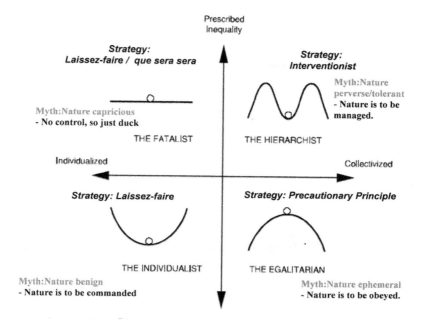

FIGURE 3.12 A typology of perceptual filters, based on four myths of nature: nature capricious, nature perverse/tolerant, nature benign, and nature ephemeral. *Source: Adapted from Adams J.,* Risk *(London: UCL Press, 1995).*

Myths of Human Nature—In John Adam's book *Risk*,[33] he discusses four myths of human nature (Figure 3.12), each of which is associated with a different management style. These myths are important in defining worldview and create a lens through which people tend to interpret risks. Broadly defined, they are:

- Nature benign: the individualist
  - Laissez-faire management style. Risks can be managed, and the world is really a pretty safe place.
- Nature ephemeral: the egalitarian
  - Use of the precautionary principle, since risks are unmanageable.
- Nature perverse/tolerant: the hierarchist
  - Interventionist management style, since risks can be managed, although you have to be careful about unintended consequences.
- Nature capricious: the fatalist
  - Laissez-faire. What will be, will be.

Members of Greenpeace would fit into the egalitarian quadrant, while members of the right-leaning Fraser Institute Think Tank fit into the individualist quadrant. Most government organizations with mandates to manage risk (such as Public Safety

Canada or FEMA) would be hierarchical. A citizen who believes that human beings have no significant control over the state of the world would be a fatalist. These different myths have a lot to do with the ways risk is perceived. A few years ago I was discussing environmental risks with an economist friend, and it quickly became clear that we had very different estimates of how these risks might play out in the future. I foresaw serious potential consequences in the next few decades (I fit into the egalitarian quadrant for many risks) and favored the use of the precautionary principle, while he saw them in terms of centuries (he fit into the individualist quadrant) and favored continued economic development and a wait-and-see approach. At the root of this issue lies the perceived ability of human beings to deterministically control complex systems such as ecosystems, economic systems, agricultural systems, or nuclear systems. This topic will be explored further in Chapter 5.

> **STUDENT EXERCISE**
> - Where would your worldview fit into Figure 3.12 for the following risks: climate change, deforestation, nuclear meltdown, airplane crashes?
> - Compare your responses to the rest of the class, and discuss the differences.

Virtual Risk—Some risks are perceived directly (e.g., falling off a cliff), some through science (e.g., germs), while others are subject to varying perspectives because scientists apparently do not or cannot agree (e.g., climate change or genetically modified foods).[34] The third category is called virtual risk. Depending on which category a risk lies in, it will be perceived differently and should be managed so. Some groups of people will perceive a risk as lying in one category, while others see it in a different category; this creates much of the variance of opinion with respect to climate change. Climate scientists overwhelmingly put it in the second category of a risk perceived through science, while much of the public consider it a virtual risk. The precautionary principle applies well to risks in the first two categories, but does not work well for virtual risks because there is so much stakeholder disagreement. It is therefore not surprising that there is so much debate over risks like climate change.

Biases in Risk Perception—Research comparing estimates of deaths versus actual deaths in the United States have shown persistent biases in people's perceptions (Figure 3.13). This experiment shows that people tend to overestimate low death rates and underestimate high ones. Additionally, while the range of actual deaths covers more than six orders of magnitude, estimates only cover about 3—a difference of a factor of 1000! These errors relate to the notion of bounded rationality[35]—the idea that people have constraints on their ability to make accurate decisions because of partial or unreliable information, the limited ability of the mind to evaluate and process information, and the finite amounts of time and resources available to analyze options. This research points toward the importance of using both data and having expert input into risk assessments; those that rely only on qualitative judgment can be very flawed.

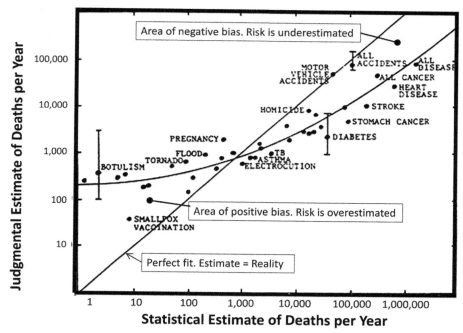

FIGURE 3.13 Biases in risk perception. Relationship between frequency estimated (perceived) by 188 subjects and actual number of deaths per year for 41 causes of death in the United States. The straight line is the perfect fit between mean estimated and actual frequencies; the curved line is a fit to the actual data. Note the large biases in perceived risk.[36] *Source: Lichtenstein S., Slovic P., Fischhoff B., Layman M., Combs B., "Judged Frequency of Lethal Events," Journal of Experimental Psychology: Human Learning and Memory 4, no. 6 (1978): 551.*

Heuristics—Work by Tversky and Kahneman (Figure 3.14) published in 1973, which led to their receiving the 2002 Nobel Prize for Economics, identified a number of rules of thumb and biases in terms of how people make decisions that deviate from a model of perfect rationality. In their words, "In general, these heuristics are quite useful, but sometimes they lead to severe and systematic errors."[37] In his Nobel Prize acceptance lecture, he stated, "Our research attempted to obtain a map of bounded rationality, by exploring the systematic biases that separate the beliefs that people have and the choices they make from the optimal beliefs and choices assumed in rational-agent models."[38]

There are many inferential rules or heuristics people tend to use to reduce complex judgments to simpler ones. Some of the more common ones are:

- Availability—This occurs when estimates of probability are selected based on the ease with which examples come to mind.
- Affect—When our feelings, such as fear or pleasure, influence decisions.
- Attribute—Substituting a more easily calculable metric for a more difficult but more accurate one.

FIGURE 3.14 Daniel Kahneman.

- Representativeness—The degree to which an event is either similar to its parent population or to the process that generated it.
- Adjustment and Anchoring—The common human tendency to rely too heavily on the first piece of information offered (the anchor) when making decisions.

There is also a tendency for people to have a desire for certainty. As a result, we are often overconfident in our judgements,[39] mostly due to having too much faith in the assumptions on which we base our evaluations. In evaluating this overconfidence, one study found that rather than 2% of true values falling outside the 98% confidence bounds, 20–50% did so.[40] Experts appear to be as subject to overconfidence as laypeople—good grounds for taking their predictions with a grain of salt. The implications of this are significant for risk assessment. First, although they are very useful we should not believe them uncritically; and second, they should be done with input from a variety of stakeholders with different backgrounds and perspectives.

The Social Amplification of Risk—The above factors contribute to what is called the social amplification of risk, a theory that was introduced in 1988.[41] This theory brought together a range of studies from disparate fields to create a common framework. The basic notion is that risk perception, and human response to it, is modulated by sets of dynamic social and psychological processes (such as blame, trust, prior attitudes, heuristics, and affective imagery) that result in some hazards becoming of increased concern within society (amplification), while others become of less of a concern (attenuation). Examples of the former are nuclear accidents and airplane crashes, while examples of the latter are cigarette smoking and automobile accidents. Figure 3.15 shows the factors involved in this process. The increasing use of social media and the internet are likely accentuating this amplification process.

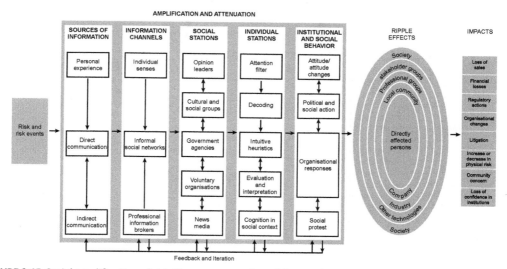

FIGURE 3.15 Social amplification of risk. There are a number of factors that can either increase or decrease perception of risk, which can also combine in complicated ways.[42] *Source: Slovic P., The Feeling of Risk: New Perspectives on Risk Perception (Routledge, 2010).*

Related to biases in risk perception and the social amplification of risk is the "white male effect" (Figure 3.16).[43] This US-based research shows that white males view a broad range of risks as being much less risky than females and nonwhite males do. The difference results from about 30% of white males who were better educated, wealthier, and politically conservative. They also tended to have more trust in institutions and authorities and to be antiegalitarian, particularly not approving of giving decision-making power to citizens in the areas of risk management. This group generally has far more power than others (in North America at least) in decisions related to risk creation[44] and also benefits more from its positive outcomes. I would expect similar findings in other parts of the world, except that instead of a white male effect, the risk differentiation would be exhibited by male elites in positions of power and wealth.

These findings are relevant to risk communication issues. Attention should be paid to what group is framing and communicating risks and who the audience is.

3.5.4 Risk as a Feeling

When one man dies it is a tragedy, when thousands die it's statistics.

Source disputed, but possibly Joseph Stalin.

The "risk as feeling" hypothesis first gained significant traction in 2001[45] and has subsequently received increasing attention as an important aspect of understanding people's perception and response to risk. There is compelling evidence that people have

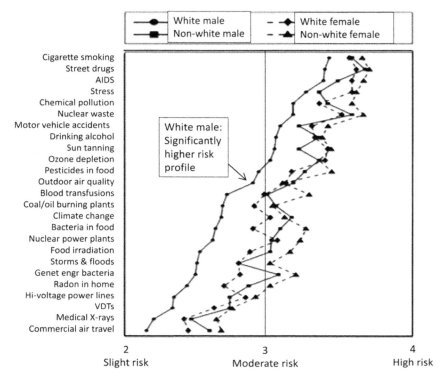

FIGURE 3.16 The white male effect. In the United States, about one-third of white males perceive risk very differently than women and nonwhite males, being much more risk tolerant. *Source: Flynn J., Slovic P., and Mertz C. K., "Gender, Race, and Perception of Environmental Health Risks,"* Risk Analysis 14, no. 6 (1994): 1101–1108.

Table 3.1 Two Modes of Thinking[47]

System 1: Experiential	System 2: Analytic
Holistic	Analytic
Affective: pleasure-pain oriented	Logical: reason oriented
Associationistic connections	Logical connections
Behavior mediated by "vibes" from past experiences	Behavior mediated by conscious appraisal of events
Encodes reality in concrete images, metaphors and narratives	Encodes reality in abstract symbols, words and numbers
More rapid processing: oriented toward immediate action	Slower processing: oriented toward delayed action
Self-evidently valid: "experiencing is believing"	Requires justification via logic and evidence

two modes of thinking and processing information,[46] one that is fast and automatic (System 1) and another that is slow and analytical (System 2). Table 3.1 summarizes the characteristics of these two systems.

These two systems are highly interactive and are both needed for rational decision making. System 1 is thought to be a much older system from an evolutionary perspective and is strongly associated with feelings of goodness or badness. System 2 is much more recent and largely differentiates *Homo sapiens* from other species (apologies to other intelligent species if I have misrepresented them).

System 1 thinking uses images, metaphors, and narratives to make risk estimations, all of which have feelings attached to them. Statistics, especially when expressed in percentages or as complex mathematics, are attached to System 2 thinking. There is evidence that people respond more strongly to System 1 than they do to System 2 in terms of how they perceive risk, which is why dread was such an important variable in Slovic's research into the psychometric paradigm. This has implications for risk communication; to be effective it should be presented using strategies that emphasize affect. One needs to be careful, however. Research by Rocky Lopes[48] has shown that too strong an affective response can lead to excessive denial and avoidance of risk, which reduce mitigative or preventative behaviors.

Slovic[49] notes that System 1 "works beautifully when our experience enables us to anticipate accurately how we will like the consequences of our decisions. It fails miserably when the consequences turn out to be much different in character than we anticipated."

STUDENT EXERCISE

- Compare your reactions to the following[50]:
  - In 2010, malaria caused an estimated 655,000 deaths (with an uncertainty range of 537,000 to 907,000), mostly among African children.
  - The death rate from malaria in Chad is 0.12%.
  - Maria, aged 2, died from malaria today; her mother was helpless to prevent it.
  - In Africa, one child dies from malaria every 30–60 s.
  - Seven jumbo jets full of children disappear every day because of malaria.

Research suggests that people respond most strongly to a single identifiable individual victim, but become increasingly insensitive to large events.[51] This is called psychosocial numbing and is the reason why nongovernmental organizations involved in disaster relief focus on individual victims, preferably children, in their fund-raising campaigns. Compare your reaction to Figures 3.17 and 3.18. In part the difference may be because our experiential memory associates emotional tags with individuals, but not with statistics.

This may also help explain our tendency to respond strongly to actual disasters while we pay little attention, relatively speaking, to the mitigation and prevention of future ones. The latter two strategies deal with events that may or may not happen to an individual and likely are not subject to strong emotional tags. A cognitive or rational approach would suggest that saving 1000 lives is a 100 times more valuable than saving 10 lives, yet that is not how we behave.

Affect is important to risk perception because it gives meaning to information, meaning that is needed for judgment and decision-making, and is therefore key to rational behavior.[52] As a student in university, I studied physics and for many years worked in a culture dominated by physical scientists. Reflecting back on that time,

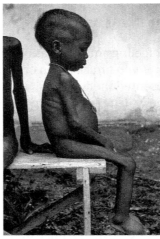

FIGURE 3.17 Starved girl, late 1960s, taken in a Nigerian relief camp. *Source: Centers for Disease Control and Prevention, Atlanta, Georgia, USA, Public Health Image Library (PHIL); ID: 6901,* http://phil.cdc.gov/.

FIGURE 3.18 Global Hunger Index for 2013.[53] *Source: von Grebmer et al. (2013). Reproduced with permission from the International Food Policy Research Institute.*

I believe the majority of physical scientists (being firmly rooted in a rational framework) tend to be bewildered as to why their arguments, based on science, facts, and statistics, may not sway others to their opinions. The dominant strategy that has been adopted by this culture to convince others is through education; that is, providing non-scientists with enough facts, statistics, and arguments to let their logical mind understand "truth" as the scientist perceives it. This is a naïve approach that generally does not work well for many risks.

3.6 The Risk Society

We are living in a world that is beyond controllability.

Ulrich Beck (Figure 3.19)

The risk society hypothesis, initially developed by Ulrich Beck[54] and Anthony Giddens, has gained traction in the field of sociology and created a very interesting debate about risk in postmodern society. To me, the most interesting notion underlying the debate is that society has been going through a transformation in terms of shared meanings, shared interests, and the construction and management of risk. It is this latter point that is most relevant to disaster theory. Beck sometimes uses the word risk in a hazard context and at other times from a vulnerability context.

People have always been exposed to a variety of hazards, mostly from natural causes or from other people (e.g., war). With the development of science, technology, and industry coupled with the juggernauts of economic growth and globalization, new and serious hazards have emerged. Some are local, but others, such as climate change, are global and ubiquitous. As a result, new risks are increasingly layered on top of the old ones. These new risks require society to reconfigure itself to deal with them; thus society is reflexive in terms of risk management. This reflexivity includes the development of new fields of study, new institutions (such as the Society for Risk Analysis or the Canadian Risk and Hazards Network), and new risk management professions such as disaster and emergency management.

Beck[55] develops a number of interesting thoughts around his thesis, including:

- The processes by which risks are generated and accumulate differ fundamentally from that of wealth, around which society has to a large degree previously organized itself. Wealth tends to accumulate at the top of society, while risks accumulate at the bottom (the two are closely related). These risks are, in Beck's words, "systematic and often cause irreversible harm."
- Risks are unevenly distributed. There are both losers and winners, and the winners are mainly big business. There is a systematic attraction between extreme poverty and extreme risk.
- Dealing with risks is fundamentally a political process and can only be meaningfully discussed within an ethical framework.

FIGURE 3.19 Ulrich Beck, author of *Risk Society: Towards a New Modernity. Source: Wikipedia.* http://en.wikipedia.org/wiki/File:Beck,Ulrich_2007.jpg.

- Risks display a "social boomerang effect"; even the rich and powerful are not safe from them, although wealth can buy various degrees of safety from some hazards.
- "Through the unrestrained production of modernization risks, a policy of making the Earth uninhabitable is being conducted in continuing leaps and bounds."

There is an irony about the risk society because it uses the same tools that cause problems to try to solve them. Albert Einstein once said that "We cannot solve our problems with the same thinking we used when we created them." Without a doubt technology can help to reduce disaster risk, but if Einstein was correct, then as an all-encompassing strategy it is based on a dangerous assumption. There are strong arguments that solutions should be largely rooted in philosophical or ethical reasoning.[56] Beck's work emphasizes the structural causes of risk as opposed to the more common tactical approach, and this is very useful. If one were to put the risk society into the PAR model, it would fit into the root causes section, where social structures contribute to increased vulnerability and the creation of hazards.

3.7 Measuring Risk

Quantifying risk is a very challenging task since aspects of it vary between communities/cultures and involve subjective and sometimes arbitrary decisions (such as the relative contributions of gender issues, poverty, and corruption to vulnerability) that are often evaluated based on expert judgment and stakeholder input. There is no single right answer because community values necessarily form the basis for many decisions. Within this framework, the notion of risk itself is ill defined, in that there is no unique set of axiomatic relationships that define it.

It is worth noting at this point that there are different risk management strategies available to societies, strategies which depend on the nature of hazard.[57] Traditional strategies based on quantitative risk estimation work well when systems are well defined, well understood, and stakeholder agreement exists in terms of management tactics. In the case of climate change, often described as a wicked problem, such strategies become problematical (depending on the specific hazard being considered). Other approaches, such as the precautionary principle or discursive methods, may be more relevant.

There are a number of methods used to estimate or measure disaster risk, ranging from techniques that are very subjective and based on limited data such as expert elicitation or the Delphi Method, to more objective data-heavy models such as those used by climatologists, geophysicists, economists and engineers. Three of the more common methodologies are discussed in the following sections.

3.7.1 Methodology 1: An Indices Approach

The first methodology is to create sets of dimensionless indices focusing on hazard and vulnerability, which can be combined in various ways to represent risk. If the same methodology is used for each hazard and location, then the results are comparable. This sort of approach has been used, for example, by Cutter[58] for vulnerability and by Carreno[59] and Cardona[60] for risk. Like the Human Development Index[a] that combines life expectancy, education, and income into a single index, this approach incorporates both hazard and vulnerability so that it can provide a useful relative measure over time or between entities. But it can also be difficult to interpret, since it involves adding apples and oranges, so to speak. An important advantage to this approach is that it can incorporate many relevant social variables for which quantitative data is very hard to obtain or can only be estimated, and where the importance of different factors must be subjectively assessed. Also, it uses a relatively straightforward mathematical process.

One type of index is based on historical records of damage. This approach suffers from several biases, as discussed in Chapter 2 on disaster data. A second type of index avoids historical events, but instead is based on various measurable physical, environmental, and social elements that are considered to be contributors to risk.[61] Examples of variables used in these two types of indices include: damaged area, number of dead people, number of injured people, damage in water mains, damage in gas network, fallen lengths of power lines, electricity substations affected, slums-squatter neighborhoods, mortality rate, delinquency rate, social disparity index, population density, number of hospital beds, health human resources, public space, rescue and firemen manpower, development level, and emergency planning.

These approaches, as discussed in Birkman,[62] are useful for examining relative risk between different groups, entities, and geographical areas and to evaluate trends. They

[a] http://hdr.undp.org.

can also be used to create profiles (as opposed to a single composite number) that can be used to identify more or most vulnerable elements.

This methodology also has limitations, such as:

- The variables/indicators represent approximations of vulnerability, resilience, and capacity and frequently are not independent of each other. They are often chosen because of data availability or for intuitive reasons with the assumption that they correlate to some aspect of vulnerability. These two issues create a fundamental limitation that makes the process mathematically unclear, particularly since levels of correlations between variables are often unknown.
- The assignment of weights to different variables is mostly subjective and in many cases varies according to local culture. It is unclear how robust they are (they are probably not at all robust).
- It can be difficult to say what averaged or summed numbers mean; output indices mainly have meaning in a comparative sense.
- In the absence of an engineering approach, risk cannot be robustly integrated over a range of hazard magnitudes, but can only be estimated for specified magnitude levels.
- Important information is lost in the averaging process. In particular, case studies have demonstrated that in many disasters the existence of a single critical variable and/or critical threshold dominates outcomes. This process is not captured by this methodology (at least the way it has been applied in studies published so far) and therefore one must question how realistically it can represent disaster risk from a conceptual viewpoint.

3.7.1.1 Examples of Indices Studies

Cutter[63] developed a social vulnerability index at a county level for the United States, based on 11 factors that accounted for 76% of the variance. The factors included various infrastructure and social factors, including race. An additive model was used to accumulate vulnerability, with equal weights being given to the various factors due to the difficulty and subjectivity of assigning weights. Some factors of relevance to the Cutter study, such as racial background, are not relevant to other cultures.

The United Nations Development Programme (UNDP) has developed a Disaster Risk Index,[64] which by setting risk values for the period 1980–2000 can be used as a reference for estimating future trends in risk. It is based on Emergency Disasters Database (EM-DAT) fatality data from droughts, earthquakes, tropical cyclones, and floods as a proxy for risk and operates at a country scale. Risk is defined as a multiplicative function: Risk = (hazard frequency) × (population) × (vulnerability). Vulnerability is quantified using an indicators approach involving 32 socioeconomic and environmental variables, with risk being calculated via a correlation analysis. This is a very restrictive proxy for risk that excludes economic and most social impacts.

A Global Urban Risk Index[65] for cities with populations more than 100,000 used a multiplicative definition of risk based on hazards (earthquakes, volcanoes, landslides, floods, and cyclones), exposed elements (city population and gross domestic product), and vulnerability (EM-DAT death and economic loss data). Empirical data are used as a proxy to estimate vulnerability, similar to the UNDP approach for risk. Note that EM-DAT data for economic loss are particularly suspect and should be used with caution.[66]

The Environmental Vulnerability Index[67] (EVI) works at a country scale and uses an additive approach based on 32 indicators of hazards, eight of resistance and 10 of damage. It is designed to be used with economic and social vulnerability indices to provide insights into the processes that can negatively influence the sustainable development of countries. The hazard indicators relate to the frequency and intensity of hazardous events. The resistance indicators refer to the inherent characteristics of a country that would tend to make it more or less able to cope with natural and anthropogenic hazards. Damage indicators relate to the vulnerability that has been acquired through the loss of ecological integrity or increasing levels of degradation of ecosystems. All of the EVI's indicators are transformed to a common scale so that they can be combined by averaging and to facilitate the setting of thresholds of vulnerability.

The Central American Probabilistic Risk Assessment[68] (CAPRA) is based on four subindices:

- "The Disaster Deficit Index measures country risk from a macroeconomic and financial perspective according to possible catastrophic events. It requires the estimation of critical impacts during a given period of exposure, as well as the country's financial ability to cope with the situation."
- "The Local Disaster Index identifies the social and environmental risks resulting from more recurrent lower level events (which are often chronic at the local and subnational levels). These events have a disproportionate impact on more socially and economically vulnerable populations, and have highly damaging impacts on national development." This index is additive in nature and is based on empirical records of deaths, people affected, and losses at a municipal level. It uses historical data to estimate the disaster index.
- "The Prevalent Vulnerability Index is made up of a series of indicators that characterize prevalent vulnerability conditions reflected in exposure in prone areas, socioeconomic weaknesses and lack of social resilience in general." This index is additive and is based on a set of weighted environmental, social, and resilience indicators.
- "The Risk Management Index brings together a group of indicators that measure a country's risk management performance. These indicators reflect the organizational, development, capacity and institutional actions taken to reduce vulnerability and losses, to prepare for crisis and to recover efficiently from disasters."

Rygel et al.[69] constructed a social vulnerability index to storm surge using a composite index of poverty, gender, race, ethnicity, age, and disabilities. They attempted to avoid issues related to assigning weights through the use of principal component analysis of census data and Pareto ranking. They ended up with three components accounting for 51% of the variance in the data in their case study. Note that this method is case specific, because different geographical regions would have different vulnerability components.

This approach requires a different set of indices for every location, using a principal component analysis.

Hahn et al.[70] discuss the CARE Livelihood Vulnerability Index as applied to climate variability and change in Mozambique using sociodemographics of livelihood, social networks, health, food and water security, natural disasters, and climate variability. From this they develop a composite vulnerability index through an additive approach in which all factors are equally weighted. The unit of analysis is household data from local surveys.

Simpson,[71] in a thorough review of issues related to the creation of indices, proposes a Disaster Resiliency/Preparedness Index that combines hazard probability and frequency with various measures of vulnerability in a multiplicative formula. The proposed methodology for constructing vulnerability is additive and uses weights. A useful list of possible indicators is listed in an appendix. Its use of the word vulnerability is very similar to my usage of the word risk.

The World Wildlife Federation[72] ranked climate vulnerability for a set of coastal cities in Asia using exposure, sensitivity, and adaptive capacity. These three categories were then averaged to calculate an overall score. "Exposure is the average of the three highlighted environmental categories including the susceptibility of the city to be impacted by 1 m sea-level rise, historical frequency of extreme weather events, including flooding and drought, and frequency of tropical storms and surges. Sensitivity is based on population, GDP, and the relative importance of that city to the national economy. Adaptive capacity is calculated by examining the overall willingness of the city to implement adaptation strategies (calculated by the available adaptation examples and/or responses to previous impacts) and the per capita GDP."

German Watch[b] publishes an annual climate risk index based on historical socioeconomic impacts of extreme weather events, not including fatalities and number affected. It only addresses direct impact and operates at a country scale. It is empirically based, using economic impact data, and is not an appropriate approach to calculating a broad based disaster risk index.

Creating a methodology that reflects disaster risk in a robust and broad-based way is complicated, perhaps impossible. An accurate, robust process that only captures a thin slice of disaster risk will misrepresent it from a social utility perspective. This is the trap in using an overly technocratic approach, and one that has been consistently critiqued in the academic literature over the past few decades.[73] The trend toward the disaster risk reduction philosophy (which emphasizes community engagement and the importance of local stakeholder participation in defining community risk) adopted by the United Nations International Strategy for Disaster Reduction reflects this concern. An all-encompassing methodology that includes the notion of risk as being socially constructed suffers from a lack of robustness. The important point to emphasize here is that it is not possible to create a broad-based methodology that is objective and without arbitrary/subjective components. There is no perfect answer to how risk should be assessed; tradeoffs are required to create an optimal strategy.

[b] http://www.germanwatch.org/klima/cri.htm.

3.7.1.2 Issues in Risk Indices

Subjectivity—There is an inherently subjective aspect to an indice-based risk assessment that is unavoidable. In part, this is because risk is socially constructed. However, it also exists because data are sometimes missing or insufficient, and therefore potentially important variables must either be excluded or estimated using expert judgment. Community input is essential to estimate local vulnerability, not just because of the local knowledge that resides there, but also since a risk estimate must reflect local values (for example, how important are issues related to poverty as compared to building damage?). Methodologies have weighting functions that can be used to assign what users consider to be appropriate levels of importance. Disagreement on exactly how variables should be weighted is inevitable.

Representativeness—Disaster risk is an amorphous concept. Beginning with the definition of disaster,[74] the very thing we are attempting to measure is not well defined. Although there is a common understanding in the literature that disaster is related to a range of hazard frequencies and intensities and to a large number of variables that are related to vulnerability (many of which are cultural), there is no universally accepted definition. Also, it is inevitable that many vulnerability variables will be correlated with each other (for example, poverty, housing quality, and gender). It becomes very difficult, therefore, to ascribe a specific meaning to a generated index.

As calculation of risk accumulates, eventually ending in a single number, information tends to be lost. The greatest usefulness of an index is in its application to relative levels of risk, between hazards and between locations. Risk therefore should be presented in several ways to provide users with options in terms of how they can use information. These include risk levels subdivided by hazard and vulnerability, spatial distribution of risk, average risk, number of elements exceeding a risk threshold, and risk as a function of a defined level of hazard (probably a critical threshold).

Robustness—Greater levels of robustness can be sought, but only at the cost of representativeness. It is a tradeoff—by reducing the number of variables and parameters (especially those that are more subjective or hard to quantify), calculations become more robust but less relevant to local community culture and characteristics and therefore less reflective of disaster risk.[75] A model that includes the many different faces of vulnerability is less robust. I expect this can be evaluated to some extent by applying the model to different types of communities.

3.7.2 Methodology 2: An Engineering Approach

The second methodology is more robust and objective in nature, but also more limited. It involves combining empirical data and hazard models to mathematically represent hazard frequency and intensity, and to then combine that information with surveys of vulnerable conditions. Such a survey would include building stock data, infrastructure type and amount, as well as their locations relative to hazard. Using functions that relate hazard intensity to damage (called damage ratio curves, loss functions, or fragility curves), one can then estimate risk by integrating the product of these two curves over the range of hazard. This is similar to what the insurance industry catastrophe (CAT) models do (see Section 6.5.6) and is part of what the CAPRA project does. In many ways it is more satisfying than the first methodology, in that it is less subjective and arbitrary and does not depend on local cultural perspectives. Various consulting firms (such as Risk Management Solutions (RMS) and EQECAT) have spent many years and large amounts of resources developing these models.

Limitations to this methodology are:

- It is more data dependent than the first methodology. Very detailed surveys of building types are required, as well as a complete set of fragility curves, which are very sensitive to type and quality of construction.
- Important social, cultural, institutional, economic, and environmental variables are impossible to include.
- These CAT models are very sensitive to initial assumptions,[76] and even small errors in measuring wind speed can result in variations of damage estimates of up to 100%.[77] Uncertainties in city scale damage estimates can have confidence intervals as much as 1–100% of probable maximum loss.[78] Unless there are very accurate scientific estimates of hazard functions, inventories of building stock, and type and quality of construction, damage outputs from these models must be considered rough estimates.

Lloyd's has produced a number of disaster scenarios[79] with estimates of insurance losses. For example, one scenario of a Florida windstorm landing in Pinellas County (Figure 3.20) results in estimated insurance losses of:

- Residential property, $88 billion US
- Commercial property, $37 billion US
- Auto, $2 billion US
- Marine, $1 billion US

FEMA has developed a methodology for estimating potential losses from earthquake, wind, and flood disasters similar to the CAT models, using a model called Hazard US (HAZUS) (Figure 3.21).[80] Natural Resources Canada is in the process of converting this model for use in Canada[81] so that in the future it may become a valuable tool for Canadians. Loss estimates from HAZUS include:

- Physical damage to residential and commercial buildings, schools, critical facilities, and infrastructure;
- Economic loss, including lost jobs, business interruptions, and repair and reconstruction costs; and
- Social impacts, including estimates of shelter requirements, displaced households, and population exposed to scenario floods, earthquakes, and hurricanes.

The results can be presented graphically using a Geographic Information System (see Figure 3.22 for an example). There is no cost for the use of HAZUS.

FIGURE 3.20 An example of a Geographic Information System-based hazard impact analysis. Footprint and damage levels for Pinellas windstorm event. *Source: Llyyd's (2012). Realistic Disaster Scenarios: Scenario Specification. January, Florida windstorm–event two. www.lloyds.com.*

3.7.3 Methodology 3: A Case Study Approach

The third approach uses case studies and scenarios. Such studies typically include some or all of the following: data on both hazard and vulnerability; descriptions of past disaster events; an analysis of the social, political, and environmental causes of the impacts; and recommendations for risk reduction. The case study approach can complement the first two methodologies and is usually based on a historical or hypothetical event. In many ways, this is the approach that provides the richest amount of information because it can include a mix of quantitative and qualitative analysis, including cultural and personal

86 DISASTER THEORY

FIGURE 3.21 HAZUS methodology. *Source: FEMA IS-922—Applications of GIS for Emergency Management Lesson 1: Introduction and Course Overview* http://emilms.fema.gov/is922/GISsummary.htm.

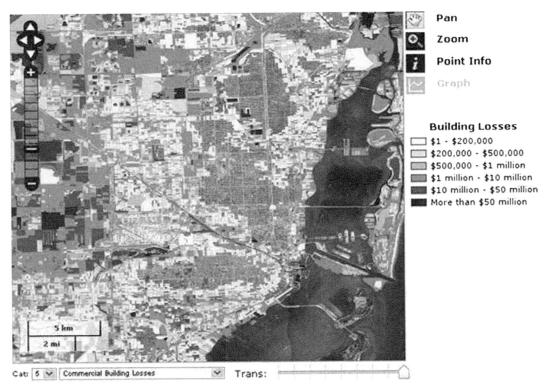

FIGURE 3.22 Screenshot of the IMMAGE HAZUS Surge Loss app, showing the distribution of forecast losses to commercial real estate in the Miami area for a Category 5 hurricane. *Source: USGS,* http://sofia.usgs.gov/proposals/2012/images/immage6x.jpg.

narratives. It is also the one that is most difficult to use for a quantitative comparison between locations and hazards.

There are many studies that use this approach. Three of my favorites that use historical perspectives are *Crucibles of Hazard: Mega-Cities and Disasters in Transition*[82] by James K. Mitchell, *A new species of trouble: explorations in disaster, trauma, and community*[83] by Kai Erikson, and *The angry earth: disaster in anthropological perspective*[84] by Anthony Oliver-Smith and Susanna Hoffman. These three books are thoughtful, insightful, and should be considered important readings for all students of disaster. There are, of course, many others; a more recent report with a technical as opposed to a social perspective is *Natural Disaster Hotspots: Case Studies.*[85] Some disasters have generated a plethora of studies, for example Hurricane Katrina, the 2004 Asian tsunami, the European 2003 heat wave, the 1998 ice storm in Canada, the 2010 Haiti earthquake, and the 2011 Japanese tsunami/nuclear disaster.

3.8 Sea Level Rise and Subsidence

Sea Level Rise—Some risks appear seemingly out of nowhere, because they were completely unanticipated. Others are within our radar but are perceived as something that is so unlikely to occur that it can be ignored. There is a category of risks, however, which are inevitable, and, like a train coming down the tracks, we can see the disaster approaching. Sea level rise, for many communities, lies within this latter category.

Global climate change is expected to affect many climate related hazards in the future. One of its most important impacts is going to be on sea level, something that has varied greatly over geological time. Sea level has not only been rising during the past century, but also is expected to continue to rise, and this poses a great problem for many low lying islands and coastal areas. The causes of sea level rise are: (1) thermal expansion of the oceans due to warming, (2) melting land ice (glaciers, Greenland, and Antarctica), and (3) changes in land water storage (snow pack, rivers, lakes, man-made reservoirs, wetland, and aquifers). From 1993 to 2007, about 30% of the rise was due to thermal expansion and about 55% due to melting land ice. Since 2007, the contribution of the latter has increased to about 80%.[86]

The most authoritative source of information on climate change is the Intergovernmental Panel on Climate Change, which publishes a summary report of peer-reviewed literature on the topic every five years. They found that from 1993 to 2010 seal levels rose by about 3.2 mm/year. These rates are expected to increase in the future, so that by the end of the 21st century sea levels may have increased by up to 1 m, depending on which population and economic growth scenario is considered (Figure 3.23).[87] Coastal sea level changes are not uniform globally, but vary regionally for a variety of reasons, such as different rates of land subsidence or uplift and changing wind patterns. Climate models, however, differ significantly in predictions of local variations, making regional generalizations extremely difficult. These projections do not include the possibility of a collapse of the Antarctic ice sheet, which would result in a rapid and catastrophic rise in sea levels. Recent studies using more rapid

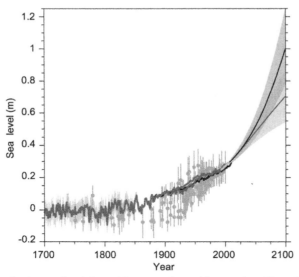

FIGURE 3.23 Compilation of paleo sea level data, tide gauge data, altimeter data (*from Figure 13.3*), and central estimates and likely ranges for projections of global mean sea level rise for RCP2.6 (blue) and RCP8.5 (red) scenarios, all relative to preindustrial values. (For interpretation of the references to color in this figure legend, the reader is referred to the online version of this book.) *Source: Stocker D. Q. (2013). Climate change 2013: the physical science basis. Working Group I Contribution to the Fifth Assessment Report of the Intergovernmental Panel on Climate Change, Summary for Policymakers, Figure 13.27.*

estimates of ice sheet melting suggest that older estimates of sea level rise may be too small[88] and that sea levels will rise by up to 2 m by 2100, although there is not scientific consensus on this issue.

Subsidence—Subsidence can occur for a number of reasons, including earthquakes, the extraction of groundwater[89] and hydrocarbons, and increasing loads due to sedimentation or urban development. Many cities have experienced subsidence over the past century, including Tokyo, Shanghai, and Bangkok. Where this occurs in coastal zones, the hazard of sea level rise is greatly amplified.

Consider the case of Bangkok, a coastal megacity with a population more than nine million. The combination of sea level rise with subsidence seems likely to create a particularly difficult situation for this city. For Bangkok, global averages of sea level rise are considered to be a good estimator, and subsidence[90] (Figures 3.24 and 3.25) has been a significant problem as well; over a 20 year period it ranged from 400 to 800 mm.

Combining the projected rates of sea level rise with an extrapolation of historical subsidence rates suggests a total drop of between 0.6 and 1.2 m by the middle of this century. Figure 3.25, shows population density within the 10 m elevation zone for Thailand. The greatest densities of people live in low elevation areas, and in Bangkok much of the city is below

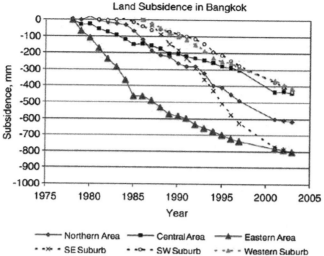

FIGURE 3.24 Land subsidence versus time in different areas of Bangkok. *Source: Phien-wej N., Giao P. H., Nutalaya P., "Land Subsidence in Bangkok, Thailand,"* Engineering Geology *82, (2006): 187–201.*

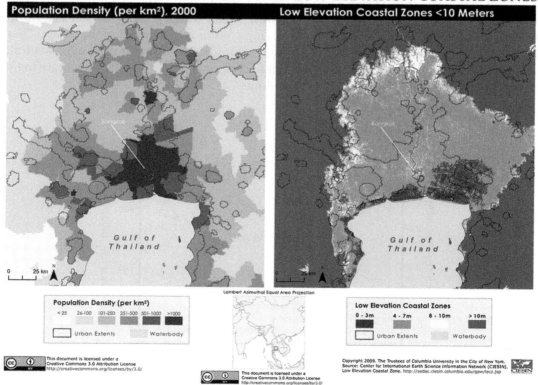

FIGURE 3.25 Population density and land elevations in Bangkok. Note the high population densities in areas below 3 m elevation. By the year 2100, sea level rise combined with subsidence may create serious flooding problems for much of the city. *Source: Center for International Earth Science Information Network, Low Elevation Coastal Zone.*

1.0 m, suggesting that this is an extremely serious long-term problem. Actual changes in elevation will greatly depend on how Bangkok deals with ground water extraction in the coming decades.

Coastal Cities—There are many low lying coastal areas with large populations and infrastructure that are going to become increasingly at risk in the future due to the combined effects of sea level rise, subsidence, and urban growth. Particularly vulnerable are some of the large river deltas in Asia such as the Ganges-Brahmaputra, the Yangtze, and the Mekong, which include some of the world's largest megacities. Significant rates of subsidence have been measured in such cities as Manila, Shanghai, and Tianjin City. In Manila (Figure 3.26), subsidence from groundwater overuse is one to two orders of magnitude greater than sea level rise.

The importance of sea level rise can be put into context by noting that 15 of the world's 20 mega cites are coastal and have rapidly growing populations. Also, although less populated, many low lying small island developing states such as the Maldives will eventually be swamped by sea level rise.[91]

Climate change seems likely to alter many of the hazards we are familiar with, such as flood, drought, and heat waves. It also seems very likely to result in a rise in sea levels, rapid on geological time scales if somewhat slow by human standards. This trend, when combined with subsidence and urban growth, will eventually create a "perfect storm" of hazards that will drastically impact low lying coastal regions on a global scale. There are several adaptation options available, such as protection through the use of sea walls (expensive, and you better hope they will not fail!), retreat, or altering city structure to live with higher water levels, such as Venice has done.

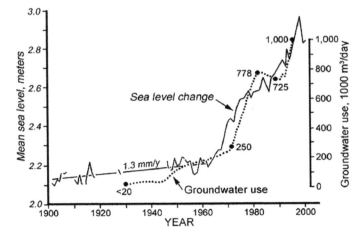

FIGURE 3.26 Sea level rise in Manila. *Source: Rodolfo K. S., and Diringan F. P., "Global Sea-Level Rise Is Recognised, but Flooding from Anthropogenic Land Subsidence Is Ignored Around Northern Manila Bay, Philippines," Disasters 30, no. 1 (2006): 118–139.*

The following data helps put the annual economic costs of climate change in perspective[92]:

- $10 billion: Current estimate of what is available annually for public funding of climate change support to developing countries (the mitigation and adaptation share is about $1 billion)
- $100 billion: The most conservative estimate of what is required for adaptation
- $1 trillion: A conservative estimate of the amounts of public funding available for harmful practices: subsidies for fossil fuels, water practices that deplete resources, fisheries and agriculture. A recent meeting at the International Monetary Fund upgraded the number to $2 trillion.

Further Reading

Fuchs RJ: Cities at risk: Asia's coastal cities in an age of climate change, *Analysis from the East-West Center* No. 96, 2010. July 2010.

Füssel H-M: An updated assessment of the risks from climate change, based on research published since the IPCC Fourth Assessment Report, *Climatic Change* 97:469–482, 2009.

Jevrejeva S, Moore JC, Grinsted A: How will sea level respond to changes in natural and anthropogenic forcings by 2100?, *Geophysical research letters* 37(L07703), 2010. http://dx.doi.org/10.1029/2010GL042947. 5 PP., 0.

Nicholls RJ, Cazenave A: Sea-level rise and its impact on coastal zones, *Science* 328(5985):1517–1520, 2010.

Phien-wej N, Giao PH, Nutalaya P: Land subsidence in Bangkok, *Thailand. Engineering Geology* 82:187–201, 2006. Scopus Exact.

Rahmstorf S: Sea-Level Rise: A Semi-Empirical Approach to Projecting Future, *Science* 315:368–370, 2007.

3.9 Summary

As an overarching framework, risk theory is very useful to the study of disasters. It has been approached in different ways, depending on the discipline in which the theory has been developed, the main ones being sociology, psychology, and the physical sciences including engineering. This has led to a historical clash of paradigms, one viewing risk as a social construct and the other viewing it as objective and measurable. This book takes a middle approach to this tension. Estimating risk can be done through quantitative and/or qualitative approaches, but each has methodological problems associated with it. Which method is best to use depends on the purpose to which it will be put and must be evaluated carefully.

3.10 Case Study: 1998 Ice Storm in Eastern Canada and Northeastern United States

…we had come to realize that there existed in this place at this time a web of support that many people thought had long since withered away from disuse. While the fruits of the power grid may make it seem as though each of us can live an autonomous life, we

learned that that is an illusion. It is family and community, not global networks, that truly sustain us.[93]

Stephen Doheny-Farina

The 1998 ice storm disaster is interesting from several perspectives. On one hand, it reveals our massive dependence on large scale critical infrastructure. An ice storm matters much more today than it did in the past, before we relied on electrical energy grids, when farmers milked their cows by hand and homes were heated with stoves and fires. At the same time, however, it shows the importance of local social networks and how family and community sustain people in times of need. It also highlights the importance of creating systems that are resilient. Especially there is risk inherent in centralized systems that lack sufficient redundancy and safe fail mechanisms. The 1998 ice storm, like so many other natural disasters that have affected people over time, shows how extreme natural occurrences can overwhelm society and destroy assumptions of normality and how people tend to band together to help each other through difficult times.

Freezing rain forms just north of warm fronts, where a temperature inversion creates an above-freezing layer in the atmosphere that is sandwiched between colder air aloft and below (Figure 3.27). Beginning as snow aloft, melting occurs in the warm layer, and the snow becomes rain. Below the warm layer, however, water droplets become supercooled in the colder air near the surface and then freeze on contact with the ground surface. Figure 3.28 shows the weather map on January 6, 1998. A quasistationary frontal zone extended from Texas toward the southern Great Lakes, and then eastward through New England. North of the front was a cold airmass, while to the south was a much warmer one. Freezing rain occurred north of the frontal zone, where warm air overlay the cold arctic air.

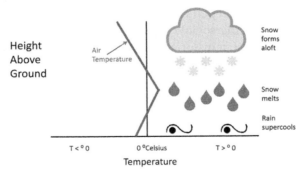

FIGURE 3.27 Formation of freezing rain. Snow forms aloft in cool and then melts as it falls into air that is above freezing. Near the surface underneath a frontal discontinuity, it falls through cold air again and supercools, subsequently freezing on contact with the ground. The reason for the temperature profile is a frontal system to the south, which creates warm air aloft while leaving colder arctic air near the surface.

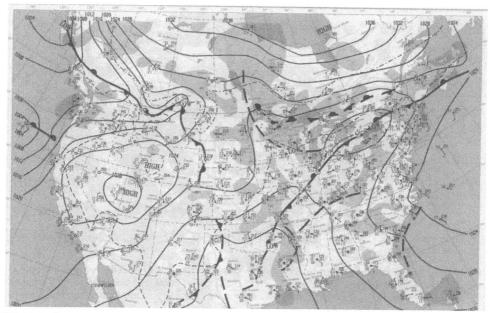

FIGURE 3.28 Surface weather map, January 6, 1998. A quasistationary front running east-west just south of Montreal results in extended periods of freezing rain. *Source: NOAA,* http://www.erh.noaa.gov/btv/events/IceStorm1998/ice98.shtml.

The 1998 ice storm began on January 4 and continued until January 10, with large amounts of freezing rain falling in several waves during that period. Normally freezing rain is brief. In this case, it was anything but brief due to a blocking pattern that had been set up in the atmosphere, a pattern that was related to an El Nino event.[94] Sometimes the atmosphere configures itself so that weather systems move rapidly; this occurs when the steering flow in the upper troposphere has a strong west to east orientation (this is called a zonal flow). At other times, the configuration has a strong south to north orientation (meridional flow) and low and high pressure systems extend up to the jet stream in what is called a blocking pattern. These patterns can remain quasistationary for weeks at a time and result in extended periods of good or bad weather in regions under their influence. This particular atmospheric configuration created a cold high pressure area in Quebec with warm overrunning moist air that originated from the Gulf of Mexico, an ideal scenario for the formation of freezing rain.

This storm, which impacted eastern Canada and the northeastern United States, was unique in historical weather records in several respects: the large amount of ice that formed on power lines over a large area for a long period, the number of people who lost power, the number of insurance claims made, and its economic impact in Canada (being Canada's costliest natural disaster until that date).

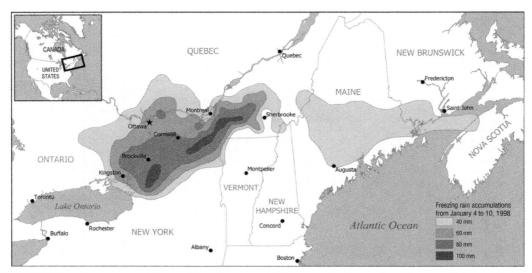

FIGURE 3.29 Very large accumulations of freezing rain, which occurred to the north of the quasistationary front. Fortunately, the highest accumulations occurred south of the large urban areas. Had the maximum amount of freezing rain been somewhat further north, the disaster that affected eastern Ontario and southwestern Quebec, including the major cities of Montreal and Ottawa, would have been much worse. *Source: Environment Canada via Wikipedia.* http://en.wikipedia.org/wiki/File:1998_Ice_Storm_map.png.

Freezing rain events are a relatively common winter occurrence in eastern Canada and northeastern United States. Using weather data from the St. Lawrence River Valley, a report by the US Army Corps of Engineers[95] identified 61 potentially damaging storm events over the period of historical observations. Significant ice storms in the valley have occurred in 1913, 1942, 1961, 1972, 1973, 1983, 1996, and 2000.

Ice accumulations from the storm are shown in Figure 3.29,[96] and when compared to the electrical grid map around the city of Montreal (Figure 3.30) show an interesting pattern. More than 75 mm of ice accumulations on electrical conductors and pylons designed to withstand 45 mm of ice resulted in 116 transmission line failures. Significantly, there was not a warming period during this event, which would have allowed the ice to melt. Wind was also a significant factor in failure, particularly when standing wave patterns on the wires are set up in the presence of moderate to strong winds; these waves resulted in large forces that contributed to failure. Only one power line remained operational, likely because it was in an area of lower freezing rain amounts compared to areas to the east and west. This seems to be because a wedge of cold air at the surface in that region caused the precipitation to fall as ice pellets instead of freezing rain, thus reducing ice accumulations on the power grid. Had this not been the case, it may well be that Montreal would have lost all power, resulting in a crisis of much greater proportions. In Maine, most of the power lines failed due to falling trees and branches. It should be noted that there were redundancies built into the high voltage power grid, but they were not designed for a threat of this level.

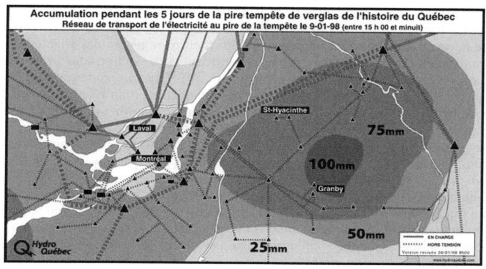

FIGURE 3.30 Freezing rain accumulations during the five days of the 1998 ice storm, in the vicinity of Montreal. The map shows the electrical grid near total collapse on January 9 (between 3 pm and midnight). The green lines are functioning electrical trunk lines; the black triangles are main transformer stations. The dotted red lines are nonfunctioning electrical trunk lines. The blue areas show the average accumulated measurements of ice in mm. (For interpretation of the references to color in this figure legend, the reader is referred to the online version of this book.) *Source: Hydro Quebec,* http://www.imiuru.com/icestormdiary/1pages/HydroQuebecMap.html.

There was a domino effect when some power lines failed, with one collapsing tower pulling others down with it. As Hydro Quebec rebuilt the lines, they made efforts to increase the resistance of the lines by strengthening pylons and power poles and increasing power supply. Every 10th tower is now a "very robust anti-cascading tower." Also, Hydro Quebec has increased redundancy in the system to maintain sources of supply in case of line failure.[97]

The impacts of the storm were massive. There were about 74 deaths from the storm overall (12 in the United States due to flooding and 16 related to other causes, and 46 in Canada from trauma, carbon monoxide (CO) poisoning, fire, hypothermia, and hazardous activities), numerous injuries, millions of trees destroyed, and damage to 1000 transmission towers, 30,000 utility poles, and tens of thousands of km of wires and cables. The ice caused numerous falls that resulted in injuries. CO poisoning was also a risk. In the United States, more than 500 people each year die from CO exposure, and in one study more than 100 cases were identified in the state of Maine. The greatest risk factor related to this was having a generator in a basement or garage. More than 4.7 million people lost power in Canada, some for up to a month, while about 500,000 lost power in the United States. Almost 800,000 insurance claims were filed in Canada and about 140,000 in the United States, resulting in a total insured loss of US$1.3 billion.[98] Also, a class action suit was filed for additional living expenses against 19 insurance companies. As of October 23, 2012, settlement had been reached with four companies, paying out an additional $12.5 million.

Those with generators were better able to deal with the power outage, although fuel was often hard to come by. Ironically, areas with more reliable power systems suffered more because they had fewer customers with their own generators, highlighting the dynamic relationship between vulnerability and resilience (Section 4.6.1). The response to the disaster required the heavy involvement of utility companies, and in Canada the military also played an important role, deploying more than 16,000 personnel (the largest deployment since the Korean War). Assistance from other provinces and states helped with the response, but it was clearly a massive job. Some communities regained power within days, but as previously mentioned, others were without power for up to a month. As a result, there was substantial of criticism of Ontario Hydro and Quebec Hydro (much of which may or may not have been warranted), particularly with respect to their communication with customers and stakeholders. In general, communication was a huge issue in this disaster. The planning assumption of most emergency management plans is that communication is intact. In this case, that assumption failed as telephone and cell networks collapsed. The usefulness of devices like hand-cranked radios quickly became apparent.

Systems are not invulnerable, but they can be designed to recover more quickly and with less effort by building in redundancy and resilience to larger degrees. For example, if towers are designed with breakaway arms, then in the event of failure only the arms and wires need to be repaired. If the transmission systems are distributed instead of centralized, then it is much less likely that a failure will be widespread. The quote at the beginning of this case study makes an important point, but it errs when it says that only family and local community sustain us. We are sustained by interactive networks across a range of scales, from personal to global. Failure in any of these networks can have drastic consequences, and their relative importance varies according to many factors. Sometimes it is the small local networks that are critical; at other times it is large-scale ones. There is no general rule, except that it is important to understand our dependencies and to ensure that they are not excessive.

End Notes

1. For example, Etkin D., and Ho E., "Climate Change: Perceptions and Discourses of Risk," *Journal of Risk Research* 10, no. 5 (2007): 623–41.
2. Blanchard W., (2008). *Guide to Emergency Management and Related Terms, Definitions, Concepts, Acronyms, Organizations, Programs, Guidance, Executive Orders & Legislation.* A Tutorial on Emergency Management, Broadly Defined, Past and Present. http://training.fema.gov/EMIWeb/edu/docs/terms%20and%20definitions/Terms%20and%20Definitions.pdf.
3. Wisner B., Cannon T., Davis I., and Blaikie P. *At Risk: Natural Hazards, People's Vulnerability and Disasters* (2nd ed.), (London, UK: Routledge, 2004).
4. At times this equation has been misquoted as risk = hazard + vulnerability.
5. Ontario Ministry of Community Safety and Correctional Services, *Hazard Identification Risk Assessment* (2012). Retrieved from http://www.emergencymanagementontario.ca/english/emcommunity/ProvincialPrograms/hira/hira.html.

6. British Columbia Ministry of Public Safety and Solicitor General Provincial Emergency Program, *Hazard, Risk and Vulnerability Analysis Tool Kit* (2004). http://www.embc.gov.bc.ca/em/hrva/toolkit.pdf.
7. Hovden J., *Risk and Uncertainty Management Strategies*. Paper Presented at the 6th International CRN Expert Workshop. Stockholm, Sweden, April 22–24, 2004.
8. Ibid.
9. Etkin D., Higuchi K., and Medayle J., "Climate Change and Natural Disasters: An Exploration of the Issues," *Journal of Climate Change* 112, no. 3–4 (2011): 585–99.
10. It is worth pointing out that many subjective and qualitative factors can be measured and quantified, although there is valid debate over methods, metrics, and values.
11. See note 7 above.
12. Adams J., *Risk* (London: University College London Press, 1995).
13. Adams J., Cars, Cholera and Cows: The Management of Risk and Uncertainty. *Policy Analysis*, no. 335 (1999), 1–49.
14. About 11 tornadoes occur in southern Ontario per year, on average.
15. Mead E., Trust and Distrust in Environmental Risk Decision-Making: The Experience of Three Communities with Low-Level Radioactive Waste Management. Ph.D. thesis, Faculty of Environmental Studies (York University, 2007).
16. Yes.
17. The answer depends on where you live. In some places, if development is controlled, the answer is yes. In many places, the answer is no, because development exceeds the extra margins of safety provided by the structural mitigation or because the dikes have the effect of increasing flood risk downstream.
18. Robertson L. S., "Reducing Death on the Road: The Effects of Minimum Safety Standards, Publicized Crash Tests, Seat Belts, and Alcohol," *American Journal of Public Health* 86, no. 1 (1996): 31–4.
19. Ontario Ministry of Transportation, *Seat Belts* (2011), http://www.mto.gov.on.ca/english/safety/seatbelt.shtml.
20. Simonet S., and Wilde G. J., "Risk: Perception, Acceptance and Homeostasis," *Applied Psychology* 46, no. 3 (1997): 235–52.
21. Wilde G. J., *Target Risk* (Toronto: PDE Publications, 1994), 124–37.
22. Handmer J., "Are Flood Warnings Futile? Risk Communication in Emergencies," *The Australasian Journal of Disaster and Trauma Studies* 2 (2000): 2000–2.
23. There are many discussions in the literature on moral hazard. See, for example, Smith V. H., and Watts M., "The New Standing Disaster Program: A SURE Invitation to Moral Hazard Behavior," *Applied Economic Perspectives and Policy* 32, no. 1 (2010): 154–69.
24. Guy A., *Floods Devastate Britain* (The Toronto Star, A11, November 27, 2012).
25. Lehmann E., *Risk: National Flood Insurance Program—a Mighty Engine that Couldn't* (E & E Publishing, January 5, 2013), http://www.eenews.net/stories/1059974778.
26. Sandink D., Kovacs P., Oulahen G., and McGillivray G., *Making Flood Insurable for Canadian Homeowners: A Discussion Paper* (Toronto: Institute for Catastrophic Loss Reduction & Swiss Reinsurance Company Ltd, 2010).
27. Hausmann P., *Floods—an Insurable Risk?* (Zurich: Swiss Re, 1998).
28. Adaptation Sub-Committee Progress Report, *Adapting to Climate Change in the UK: Measuring Progress* (UK: Committee on Climate Change, 2011), http://archive.theccc.org.uk/aws/ASC/ASC%20Adaptation%20Report_print_spreads.pdf.

29. Slovic P., *The Feeling of Risk: New Perspectives on Risk Perception* (New York, NY: Earthscan, 2010).
30. Ibid.
31. Slovic P. E., *The Perception of Risk* (New York, NY: Earthscan Publications, 2000).
32. See note 7 above.
33. See note 12 above.
34. See note 12 above.
35. Simon H.A., "Bounded Rationality and Organizational Learning," *Organization Science* 2, no. 1 (1991): 125–34.
36. Lichtenstein S., Slovic P., Fischhoff B., Layman M., and Combs B., "Judged Frequency of Lethal Events," *Journal of Experimental Psychology: Human Learning and Memory* 4, no. 6 (1978): 551.
37. Tversky, and Kahneman, "Judgment Under Uncertainty: Heuristics and Biases," *Amos Tversky; Daniel Kahneman Science*, New Series 185, no. 4157 (September 27, 1974).
38. Kahneman Nobel Lecture.
39. See note 31 above.
40. See note 36 above, 551–78.
41. See note 29 above.
42. See note 29 above, 319.
43. Finucane M. L., Slovic P., Mertz C. K., Flynn J., and Satterfield T. A., "Gender, Race, and Perceived Risk: The 'White Male' Effect," *Health, Risk & Society* 2, no. 2 (2000): 159–72.
44. Related to this is that acceptable levels of risk are greater by about a factor of 1000, when they are voluntarily chosen, as compared to when they are imposed.
45. Loewenstein G. F., Weber E. U., Hsee C. K., and Welch N., "Risk as Feelings," *Psychological Bulletin* 127, no. 2 (2001): 267.
46. Stanovich K. E., and West R. F., "Individual Differences in Reasoning: Implications for the Rationality Debate?," *Behavioral and Brain Sciences* 23, no. 5 (2000): 645–65.
47. See note 29 above.
48. Lopes R., *Public Perception of Disaster Preparedness Presentations Using Disaster Damage Images* (Natural Hazards Research and Applications Information Center, Institute of Behavioral Science, University of Colorado, 1992).
49. Slovic P., Finucane M. L., Peters E., and MacGregor D. G., "Risk as Analysis and Risk as Feelings: Some Thoughts about Affect, Reason, Risk, and Rationality," *Risk Analysis* 24, no. 2 (2004): 311–22.
50. Sources: WMO Malaria fact sheet; www.worldlifeexpectancy.com, www.againstmalaria.com.
51. Slovic P., "The More Who Die, the Less We Care," in Michel-Kerjan E., and Slovic P., eds., *The Irrational Economist: Making Decisions in a Dangerous World* (New York: Public Affairs Press, 2009).
52. Slovic S., and Slovic P., "Numbers and Nerves: Toward an Affective Apprehension of Environmental Risk," *Whole Terrain* 13 (2004): 14–8.
53. von Grebmer, K., D. Headey, T. Olofinbiyi, D. Wiesmann, H. Fritschel, S. Yin, Y. Yohannes, C. Foley, C. von Oppeln, B. Iseli, C. Béné, and L. Haddad. 2013. 2013 Global Hunger Index — The Challenge of Hunger: Building Resilience to Achieve Food and Nutrition Security. Figure 2.4, "Global Hunger Index by Severity" map. Bonn, Germany: Welthungerhilfe; Washington, DC: International Food Policy Research Institute; Dublin, Ireland: Concern Worldwide.
54. Beck U., *World at Risk* (Cambridge, UK: Polity Press, 2007).
55. Ibid.

56. Mileti D., *Disasters by Design: A Reassessment of Natural Hazards in the United States* (Washington, DC: National Academies Press, 1999).
57. See note 7 above.
58. Cutter S. L., Boruff B. J., and Shirley W. L., "Social Vulnerability to Environmental Hazards," *Social Science Quarterly* 84, no. 2 (2003): 242–61.
59. Carreno M. L., Cardona O. D., and Barbat A. H., "A Disaster Risk Management Performance Index," *Natural Hazards* 41, no. 1 (2007): 1–20.
60. Cardona O. D., *Indicators of Disaster Risk and Risk Management.* Summary Report. IDB/IDEA Program on Indicators for Disaster Risk Management (Washington DC: Inter-American Development Bank, Sustainable Development Department Environment Division, 2005); Carreno M. L., Cardona O. D., and Barbat A. H., *Application and Robustness of the Holistic Approach for the Seismic Risk Evaluation of Megacities* (2008), in Innovation Practice Safety: Proceedings, The 14th World Conference on Earthquake Engineering October 12–17, 2008, Beijing, China.
61. These data suffer from various problems that can make it unreliable as an indicator of risk, including observational biases, lack of data, lack of common methodologies, and sampling error.
62. Birkmann J., "Indicators and Criteria for Measuring Vulnerability: Theoretical Bases and Requirements," in Birkmann J., ed., *Measuring Vulnerability to Natural Hazards: Towards Disaster Resilient Societies* (New York: United Nations Press, 2006), 55–78.
63. See note 57 above.
64. Peduzzi P., Dao H., Herold C., and Mouton F., "Assessing Global Exposure and Vulnerability towards Natural Hazards: The Disaster Risk Index," *Natural Hazards Earth System Science* 9 (2009): 1149–59.
65. Brecht H., Deichmann, and Wang H. G., *A Global Urban Risk Index*, Policy Research Working Paper 6506 (World Bank Urban and Disaster Risk Management Department, 2013), http://elibrary.worldbank.org/doi/pdf/10.1596/1813-9450-6506.
66. Guha-Sapir B., and Below R., *The Quality and Accuracy of Disaster Data: A Comparative Analyses of Three Global Data Sets* (Brussels, Belgium: The ProVention Consortium, The Disaster Management Facility, The World Bank, 2005).
67. Kaly U. L., Pratt C., and Mitchell J., *Building Resilience in SIDS: The Environmental Vulnerability Index*, Final Report (SOPAC, UNEP, 2005), http://www.vulnerabilityindex.net/EVI_Library.htm.
68. Cardona O. D., *Indicators of Disaster Risk and Risk Management.* Summary Report. IDB/IDEA Program on Indicators for Disaster Risk Management (Washington DC: Inter-American Development Bank, Sustainable Development Department Environment Division, 2005).
69. Rygel L., O'sullivan D., and Yarnal B., "A Method of Constructing a Social Vulnerability Index: An Application to Hurricane Storm Surges in a Developed Country," *Mitigation and Adaptation Strategies for Global Change* 11 (2006): 741–64.
70. Hahn M. B., Riederer A. M., and Foster S., "The Livelihood Vulnerability Index: Apragmatic Approach to Assessing Risks from Climate Variability and Change—A Case Study in Mozambique," *Global Environmental Change* 19, no. 1 (2009): 74–88.
71. Simpson D. M., (2006), Indicator Issues and Proposed Framework for a Disaster Preparedness Index (DPi), Draft Report to the: Fritz Institute Disaster Preparedness Assessment Project by the Center for Hazards Research and Policy Development University of Louisville.
72. WWF, *Mega-Stress for Mega-Cities: A Climate Vulnerability Ranking of Major Coastal Cities in Asia, Dhaka, Calcutta and Minh, H.C.* (2009).
73. Hewitt K., *Regions of Risk—A Geographical Introduction to Disasters* (London, UK: Longman, 1997), 389.

74. Perry R. W., and Qarantelli E. L., *What is a Disaster? New Answers to Old Questions* (Xlibris Corporation, 2005).
75. See for example Figure 2.2 in Birkmann J., "Indicators and Criteria for Measuring Vulnerability: Theoretical Bases and Requirements," in Birkmann J., ed., *Measuring Vulnerability to Natural Hazards: Towards Disaster Resilient Societies* (New York: United Nations Press, 2006), 55–78. This figure contrasts objective versus normative information.
76. Simpson A. G., "P/C Industry Depends Too Much on Catastrophe Models, Says Pioneer Clark," *Insurance Journal* (April 14, 2011), http://www.insurancejournal.com/news/national/2011/04/14/194464.htm.
77. Air Worldwide, *Anatomy of a Damage Function: Dispelling the Myths* (March 16, 2010), http://www.air-worldwide.com/PublicationsItem.aspx?id=19052.
78. Guy Carpenter Briefing, *Managing Catastrophe Model Uncertainty: Issues and Challenges*. Guy Carpenter, (March 2011), http://www.guycarp.com/portal/extranet/insights/briefingsPDF/2011/Catastrophe%20Model%20Change%20Issues%20and%20Challenges.pdf;jsessionid=NS3Kdh69zjY3zH7kp2kccpfGz5c4PTDyLQJdyGvQQ8Q2RNJnCMTG!-783285485?vid=1.
79. Lloyd's, *Realistic Disaster Scenarios, Scenario Specification*, January 2012, http://www.lloyds.com/~/media/Files/The%20Market/Tools%20and%20resources/Exposure%20management/RDS%20%20Scenario%20Specification%20%20January%202012.pdf.
80. FEMA, *The Federal Emergency Management Agency's (FEMA's) Methodology for Estimating Potential Losses from Disasters* (2013), http://www.fema.gov/hazus.
81. Canadian HAZUS User Group (n.d.). What we do, http://www.usehazus.com/canadianhug/.
82. Mitchell J. K., ed., *Crucibles of Hazard: Mega-Cities and Disasters in Transition* (New York: United Nations University Press, 1999).
83. Erikson K. T., *A New Species of Trouble: Explorations in Disaster, Trauma, and Community*, vol. 7 (New York: W.W. Norton and Co., 1994).
84. Oliver-Smith A., and Hoffman S. M., eds., *The Angry Earth: Disaster in Anthropological Perspective* (New York: Routledge, 1999).
85. Arnold M., Chen R. S., Deichmann U., Dilley M., Lerner-Lam A., Pullen R. E., and Trohanis Z., *Natural Disaster Hotspots: Case Studies*. Disaster Risk Management Series No. 6, (Washington, DC: World Bank, Hazard Management Unit, 2006), http://siteresources.worldbank.org/INTDISMGMT/Resources/0821363328.pdf?resourceurlname=0821363328.pdf.
86. Cazenave A., and Llovel W., "Contemporary Sea Level Rise," *Annual Review of Marine Science* 2 (2010): 145–73.
87. Stocker D. Q., Climate Change 2013: The Physical Science Basis. *Working Group I Contribution to the Fifth Assessment Report of the Intergovernmental Panel on Climate Change, Summary for Policymakers*, (IPCC, 2013).
88. Nicholls R. J., and Cazenave A., "Sea-Level Rise and Its Impact on Coastal Zones," *Science* 328, no. 5985 (2010): 1517–20.
89. As well as contributing to flooding, excessive withdrawal of groundwater can also draw salty water inland, poisoning aquifers.
90. Phien-wej N., Giao P. H., and Nutalaya P., "Land Subsidence in Bangkok, Thailand," *Engineering Geology* 82, no. 4, (2006): 187–201.
91. Kelman I., and West J. J., "Climate Change and Small Island Developing States: A Critical Review," *Ecological and Environmental Anthropology* 5, no. 1, (2009): 1–16.
92. Source: Inclusive Green Growth World Bank 2012 and Rob van den Berg (Global Environment Facility). See Also Fifth Overall Performance Study of the GEF: Cumulative Evidence on the Challenging Pathways to Impact, www.gefeo.org.

93. Doheny-Farina S., The Grid and the Village (2001). *Orion Magazine*, Autumn 2001, http://www.orionmagazine.org/index.php/articles/article/90/.
94. Higuchi K., Yuen C. W., and Shabbar A., "Ice Storm'98 in Southcentral Canada and Northeastern United States: A Climatological Perspective," *Theoretical and Applied Climatology* 66, no. 1–2 (2000): 61–79.
95. Jones K. F., *Ice Storms in the St. Lawrence Valley Region* (No. ERDC/CRREL TR-03-1), Engineer Research and Development Center, Hanover, NH, Cold Regions Research and Engineering Lab (2003).
96. This region, although larger than most ice storms, is typically where they occur due to the role local topography has in the movement of cold air near the surface.
97. Hydro Quebec (2013), *Ice Storm 1998: 15 Years later,* January 7, 2013, http://news.hydroquebec.com/en/news/116/ice-storm-1998-15-years-later/#.Ui4RKV_D_CI.
98. RMS, *The 1998 Ice Storm: 10-Year Eetrospective*, RMS Special Report, (Risk Management Solutions Inc., 2008).

4 Hazard, Vulnerability, and Resilience

To be alive is to be vulnerable.
Madeleine L'Engle

Bend and be straight;

Empty and be full;

Wear out and be new;

Have little and gain;

Have much and be confused.

Tao Te Ching, XXII

FIGURE 4.1 A plant pushing its way through an asphalt surface displays amazing resilience.

CHAPTER OUTLINE

- 4.1 Why This Topic Matters 105
- 4.2 Recommended Readings 105
- 4.3 Hazard 106
- 4.4 Introduction to Vulnerability and Resilience 111
- 4.5 Vulnerability 113
  - 4.5.1 Creative Tensions and Debates 113
  - 4.5.2 Scales and Causation 115
  - 4.5.3 Vulnerability Assessment 119
- 4.6 Resilience 122
  - 4.6.1 The Shift toward Resilience Thinking 123
  - 4.6.2 Engineering Resilience 124
  - 4.6.3 Ecological Resilience 126
  - 4.6.4 Psychological Resilience 128
  - 4.6.5 Community Resilience 130
  - 4.6.6 Religion and Spirituality 134
- 4.7 Grassy Narrows 138
- Further Reading 142
- 4.8 Responsibility and Response Ability—Comments on Vulnerability and Community by John (Jack) Lindsay 142
- End Notes 145

CHAPTER OVERVIEW

The concepts of hazard, vulnerability, and resilience are inseparable from one another. This chapter is an overview of each area and how they interrelate. In particular, some debates within vulnerability literature are identified, and different approaches to understanding resilience. The concept of resilience is gaining increasing traction within the practice of disaster management but has great challenges in terms of being operationalized. This is an enormous topic that one chapter cannot possibly do justice to, and the reader is encouraged to do further reading.

KEYWORDS

- Adaptation
- Assessment
- Community

- Critical threshold
- Environment
- Grassy narrows
- Hazard
- Mapping
- Resilience
- Robustness
- Spirituality
- Vulnerability and Capacity Assessment
- Vulnerability

4.1 Why This Topic Matters

The concepts of hazard, vulnerability, and resilience have become core to the study and practice of disaster management. People have always been exposed to a variety of hazards, have gotten much better at reducing their vulnerability to some of these hazards, and are generally very resilient. Yet the changing world we are living in exposes us to new and different risks, and sometimes the old strategies are insufficient. New layers of coping are required in addition to the historical ones, if societies are going to continue to be effective in reducing disaster risk.

Virtually every book in this field, and innumerable articles from a plethora of other disciplines, discuss these concepts; nevertheless, as we have seen before with other topics that there is a lack of general agreement on the precise meaning of the terms. Definitions become difficult partly because "the concepts of resilience, robustness and vulnerability can only be understood in relation to one another."[1] This chapter will discuss various approaches to understanding and measuring hazard, vulnerability, and resilience, with the purpose of clarifying meaning, usage, and measurement.

■ ■ 4.2 Recommended Readings ■

- Bankoff, G., Frerks, G., and Hilhorst, D. (2004). *Mapping vulnerability: disasters, development, and people.* Earthscan/James & James.
- Birkmann, J. and Wisner, B. (2006). *Measuring the un-measurable: the challenge of vulnerability.* United Nations University, Institute for Environment and Human Security. No. 5/2006.
- Erikson, K.T. (1994). *A new species of trouble: explorations in disaster, trauma, and community* Vol. 7. Norton, New York.
- Hazard, risk and vulnerability analysis tool kit http://embc.gov.bc.ca/em/hrva/toolkit.html.

- Kirmayer, L.J., Sehdev, M., Whitley, R., Dandeneau, S.F., and Issac, C. (2009). Community resilience: models, metaphors and measures. *Journal of Aboriginal Health*, November. National Aboriginal Health Organization.
- Lawson, E.J. and Thomas, C. (2007).Wading in the waters: spirituality and older black Katrina survivors. *Journal of Health Care for the Poor and Underserved* 18, 341–354.
- Mitchell, J.K. (1999). *Crucibles of hazard: mega-cities and disasters in transition.* United Nations University Press, New York.
- Oliver-Smith, A. (1999). Peru's five hundred-year earthquake: vulnerability in historical context. *The Angry Earth*, 74–88.
- Tuohy, R. and Stephens, C. (2012). Older adults' narratives about a flood disaster: resilience, coherence, and personal identity. *Journal of Aging Studies* 26, 26–34.
- Walker, B., Holling, C. S., Carpenter, S. R., and Kinzig, A. (2004). Resilience, adaptability and transformability in social--ecological systems. *Ecology and Society* 9(2), 5.
- Wisner, B. (2003). *At risk: natural hazards, people's vulnerability and disasters.* Routledge.

■ ■ Questions to Ponder ■

- Are vulnerability and resilience states or processes (or both)?
- Are they two ends of a spectrum, or different kinds of things?

STUDENT EXERCISE

Make a list of what aspects of your life, personality, or community make you vulnerable or resilient.

4.3 Hazard

It is not possible to disentangle the notion of hazard from vulnerability and adaptation, and it is closely linked to how we engage with resources. For example, rain is an important resource, but when there is too much or too little of it, then it becomes a hazard. What determines too much or too little depends on how we have adapted to the natural environment and varies with culture, agricultural practices, building codes, land use planning, etc. Similar issues exist for technological resources, which are simultaneously beneficial to society but also a source of risk. Examples are nuclear power plants, most chemicals, and automobiles.

An early definition of a natural hazard was "those elements in the physical environment, harmful to man and caused by forces extraneous to him."[2] The notion of hazard as an external threat is the basis of what has been called the hazard paradigm, first notably critiqued by

Ken Hewitt.[3] The hazards paradigm leads to management approaches rooted in controlling nature using top-down and engineering solutions. Nature becomes, in part, an enemy that needs to be defeated, and war-like metaphors become common when referring to disaster response. Illustrations abound in media headlines, for example, "German, Hungarian Towns Battle Floods"[4] and "Wealthy US Coastal Enclaves Battle Surging Seas."[5] The problem with the hazards paradigm is that it places responsibility for disaster outside of human influence. In contrast, the development of vulnerability and resilience theory and the use of a human ecological perspective[6] shift cause and effect away from nature and firmly into the human-social context. An expanded definition of hazard would include nonphysical and internal threats, perhaps including social constructs such as overly rigid institutions and corruption.

> **STUDENT EXERCISE**
>
> Define the word hazard. Afterwards, compare it to the United Nations International Strategy for Disaster Reduction (UN ISDR) definition.[7]

Hazards have typically been classified as natural, technological, or human caused. These divisions are not well defined since many disasters have characteristics of two, or even all three, categories. For example, a storm might trigger a failure in an industrial site, but poor planning and inappropriate use of technology may be more important causes of damage (see Chapter 5.5 for more discussion about this). If you build in a flood plain and eventually a flood causes a disaster, how appropriate is it to call this a natural disaster? In *The Environment as Hazard*,[8] the authors further classify hazard by the variables frequency, duration, areal extent, speed of onset, spatial dispersion, and temporal spacing (Figure 4.2). This leads to hazard profiles that can be very different from each other; these differences are important in terms of developing effective disaster management strategies.

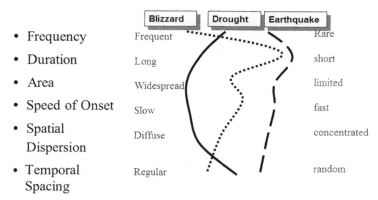

FIGURE 4.2 Hazard Event Profiles. *Source: Adapted from Burton, I., Kates, R.W., and White, G. (1993). The environment as hazard. Guilford Press, New York.*

> **STUDENT EXERCISE**
>
> Consider the variable " speed of onset."
>
> - How might emergency managers use different planning strategies for hazards at opposite ends of this spectrum?

There are other ways that hazards can be differentiated as well, including risk perception variables (Figure 3.11) such as documented by the psychometric paradigm[9]; dread is particularly important in this regard (see Chapter 3.5). An example of its effect can be seen on perceptions of wildfire risk; in one study it was noted that wildfires rank low on people's risk perceptions because of a feeling of control, limited areal extent, and high level of knowledge.[10] As a result, there is a tendency to develop areas at risk of wildfires, beyond what many experts would consider reasonable.

Context is important to understanding hazard. Contexts include physical processes that can alter the probability of extreme events (i.e., climate change and deforestation), changing demographics that increase exposure (i.e., development in more risky locations), and people's adjustments to risk that alter adaptive ranges (i.e., changing land use practices or better warning systems). These contexts can be cultural, political, environmental, economic, or organizational. Hazard characteristics are therefore dynamic, not static, and are affected by a complicated set of feedback between various factors. Figure 4.3 shows the model used by Mitchell et al.[11] to illustrate this perspective.[12]

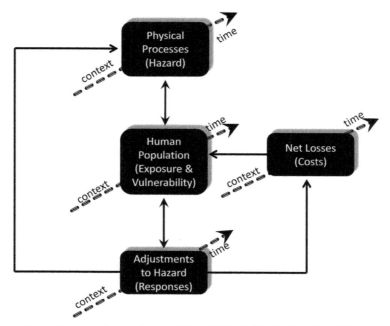

FIGURE 4.3 A natural hazard system. *Source: Adapted from Mitchell, J.K., Devine, N., and Jagger, K. (1989). A contextual model of natural hazard. Geographical Review, 391–409.*

Statistics on hazards are usually available from government agencies and are often incorporated into building codes. Examples include seismic hazard values from the National Building Code of Canada that can be accessed online at Natural Resources Canada[13] and snow load calculations from the Canadian Mortgagee and Housing Corporation.[14]

Hazard Mapping: A great deal of attention has been given to mapping hazards, for the good reason that they are extraordinarily useful for risk assessments. These maps, which include histories of hazardous events and/or probabilities of occurrence of events of varying magnitudes, are available on a variety of scales from local to global (i.e., the Natural Disaster Hotspots report by the World Bank[15]). Mapping is also available on the Web on a real-time basis, for example, by the World Meteorological Organization on their Severe Weather Information Centre Web site,[16] by the United States Geological Survey,[17] or by the European-Mediterranean Seismological Centre.[18] Most governments have departments of meteorology and geology that publish hazard maps. The United States Geological Survey has a number of seismic hazard maps available on their Web site,[19] some of which are interactive (Figure 4.4). The U.S. Federal Emergency Management Agency (FEMA) also has a mapping information platform available to the public,[20] and has Geographical Information System (GIS) software that can be used to not only map selected hazards but also to estimate potential losses from disasters (HAZUS; see Chapter 6.5.6).[21]

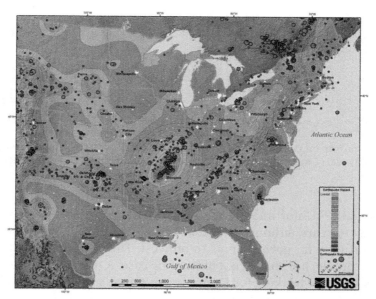

FIGURE 4.4 Earthquake hazard in Eastern United States and Southeastern Ontario. Red dots indicate historical earthquakes, while the colored contours indicate relative hazard magnitude. *Source: United States Geological Survey,* http://www.usgs.gov/public_lecture_series/images/hazards_map.jpg. (For interpretation of the references to color in this figure legend, the reader is referred to the online version of this book.)

Given how much damage floods cause, flood risk maps (e.g., Figure 4.5) are one of the most important hazard maps; they are available in Canada through the Flood Damage

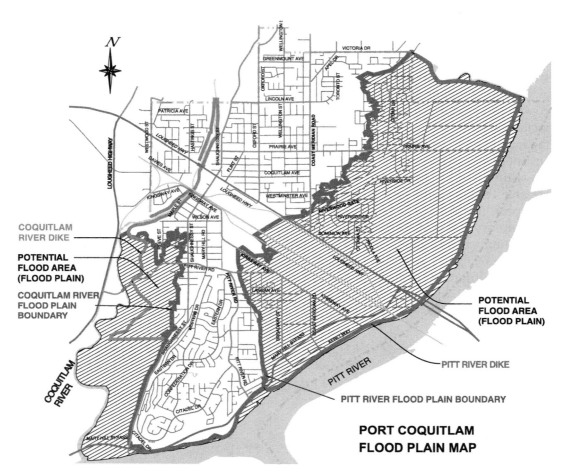

FIGURE 4.5 Flood plain map for Coquitlam, British Columbia, Canada. *Source: City of Coquitlam, BC,* http://www.portcoquitlam.ca/City_Hall/City_Departments/Fire_Emergency_Services/Emergency_Preparedness/Flood_Preparedness.htm.

Reduction Program.[22] Unfortunately, this program has had greatly reduced funding over the past number of years, although the recent floods in Calgary and Toronto in 2013 may create an opportunity for governments to reinvest. These maps should be made easily available to the public, but such is not always the case. A number of years ago I was told a story by a hydrologist working for Environment Canada of how they distributed flood risk maps to dentists' offices as part of their public information program, only to have them removed by real estate agents who viewed them as detrimental to house values and sales. From a risk management perspective, home owners should know before they purchase a property where it lies relative to flood plains, but competing interests make this a challenging process to implement. Following the 2013 floods, the Alberta government had planned to put warnings on all land titles for properties in flood risk zones, but backed away from this for properties not directly in the floodway, due to pressure from homeowners.[23]

Hazards evolve. In the past, (hostile people aside) they were either environmental or biological. New hazards, however, that are technological and institutional have emerged, to which society tries to adjust as best it can, given its limitations and constraints. It has become an enormously complex system that some, such as Thomas Homer-Dixon,[24] believe may be increasingly operating beyond our control. Some hazards are relatively easy to understand and manage, while others, such as global warming, deforestation, species extinction, pollution, and terrorism, are wicked problems, sometimes unexpected and highly resistant to management.

4.4 Introduction to Vulnerability and Resilience

There are many definitions of the terms *vulnerability* and *resilience*. They are neither "correct" nor "incorrect"; usage depends on common agreement and what is most useful for the purposes of discussion. Some definitions subsume resilience under vulnerability, while others do the reverse. For the purpose of this chapter, I will initially consider the terms as (somewhat simplistically) portrayed in Figure 4.6, which displays the theoretical impact of some harmful event on a community.

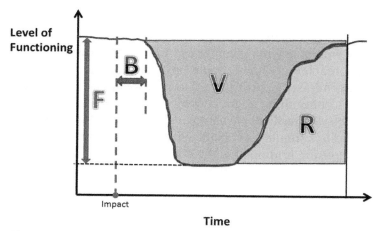

FIGURE 4.6 A simplified but useful view of vulnerability and resilience. F = fragility, B = buffer, V = vulnerability, R = resilience.

I separate the notions of vulnerability and resilience because they are both so useful; although they are linked, it seems to me that making one of them a subset of the other is not helpful. In this figure, F = fragility, which is the maximum loss of function (or harm incurred) as a result of some event; B = buffer, which reflects the ability of the system to absorb a harmful event without loss of function; V = vulnerability, which is the cumulative harmful impact over time until recovery occurs; and R = resilience, which represents the system's ability to recover functionality over time.

There are some problems with this depiction. First, it displays vulnerability and resilience as part of a dynamic process when many consider them to be state variables (i.e., being in a state of vulnerability). In other words, even in the absence of a damaging event, people or communities may possess characteristics of vulnerability and resilience. Second, any given event would produce a different calculation of the variables, which complicates interpretation. Third, it implies that resilience and vulnerability are two ends of a spectrum, which is a debatable proposition. Although they are certainly related, they are, in part, different kinds of things. Fourth, the important quality of robustness (the capacity to resist harm) is not explicitly represented in this figure, though it is implied. Fifth, if R is very small because of an event that causes massive destruction, then V→∞ as time→∞, which is not helpful; the figure only has a useful meaning if a recovery process exists or if V and R are calculated over finite periods of time. Next, it ignores many of the subtleties involved in understanding disaster recovery, particularly that some aspects are not objective or easily quantifiable, are culturally defined, or involve trade-offs. Finally, this depiction assumes that human-environmental systems exist in a state of equilibrium and that the purpose of recovery is to return to a previous, better state. If one assumes a chaotic system, then this notion is debatable as there may be more than one stable equilibrium state (see Chapter 5). This final critique is very much a function of scale since a system might remain in a particular equilibrium state for a period of time (such as an interglacial climate period) but be in a nonequilibrium state at other times. In spite of the above caveats, this diagram is helpful for visualizing the meaning of the terms.

Similar to how emergency management can be viewed through an all-hazards and/or a risk-based approach (note: they should be combined), some aspects of vulnerability and resilience are generic (e.g., poverty and level of health), while others are hazard specific. For example, a community might be very vulnerable to a flood but not to a blizzard, or a person might be very resilient psychologically but not physically. For this reason, discussions or calculations should be specific with regard to sources of risk. It is also worth mentioning that exposure is implicit in the discussion of vulnerability. Without exposure, the notion has no utility. Also, vulnerability and resilience vary over a range of hazard magnitudes; a community might have low vulnerability to flood hazard up to some critical threshold, but be very vulnerable beyond it. This is the purpose of fragility curves (Chapter 6.5.6)—to quantify such thresholds. *The importance of knowing the location of critical thresholds cannot be overstated.*

The notion of scale is important to both vulnerability and resilience[25]; each operates at scales from individual to global and is linked in time and in space. An individual's vulnerability and resilience depends not only on their own resources but also on those of the community in which they live, the critical infrastructure (regional, national, and international) that they depend on, how risk is constructed in distant boardrooms that do not recognize political boundaries, and many other factors. Climate change, toxic pollution, burgeoning national debts, terrorism, and mortgage-lending practices all have global implications and affect the vulnerability and resilience of much of the world's population. This makes the process of assessment extraordinarily complicated, especially since different factors and scales have varying levels of importance depending on place and time. An illustration is vulnerability to heat stress, which varies by geography and season, as well as in the long term if climate change theories are correct. Social vulnerability varies with demographics and immigration patterns, while economic vulnerability varies depending on whether economies are or are not in recession. No measure of vulnerability or resilience is static or universal.

The notions of vulnerability and resilience are enormously useful. Using them in an operational way, however, is challenging and suffers from several difficulties[26]: (1) lack of a common definition, (2) difficulties in finding metrics that represent them, and (3) problems related to how they function at different scales.

4.5 Vulnerability

I understand now that the vulnerability I've always felt is the greatest strength a person can have. You can't experience life without feeling life. What I've learned is that being vulnerable… is not a weakness, it's a strength.

Elisabeth Shue

The UN ISDR defines vulnerability as "The characteristics and circumstances of a community, system or asset that make it susceptible to the damaging effects of a hazard" with the following added comment:

There are many aspects of vulnerability, arising from various physical, social, economic, and environmental factors. Examples may include poor design and construction of buildings, inadequate protection of assets, lack of public information and awareness, limited official recognition of risks and preparedness measures, and disregard for wise environmental management. Vulnerability varies significantly within a community and over time. This definition identifies vulnerability as a characteristic of the element of interest (community, system or asset) which is independent of its exposure. However, in common use the word is often used more broadly to include the element's exposure.

4.5.1 Creative Tensions and Debates

How to approach analyzing and measuring vulnerability is a subject of much debate, as illustrated in the following list, which is from a paper written by Jörn Birkmann and Ben Wisner[27] on challenges in measuring vulnerability. The various tensions described here are ongoing and reflect the dualism that exists between its complexity and the potential usefulness of an application.

"*Complexity versus simplicity*

- Cultural, livelihood, situational, institutional/political complexity vs. desire for mathematical parsimony
- Complex, dynamic, even chaotic process versus product

Understanding versus implementation

- Utility/necessity of basic starting points: e.g., exposure maps in Sri Lanka; e.g., district/village surveys in Tanzania;
- But how does one get from basic description to understanding?

Natural science versus social science epistemology

- Are controlled experiments possible when groups of people are involved?
- When local people themselves are researching their own hazards and vulnerabilities/capacities and implementing action plans, does this constitute an experiment?

Nomothetic versus ideographic goals

- Are we trying to establish law-like statements and the ability to predict outcomes or are we providing narratives and descriptions of situations that raise consciousness of risks and mobilize local and outside action, vigilance, and preparedness?

Ethical issues

- Are people research objects or subjects?
- Are there not winners and losers in any intervention into the risk-scape of a locality?

Cacophony versus polyphony

- Full understanding of vulnerability may involve a large team, but can they speak with one voice when a common language or metric of vulnerability may not exist?
- Decision makers want clear options, not nuanced understandings."

Vulnerability is generally considered to be a negative thing, to be reduced as much as possible. Yet there is a positive side to it. The quote above by the American actress Elizabeth Shue expresses the notion of vulnerability (within the context of personal relationships) as strength. To be vulnerable means having the capacity to be affected by others or your environment, which is important not only in relationships but also within the context of learning and adaptive management.[28] To coevolve in an ecosystem sense requires vulnerability (or else there is no feedback), and from this perspective it should be simultaneously viewed as both a weakness and a strength. Rabbits are vulnerable to many predators, but have become a fast and resilient species because of this. It is a case of vulnerability forcing adaptation. I view vulnerability as being highly contextual; in some contexts a degree of vulnerability is desirable. Too little vulnerability implies a level of disconnectedness that may be pathological while too much is suicidal. The disconnect between modern urban technological society and the nonhuman natural environment has been discussed in these terms.[29] This notion connects vulnerability to resilience in a more dynamic way than shown in Figure 4.6 in the sense that vulnerability creates system feedback that result in increased resilience. Figure 4.7, in a highly simplified way, illustrates how these two possible relationships might look. The dynamic interpretation suggests a target level of vulnerability that is not minimized but instead optimized.

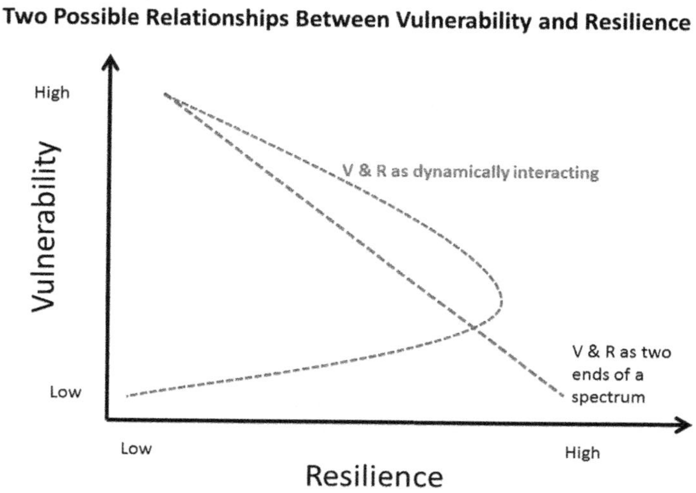

Two Possible Relationships Between Vulnerability and Resilience

FIGURE 4.7 Two ways of looking at the relationship between vulnerability and resilience. The straight line views the two variables as opposite ends of a spectrum, while the curved line portrays them as being dynamically interactive. In the latter case, low vulnerability is paired with low resilience within the context of a complex adaptive system, which requires system feedback in order to sustain resilience.

> **STUDENT EXERCISE**
>
> Describe two scenarios: (1) one related to the hypothesis that too little vulnerability creates a pathological relationship that limits resilience, and (2) one that shows them as two ends of a spectrum.

4.5.2 Scales and Causation

The multidimensional aspect of vulnerability is shown in Figure 4.8.[30] Each sphere is a useful perspective, depending on the purpose for which it is being used.[31]

Vulnerability as it relates to physical hazard is normally examined with respect to extreme events, particularly from an engineering perspective. Flood hazard, for example, typically uses a 1/100-year event while building codes might consider 1/30- or 1/50-year occurrences of wind or snow loads (there will be many exceptions to these levels); which level is chosen is a function of how risk is constructed and perceptions of acceptable level of risk in society. If a community is designed to manage a 1/30-year snowfall, then risk analyses should focus on that level of event. This issue depends on critical thresholds, as illustrated in Figure 4.9, which shows the typical "S"-shaped curve associated with hazard impacts. Extreme value/critical threshold analysis is important when analyzing potential impacts to people and communities.

Disaster, when viewed as a chain of cause and effect beginning with the hazard, has been described by Kenneth Hewitt as part of the dominant paradigm of disaster research.[32] While

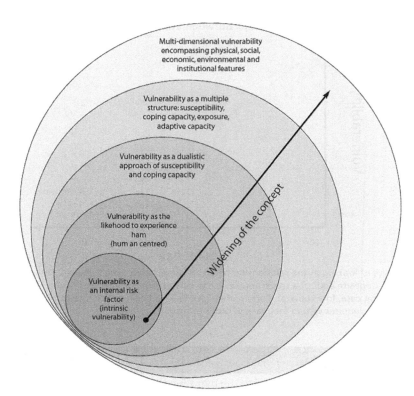

FIGURE 4.8 Key spheres of the concept of vulnerability. *Source: Birkmann, J. and Wisner, B. (2006). Measuring the un-measurable: the challenge of vulnerability. United Nations University, Institute for Environment and Human Security. No. 5/2006.*

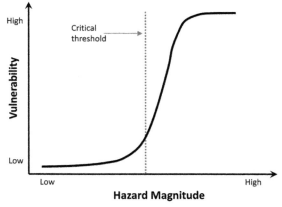

FIGURE 4.9 "S"-shaped vulnerability curve and the associated critical threshold. Below the critical threshold, systems are very resistant to damage; but near and beyond the threshold, vulnerability increases exponentially, until a point of saturation is reached.

acknowledging the usefulness of this approach, he suggests an alternate paradigm that better reflects the complexity of human–natural interactions (please forgive my dualism), where:

- "Most natural disasters, or most damages in them, are characteristic rather than accidental features of the places and societies where they occur.
- The risks, pressures, uncertainties that bear on awareness of and preparedness for natural fluctuations flow mainly from what is called 'ordinary life,' rather than from the rareness and scale of those fluctuations.
- The natural extremes involved are, in a human ecological sense, more expected and knowable than many of the contemporary social developments that pervade everyday life."[33]

Risk is constructed in the everyday decisions made by people and organizations, and understanding it requires an examination of those processes. Particularly important is the distribution of power in society and the resulting marginalization of some groups. Aboriginals typically suffer because of this; certainly in North America, the history of their marginalization has resulted in terribly vulnerable communities with high rates of poverty, health issues, and suicides (Figure 4.10). Also, see the case study on Grassy Narrows at the end of this chapter.

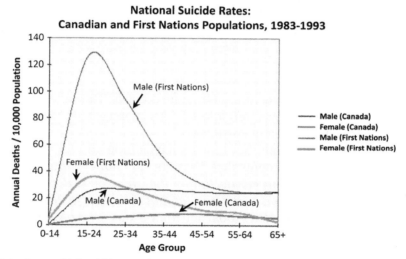

FIGURE 4.10 Note the very high suicide rates for First Nations people (particularly males), as compared to the general population. *Adapted from Government of Manitoba,* http://www.gov.mb.ca/ana/apm2000/2/d.html.

Corruption also has a large impact. Figure 4.11[34] shows a strong relationship between higher levels of corruption and greater earthquake fatalities. In a similar study in the United States,[35] a positive relationship was found between state corruption and natural disasters, with each additional $100 per capita increase in FEMA relief corresponding to a 102% increase in average state corruption. The article raises the interesting issue of cause and effect, suggesting that not only does corruption contribute to disasters but also that

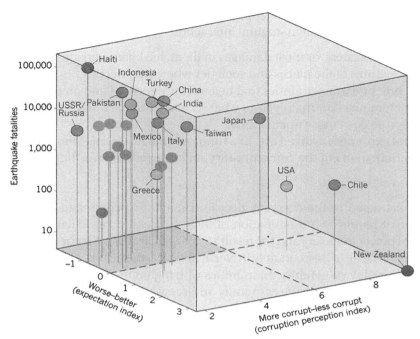

FIGURE 4.11 Corruption's toll. Corruption versus the level of corruption that might be expected from per capita income. Of all earthquake fatalities attributable to building collapse in the past three decades, 82.6% occur in societies that are anomalously corrupt (left-hand corner of the plot). *Source: Ambraseys, N. and Bilham, R. (2011). Corruption kills. Nature 469, 153–155.*

increased aid results in greater corruption by state officials. This latter issue has resulted in serious ethical critiques of how international aid is managed, particularly with nations that are known to have high levels of corruption.[36]

Historical context is important, both for understanding and for creating risk reduction strategies. Ken Hewitt comments that disasters ***"are not explained by conditions or behaviour peculiar to calamitous events. Rather they are seen to depend on the ongoing social order, its everyday relations to the habitat and the larger historical circumstances that shape or frustrate these matters"*** (bold added). Various case studies have emphasized the importance of history. For example, the root causes of vulnerability to earthquakes in Peru have been shown to have begun almost 500 years ago,[37] as a result of conquest and colonization by European powers (I highly recommend the book *The Angry Earth* edited by Anthony Oliver-Smith and Susanna Hoffman). Similarly, the extreme vulnerability of Haiti is a function of centuries of political and economic relationships with other countries, including a large debt that had to be repaid to France for its loss of men and slaves as a result of Haiti gaining independence in 1804; this debt took 122 years to pay off. Having to pay off this debt undermined Haiti's export base, which in the presence of failed local development led to widespread environmental degradation and highly vulnerable communities.[38]

Vulnerability is paradoxical. Intuitively obvious as a concept to most people, its complexity is enormous and it resists attempts to quantify it.

4.5.3 Vulnerability Assessment

As interesting as the study of vulnerability is from a conceptual perspective, it only becomes useful if it can be assessed. It can be examined from a variety of sectors (e.g., engineering, economic, health, communications, social, environmental) and perspectives. What the different definitions and approaches have in common is a consideration of actual or potential for harm, or disruption of norms. Ben Wisner describes four main clusters of approaches to examining social vulnerability[39]:

- *Laundry List Approach*: Lists of elements at risk, including structural fragility of buildings, bridges, health care systems, people, etc. His critique of this approach is that people "get lost in the process of conceptualizing whole systems."
- *Taxonomy Approach*: The second approach examines causes of social vulnerability and uses empirically based taxonomies of groups of people who are especially vulnerable, such as the elderly, sick, disabled, poor, women, etc. Such taxonomies add much value to the previous approach but can still be blunt tools. For example, depending on a specific situation, being female may or may not contribute toward being more vulnerable.
- *Situational Approach*: This addresses specific situations, such as the nature of a household's daily life, and include recognition that situations are dynamic and shift over time. Different factors may overlap, but a great deal of depth is added by using this approach. "Situational analysis breaks out human beings in their complexity and also groups of humans from the heterogeneous mass of things and systems said by mainstream planners to be 'vulnerable'."

These first three approaches are structural in nature and reflect a Western technocratic discourse, with reliance on experts for a meaningful assessment. Each is progressively more sophisticated but also requires more data and effort. They are not exclusionary, and they provide a great deal of useful information needed in any vulnerability assessment. The fourth approach below is somewhat different.

- *Community-Based Approach*: When communities move beyond a victim stance and become empowered to define and assess their own vulnerabilities and capacities, then the logic of the process changes. Dynamics become internalized within affected communities, thereby increasing social capital. Social space then becomes available for alternative narratives and voices. This approach does not negate the possibility of using the first three approaches; in most cases, they would be essential components given the complexity of many of the risks in today's world.

The complexity of vulnerability assessment is captured in Figure 4.12; it is easy to see that a holistic evaluation is challenging indeed![40]

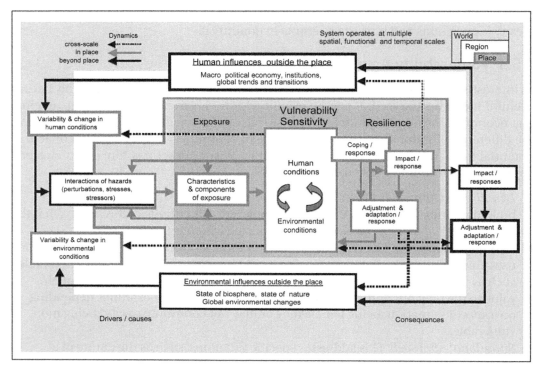

FIGURE 4.12 A vulnerability framework. *Source: Turner, B.L., Kasperson, R.E., Matson, P.A., McCarthy, J.J., Corell, R.W., Christensen, L., Eckley, N., Kasperson, J.X., Luers, A., Martello, M.L., Polsky, C., Pulsipher, A., Schiller, A. (2003). A framework for vulnerability analysis in sustainability science.* Proceedings of the National Academy of Sciences *100(14), 8074–8079.*

Approaches may be quantitative, qualitative or narrative based. Quantitative assessments of vulnerability rely on statistical and indice-based methods, as discussed in Chapter 3.7. These sorts of analyses are interesting and useful, can be robust, but must only capture a narrow slice of vulnerability. Susan Cutter[41] notes that "there is a general consensus within the social science community about some of the major factors that influence social vulnerability. These include: lack of access to resources (including information, knowledge, and technology); limited access to political power and representation; social capital, including social networks and connections; beliefs and customs; building stock and age; frail and physically limited individuals; and type and density of infrastructure and lifelines." The methodology of how to actually measure these factors is often unclear and subject to different opinions. Particularly, what relative levels of importance should be assigned to each factor is highly subjective. Her paper lists a number of variables and dimensions, the most important of which are (starting with highest-ranked variables): personal wealth, age, density of the built environment, single-sector economic dependence, housing stock and tenancy, and race/ethnicity. This list was developed based on research in the United States, is culturally relative, and may not be representative of other countries or cultures.

Qualitative analyses and narratives add insight and provide a historical and cultural context that is essential in a comprehensive vulnerability assessment. For further study, I recommend the following books: *At Risk* 2nd ed. by Ben Wisner, Piers Blaikie, Terry Cannon, and Ian Davis; *Mapping Vulnerability: Disasters Development and People* edited by Greg Bankoff, Georg Frerks, and Dorothea Hilhorst; *A New Species of Trouble* by Kai Erikson; *Crucibles of Hazard: Mega-Cities and Disasters in Transition* edited by James Mitchell; *Measuring Vulnerability to Natural Hazards: Towards Disaster Resilient Societies* edited by Jorn Birkmann; *Disaster Risk and Vulnerability: Mitigation through Mobilizing Communities and Partnerships* edited by C. Emdad Haque and David Etkin.

Operational tools or guidelines have been developed to carry out vulnerability and capacity analyses (VCA). Generally, they include the following steps:

1. Select a framework for analysis to establish clear and shared understanding of what is to be analyzed, as well as the role of the VCA
2. Select unit/level of analysis to facilitate planning the scope and focus of the VCA and selection of the methodology
3. Identify stakeholders to provide expert knowledge and ensure ownership of findings
4. Select approach for data collection and analysis appropriate to the scale, scope, and purpose of the VCA
5. Collect data using a series of data-gathering methods to build up evidence
6. Analyze data in order to link different dimensions of vulnerability to present a full picture and reveal cause/effect linkages
7. Decision making and action feed findings into risk assessment and project design and make appropriate modifications to reduce vulnerability

For a more in-depth description of VCA, refer to *Tools for Mainstreaming Disaster Risk Reduction: Guidance Notes for Development Organizations*[42] by Charlotte Benson and John Twigg; the British Columbia *Hazard, Risk and Vulnerability Analysis Tool Kit*[43]; or *Community-Wide Vulnerability and Capacity Assessment*[44] by Ron Kuban and Heather MacKenzie-Carey.

STUDENT EXERCISE

Make a detailed list of measurable factors that you would include in a vulnerability assessment. Rank them for a community you know well, and for each one comment on its degree of universality. Are there factors that cannot be measured that you consider significant?

STUDENT RESEARCH PROJECT

Select a community you know well and for which there are good datasets, and do a VCA. This is a major project that is best done in groups.

In 2006, Jörn Birkmann and Ben Wisner published a report for the United Nations University called *Measuring the Un-measurable: The Challenge of Vulnerability*.[45] Much of the material in this section of Chapter 4 was taken from it, and I recommend it as further reading. The title says a lot about this topic and is well worth keeping in mind.

4.6 Resilience[46]

An Oak that grew on the bank of a river was uprooted by a severe gale of wind, and thrown across the stream. It fell among some reeds growing by the water, and said to them,

"How is it that you, who are so frail and slender, have managed to weather the storm, whereas I, with all my strength, have been torn up by the roots and hurled into the river?"

"You were stubborn," came the reply, *"and fought against the storm, which proved stronger than you: but we bow and yield to every breeze, and thus the gale passed harmlessly over our heads."*

The Oak Tree and the Reeds: Aesop's Fable (Figure 4.13)

FIGURE 4.13 Oak tree and the reed.

Sometimes resilience is conceptualized as a metaphor taken from the physical sciences, other times as a theory, a set of capacities or as a strategy. The UN ISDR defines resilience as "The ability of a system, community or society exposed to hazards to resist, absorb, accommodate to and recover from the effects of a hazard in a timely and efficient manner, including through the preservation and restoration of its essential basic structures and functions... Resilience means the ability to 'resile from' or 'spring back from' a shock. The resilience of a community in respect to potential hazard events is determined by the degree to which the community has the necessary resources and is capable of organizing itself both prior to and during times of need."

The UN definition differs from how I am using the word, since I separate out resistance (or robustness) and absorption from the process of recovery. I think that if you try to "stuff" too much into a single definition it loses usefulness. There are many factors involved in determining resilience, most of which are in common with vulnerability. For example, Castleden[47] in a literature review on resilience identified the following components: communication, learning, adaptation, risk awareness, social capital, good governance, planning and preparedness, redundancy, economic capacity and diversification, and population physical and mental health. One problem with the UN definition is that it does not include the notion of reorganization and adapting to a different state, which is fundamental to ecological approaches. An alternate definition to the UN one might be "a process linking a set of adaptive capacities to a positive trajectory of functioning and adaptation after a disturbance."[48]

Critiques of the concept of resilience suggests that (1) it is a concept rooted in the neoliberal discourse of choice, agency, and flexibility, (2) it focuses on the response to adversity but may well be evident in more normal conditions during everyday life, and (3) it may not sufficiently take into account "individual agency, situational context and processes of improvisation in everyday life."[49] As well, it has been critiqued for being imprecise, possibly inappropriate, and for reproducing some of the same biases and stereotypes that occur with discussions of risk and protection, particularly that of blaming victims.[50]

A further complicating factor is that a process or factor may contribute to resilience at smaller scales but detract from it at larger scales (e.g., The Tragedy of the Commons[51]). What is good for an individual may harm the larger community. Similarly, one could argue that some resilience variables interact with others in perverse ways; one example is economic development at the expense of environmental quality (e.g., developing the Tar Sands). Because of these factors, many decisions around resilience (like vulnerability) involve trade-offs.

4.6.1 The Shift toward Resilience Thinking

There are ancient references to the concept of resilience, such as Aesop's Fable of *The Oak Tree and the Reeds* and the Japanese proverb "The bamboo that bends is stronger than the oak that resists." Notions of vulnerability and resilience both address response to adversity but are different in that resilience, particularly from engineering and ecological perspectives, tend to be underlain by a positivist approach. Vulnerability theory, on the other hand, is more influenced by social constructionism. Resilience thinking is consistent with an all-hazards approach to disaster management and integrates social and physical systems. Incorporating notions of uncertainty and change, it is a very useful way of looking at how systems respond to crisis.

One reason why there has been a shift toward resilience in the past decade is that it emphasizes strengths and capacities within communities, whereas vulnerability emphasizes weaknesses. This shift is also the result of an explosion of literature on the subject and its widespread adoption by the social sciences over the past few decades. Coupling resistance, vulnerability, and resilience approaches forms a much more holistic approach to disaster risk reduction than any approach by itself.

People, communities, and systems are both vulnerable and resilient simultaneously. One interesting example of this is the elderly, frequently cited as a vulnerable group in vulnerability assessments. Elders, though, are considered a source of wisdom by First Nations and are explicitly part of their Emergency Management Circle (Chapter 6.5.9). The importance of elders has been noted elsewhere as well. In a narrative-based study of the coping of older adults to a flood disaster in New Zealand,[52] the authors highlight their resilience: "life wisdom and the development of a life narrative is an important aspect of dealing with traumatic experience." They emphasize the importance of remembering the past as a resource for the present and future. Because of the lives they have led and their greater experience, elderly people are usually more able to place events into a larger context and can be better at (re)creating narratives, an important process in providing meaningfulness to crisis. The authors of this study note that "Within each story the narrator drew on a coherent ontology of self, which operated to maintain an ongoing personal identity in the face of disruption and discontinuity, and to make sense of the disruptive even." and that "These findings suggest that a lifetime of experience provides resources for psychological resilience and strength rather than vulnerability in the face of disaster..."

4.6.2 Engineering Resilience

The first scientific concept of resilience originated from physics and the observation that some bodies, described as being elastic, regain their shape after being deformed. Hooke's law, first stated in 1678, is a statement of engineering resilience. It is an observation about the elastic properties of some materials and states that the force needed to compress or deform a spring varies directly with the amount of compression or deformation: $F = kX$ where F is the force applied, k is the coefficient of stiffness, and X is the amount of stretching. Beyond some point, even elastic bodies reach their limit, as illustrated in Figure 4.14, and subsequently show plastic properties, which unlike elastic bodies do not regain their original shape after deformation. Every object, person, community, and system has its breaking point. Engineering resilience addresses properties of efficiency, constancy, and predictability, all of which are needed for good fail-safe designs, and assumes the presence of a single stable steady-state system. Fail-safe designs (in the sense that they are not supposed to fail) can be contrasted with safe-fail designs (fuses are an example), where failure occurs with a minimum of harm. Safe-fail is an important component of resilience and vulnerability.

The elastic property of materials has been transferred to other fields of study to describe the ability of people, communities, or systems to bounce back to their original

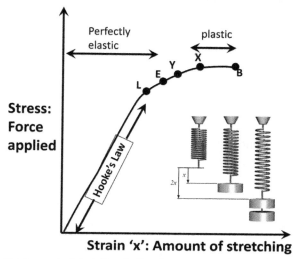

FIGURE 4.14 Hooke's law. The amount of stretching is directly proportional to the force applied, up to a limit, after which stretching is nonlinear and the object exhibits plastic properties. L, limit of proportionality; E, elastic limit; Y, yield point (plastic deformation begins); X, body has permanent strain; B, breaking point. Spring illustration. *Source: Wikipedia,* http://en.wikipedia.org/wiki/File:Hookes-law-springs.png.

state following a disturbance. Because of this, resilience is often described as a metaphor when applied to the social sciences.

Engineering approaches to resilience have evolved considerably since 1678, as illustrated by the following quote from the Resilience Engineering Network[53]:

The term Resilience Engineering represents a new way of thinking about safety. Whereas conventional risk management approaches are based on hindsight and emphasise error tabulation and calculation of failure probabilities, Resilience Engineering looks for ways to enhance the ability of organisations to create processes that are robust yet flexible, to monitor and revise risk models, and to use resources proactively in the face of disruptions or ongoing production and economic pressures. In Resilience Engineering failures do not stand for a breakdown or malfunctioning of normal system functions, but rather represent the converse of the adaptations necessary to cope with the real world complexity. Individuals and organisations must always adjust their performance to the current conditions; and because resources and time are finite it is inevitable that such adjustments are approximate. Success has been ascribed to the ability of groups, individuals, and organisations to anticipate the changing shape of risk before damage occurs; failure is simply the temporary or permanent absence of that.

Traditional engineering resilience, which emphasizes efficiency, control, constancy, and the return to a previous stable state, when transferred to psychosocial systems has been critiqued as being too simplistic; models of resilience developed from ecological theory are better at addressing several aspects of human systems, including adaptive management. After all, people (perhaps not often enough) learn from their environment, whereas springs do not.

4.6.3 Ecological Resilience

The first literature on ecological resilience emerged in the 1970s[54] and was further developed with the notion of Panarchy (see Chapters 5 and 6). This approach considers systems existing far from stable steady-state conditions and acknowledges the presence of multiple stable states where instabilities in the system can flip it from one state into another. "In this case, resilience is measured by the magnitude of disturbance that can be absorbed before the system redefines its structure by changing the variables and processes that control behavior."[55] Whether or not there exists a single or multiple stable states of a system is the determining factor as to which model (engineering or ecological) of resilience is appropriate. We need to know when conditions result in nonlinear responses, particularly threshold effects or runaway feedbacks. However, this level of knowledge is not always attainable.

Building ecological resilience emphasizes ongoing renewal and reorganization and can be addressed through the following broad mechanisms that address the variables of flexibility, redundancy, diversity, and connectivity[56] by:

- buffering disturbances to reduce their impact
- learning to live with change and uncertainty
- nurturing diversity
- creating opportunities for self-organization and cross-scale linkages to maintain crucial system functions
- learning and adaptation.

Some of the above mechanisms rely on the interaction of processes that occur at different scales, in the sense that slower larger-scale processes provide a level of stability for faster smaller systems, which can then engage in experimentation and learning. This is common in ecosystems where species adapt to local conditions, but larger ecosystem functions are maintained through, for example, the accumulation and storage of nutrients. A social example might be different communities experimenting with various recovery strategies, while the national system of governance provides the support and stability required for the experiments.

Similar to ecological systems are psychosocial systems, and theories of resilience coming from the former have been increasingly adopted by the latter. There are strong connections between ecological resilience and the material in Chapter 5 on complexity.

In a very useful discussion of resilience in socioecological systems[57] (SES), the authors Walker, Holling, Carpenter, and Kinzig make a distinction between resilience, adaptability, and transformability, linked system characteristics that determine stability dynamics.[58] The first, resilience, addresses "the capacity of a system to absorb disturbance and reorganize while undergoing change so as to still retain essentially the same function, structure, identity, and feedbacks" and is determined by four factors:

- Latitude: the maximum amount a system can be changed before losing its ability to recover (before crossing a threshold that, if breached, makes recovery difficult or impossible).
- Resistance: the ease or difficulty of changing the system; how "resistant" it is to being changed.
- Precariousness: how close the current state of the system is to a limit or "threshold," which if breached makes recovery difficult or impossible.
- Panarchy: because of cross-scale interactions, the resilience of a system at a particular focal scale will depend on the influences from states and dynamics at scales above and below. For example, external oppressive politics, invasions, market shifts, or global climate change can trigger local surprises and regime shifts.

The second characteristic, adaptability, refers to the ability of social systems to intentionally manage vulnerability and resilience; after all, we do believe that people and societies can deliberately manage their affairs. The social learning required to do this leads to an interesting conundrum since it requires smaller-scale events (i.e., disasters) that can act as learning opportunities that can be applied at larger scales. This is sometimes acknowledged by disaster managers, who ironically note that the best way to prepare for a large disaster is to have a few smaller ones. This is a tricky business; even small disasters have the capacity to either strengthen or weaken people and communities and tend not to be socially acceptable events. Adaptable SES can intentionally alter latitude, resistance, precariousness, or Panarchy. There are, however, social traps that can prevent this from occurring; in these cases, an SES has become locked into a fragile state.

Systems become transformed when they configure themselves in an entirely new and stable landscape. An example of this is a forest converted to agricultural use, or a change in government from a monarchy to a republic. Transformability emphasizes the attributes of novelty, diversity, effective institutions, and strong cross-scale communications. These issues are important to the discussion of environmental ethics in Chapter 9.4.4.

Applying notions of resilience to global dynamics is an interesting exercise. Recent global trends include increased connectedness, greater densification (such as urbanization), declining diversity, and more rapid interactions between systems and subsystems. As well, the usual process where large-scale events move slowly and provide stability to smaller-scale events (which move quickly and can be short lasting) is changing such that the large-scale events are losing their ability to create stability and can act as a disturbance to more local scales[59] (consider the effect of the 2008 recession that began in the United States). These trends have important implications that are yet to be fully understood but suggest that the more traditional ways of assessing risk (such as stability and control) need to shift toward a paradigm of resilience, adaptability, and transformability.

4.6.4 Psychological Resilience

The adoption of resilience by psychologists traces its roots back to studies of concentration camp survivors during World War II, where some individuals continued to function in the face of unspeakable adversity while others did not.[60] Other long-term research on children at risk showed large variations in outcomes under similarly adverse conditions.[61] But this is not just about individuals. The processes that determine resilience include genetics, neural systems, and an individual's interactions with family, workplace, and community. These critical relationships are shown graphically from a child's perspective in Figure 4.15. One example of the importance of interactions is the phenomenon of "social referencing" (the contagious effect of feelings and behaviors), which can be very significant in terms of how people frame and respond to crisis. This particularly applies to the resilience of children, who largely depend on their relationships with parents to provide a place of security and reassurance in threatening situations. Along these lines, Masten notes that "All planning for disaster must account for the attachment system and how such relationships are likely to motivate behavior and provide for a sense of security." Attachment to place can also be extraordinarily strong and often motivate people to reconstruct their lives in hazardous areas.[62] For example, an article from the *Daily News* notes that "Repair, rebuild and reopen—that's the mantra for the post-Sandy Jersey Shore."[63]

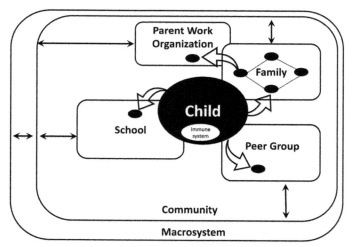

FIGURE 4.15 Multiple-level embedded systems that interact to influence a child's life. The child as a living system is shown embedded in three microsystems: family, peer group, and school. These microsystems are in turn embedded in larger-scale systems: the community and the macrosystem. *Source: Adapted from Masten, A.S. and Obradovic, J. (2007). Disaster preparation and recovery: lessons from research on resilience in human development. Ecology and Society 13(1), 9.*

Psychological literature discusses resilience in terms of functionality and structure, as do other fields, and emphasizes the importance of flexibility and adaptability rather than a return to an original state. Unique aspects such as identity, faith, and spirituality not present in engineering or ecological paradigms are also important. A disaster can result in a shift to different patterns of behavior, and it is common for victims of disaster to comment on newfound insights and values. In complexity terms, this would be viewed as a shift to a different strange attractor, while in ecological terms it would be described as an alternate functional state. The discussion in Chapter 7.5 on the myths at the core of our internal world provides one possible explanation for such a shift.

I found a comment by Ann Masten[64] to be rather interesting—that resilience depends on "ordinary magic," or the everyday adaptive systems that have evolved over time. This parallels the thoughts of Ken Hewitt who also emphasizes the importance of everyday patterns in the construction of risk. Thus, the notion of both social capital (adaptive capacity that results from social relationships) and human capital (adaptive capacity that is internal to individuals) become crucial to the resilience paradigm.[65]

Psychological resilience plays a large factor in recovery. For example, about one-third of survivors of the Orissa cyclone of 1999 had disabling symptoms one year after the event.[66] Nevertheless, studies have shown that "Mental health aspects of relief and rehabilitation have been frequently neglected in disaster initiatives."[67] While responding to the 2004 tsunami in the Indian Ocean, a team of mental health providers from the National Institute of Mental Health and Neurosciences in Bangalore, India, used a variety of strategies to help victims recover, including active listening, empathy, ventilation, relaxation, play therapy, storytelling, and spirituality.[68] A particular emphasis was placed on reuniting families and restoring normalcy to their lives. Adult survivors commonly experienced guilt and anger, as well as fear, depression, anxiety (specifically hypervigilance, hyperarousal, sleeplessness, depersonalization, and panic attacks) and posttraumatic stress syndrome. Gender differences were present, with men more likely to experience alcoholism and women more likely to experience depression. Self-guilt and mutual blame were common. The reactions also included resentment of the invasion of privacy by media and fear of being dependent on aid. The study found that involving victims in planning and recovery activities reinforced their needs for dignity and independence. The psychosocial care that was provided to adults was based on the strategies of ventilation, empathy, relaxation techniques, spirituality, active listening, social support, and the externalization of interests and recreation to distract thoughts from the disaster. Other research has shown the usefulness of civic ecological activities in the healing process, such as gardening, developing urban forests, and stewardship.[69] In the words of one survivor, "even one small little garden can create a sense of peace and hope."

A person's resilience is a function of their creativity, competence, problem solving ability, persistence in the face of adversity, level of self-control, and their belief in their own capabilities—all of which are positively correlated with confidence and positive views of self-worth and effectiveness. External factors that affect resilience during disaster include the amount and quality of available information, the strength of personal networks (especially the attachment network of family and friends), and the community competences and skills that are present to enable response and adaptation.

Until recently, critical incident stress debriefing (CISD) was a favored technique to assist people after a disaster. Multiple studies have shown, however, that CISD is not only ineffective but in some cases can actually be psychologically harmful.[70] Other less invasive and more practical forms of immediate intervention have been developed for use with both children and adults.

> What doesn't break us makes us stronger. An article written by Diane Gail Colina called *Reflections from a Teacher and Survivor*[71] said the following:
>
> "But the frustrations brought tremendous change in me. The heat, the workload, the sorrow at seeing such destruction began to release in me an adrenaline that forced me to stand on my own two feet. If once I had been the woman who never said no, now I was developing my own voice and it was plenty pissed off. I began to argue with males or authority figures for the first time in my life. I asserted my rights with contractors, roofers, adjusters, even my bosses at work. The more I practiced, the better it felt. No longer would I sit back and wait. When the insurance office called a security guard to stand by as I ranted and raved about the promised check that could not be found once again, I realized someone was actually afraid of me. This was an incredible feeling of empowerment—the first time I ever felt that way.
>
> Perhaps it was seeing my students' difficulties that gave me the strength to get up every morning, take a cold shower by candlelight, and report to work. Perhaps it was seeing their courage to cross the seas on a small boat that diminished my own troubles. In the aftermath of the hurricane I found a source of strength and power of survival that I hold within myself and because of that I am a stronger person today."

4.6.5 Community Resilience

The word *community* can be interpreted in many ways. Traditionally, it refers to a group of people geographically tied together (such as living in the same town), although it can also refer to any group of people connected through some common bond, such as a club, religious/cultural group, or professional association. More recently, new kinds of communities (such as gaming or other special interest groups) have evolved as a result of the Internet.

Community resilience assessments either can examine how people use community resources to enhance resilience through social, workplace, cultural, and other networks, or can look at how communities themselves function. From this perspective, community resilience can be seen as "a set of capacities that can be fostered through interventions and policies, which in turn help build and enhance a community's ability to respond and recover from disasters."[72] Using this approach, Susan Cutter developed a composite Resilience Index (similar to the Social Vulnerability Index) that can be used to measure baseline factors that contribute to community resilience. While recognizing that there is no single set of indicators that have been universally accepted for quantifying resilience, she selected the variables listed in Table 4.1 using the categories of social, economic, institutional, infrastructure resilience, and community capital.

Table 4.1 Variables Used to Measure Disaster Resilience (Cutter[74])

Component	Category	Variable	Effect on Resilience
Community capital	Place attachment	Net international migration	Negative
Institutional resilience	Political fragmentation	No. of governments and special districts	Negative
Social resilience	Educational equity	% Population with college education: % population without high school	Negative
Community capital	Innovation	% Employed in creative class occupations	Positive
Community capital	Place attachment	% Born and still residing in state	Positive
Community capital	Political engagement	% Voter participation	Positive
Community capital	Social capital—advocacy	No. of social advocacy organizations/10,000 population	Positive
Community capital	Social capital—civic involvement	No. civic organizations/10,000 population	Positive
Community capital	Social capital—religion	No. religious adherents/10,000 population	Positive
Economic resilience	Business size	Ratio of large to small businesses	Positive
Economic resilience	Employment	% Employed	Positive
Economic resilience	Employment	% Female labor force participation	Positive
Economic resilience	Health Access	No. physicians/10,000 population	Positive
Economic resilience	Housing capital	% Home ownership	Positive
Economic resilience	Income and equality	GINI coefficient	Positive
Economic resilience	Single sector employment dependence	% Not employed in farming, fishing, forestry and extractive industries	Positive
Infrastructure resilience	Access/evacuation potential	Principle arterial miles/square mile	Positive
Infrastructure resilience	Housing age	% Housing units from 1971 to 1994	Positive
Infrastructure resilience	Housing type	% Housing units that are not mobile homes	Positive
Infrastructure resilience	Medical capacity	Hospital beds/10,000 population	Positive
Infrastructure resilience	Recovery	No. public schools/square mile	Positive
Infrastructure resilience	Shelter capacity	% Vacant rental units	Positive
Infrastructure resilience	Sheltering needs	No. hotels and motels/square mile	Positive
Institutional resilience	Flood coverage	% Housing covered by NFIP	Positive
Institutional resilience	Mitigation	% Covered by a recent hazard mitigation plan	Positive
Institutional resilience	Mitigation	% Participating in a community rating system for flood	Positive
Institutional resilience	Mitigation	% In storm-ready communities	Positive
Institutional resilience	Mitigation and social connectivity	% Covered by citizen corps programs	Positive
Institutional resilience	Municipal services	% Expenditures for fire, police and EMS	Positive
Institutional resilience	Previous disaster experience	No. of paid disaster declarations	Positive
Social resilience	Access to transportation	% Population with a vehicle	Positive
Social resilience	Age	% Nonelderly	Positive
Social resilience	Communication capacity	% Population with a telephone	Positive
Social resilience	Health coverage	% Population with health insurance	Positive
Social resilience	Language competency	% Population not speaking English as a second language	Positive
Social resilience	Special needs	% Population without a disability	Positive

Cutter and her fellow researchers acknowledge that this is a "broad brush" approach and that analyses that are more detailed are required to better understand communities. Particularly, an argument can be made that resilience is culturally relative and depends on local norms and social roles. There are other problems with indice-based models as well, including the assumption that different factors are independent of each other and can be added. Additionally, resilience must be viewed as a dynamic process that varies over time and depends on changes within individuals, the communities within which they live, and the shifting larger social, economic, and environmental context.

Social capital is a concept that is essential to understanding community resilience. It includes social relationships, networks and reciprocity, shared norms and values, a culture of trust, collective participation, and access to resources,[73] all of which contribute to the strength and effectiveness of individual and community relationships. In some places these factors may rely heavily on family and informal relationships, while in others they may be more institutionalized.

Community resilience is both a function of the collection of individuals who comprise it and emergent properties that are dependent on social capital, social/cultural values, and practices,[75] including duties to family members, the role of authorities within the community, trust, level of engagement and ethical norms. An adaptive management perspective would emphasize the degree to which experimentation and social learning is encouraged by society. Some societies are very rigid, particularly ones that are strongly orthodox or fundamentalist.

The UN ISDR developed a guidance note based on the principles of disaster risk reduction (DRR), on the characteristics of a disaster resilient community.[76] They emphasize the generality of the variables, noting that each community is unique and should therefore apply it appropriately within their own context. Their approach to resilience includes the following:

- capacity to absorb stress or destructive forces through resistance or adaptation
- capacity to manage, or maintain certain basic functions and structures, during disastrous events
- capacity to recover or "bounce back" after an event.

Variables are evaluated according to the following scale:

- Level 1. Little awareness of the issue(s) or motivation to address them. Actions limited to crisis response.
- Level 2. Awareness of the issue(s) and willingness to address them. Capacity to act (knowledge and skills, human, material. and other resources) remains limited. Interventions tend to be one-off, piecemeal, and short term.
- Level 3. Development and implementation of solutions. Capacity to act is improved and substantial. Interventions are more numerous and long term.
- Level 4. Coherence and integration. Interventions are extensive, covering all main aspects of the problem, and they are linked within a coherent long-term strategy.
- Level 5. A "culture of safety" exists among all stakeholders, where DRR is embedded in all relevant policy, planning, practice, attitudes and behavior.

The note addresses five thematic areas based on the Hyogo Framework for Action:[77]

- Thematic Area 1: governance
- Thematic Area 2: risk assessment
- Thematic Area 3: knowledge and education
- Thematic Area 4: risk management and vulnerability reduction
- Thematic Area 5: disaster preparedness and response.

Each thematic area is divided into three sections:

- Components of resilience
- Characteristics of a resilient community
- Characteristics of an enabling environment.

Table 4.2 lists the specific components of the first factor for the thematic areas.

Table 4.2 Specific Components of the First Factor for the Thematic Areas

	Thematic Area	Components of Resilience
1	Governance	• Policy, planning, priorities, and political commitment • Legal and regulatory systems • Integration with development policies and planning • Integration with emergency response and recovery • Institutional mechanisms, capacities, and structures; allocation of responsibilities • Partnerships • Accountability and community participation
2	Risk assessment	• Hazards/risk data and analysis • Vulnerability and impact data/indicators • Scientific and technical capacities and innovation
3	Knowledge and education	• Public awareness, knowledge, and skills • Information management and sharing • Education and training • Cultures, attitudes, motivation • Learning and research
4	Risk management and vulnerability reduction	• Environmental and natural resource management • Health and well-being • Sustainable livelihoods • Social protection • Financial instruments • Physical protection; structural and technical measures • Planning regimes
5	Disaster preparedness and response	• Organizational capacities and coordination • Early warning systems • Preparedness and contingency planning • Emergency resources and infrastructure • Emergency response and recovery • Participation, voluntarism, accountability

For more detailed information, I recommend reading the source document.[78]

The Canadian National Platform on DRR has a number of working groups. one of which, as part of the World Disaster Reduction Campaign "Making Cities Resilient," focuses on the development of resilient communities.[79] The Resilient Communities Working Group supports the Canadian Platform by working to provide information to Canadians, communities, and regions on actions that they can take to increase their resilience to disasters. Group members work to bring the idea and importance of disaster risk reduction to communities and regions, and encourage communities to engage in activities that can make them more resilient.

4.6.6 Religion and Spirituality

Too often, development programs are done with a secular worldview that can privatize or problematize religion, thereby overlooking the religious sources of resilience.

Margarita A. Mooney

Religion and spirituality might best be considered as sources that contribute to resilience, as opposed to a kind or type of resilience. Their role with respect to individual and community recovery has been examined in several studies and has been shown to be helpful,[80] although different studies show varying levels of long-term effect.[81] This issue varies greatly by culture and worldview. Religion and spirituality have the ability to: allow people to create a vision of a better postdisaster world (Box 4.1 and 4.2); provide meaning and purpose; facilitate psychological integration, hope and motivation, personal empowerment, a sense of control; provide role models for suffering, guidance for decision making, social support; and to create feelings of self-efficacy.[82] The effect varies by culture, being more pronounced in religious countries such as Haiti and much less so in less religious countries such as Norway.

BOX 4.1　COMMENTS BY SURVIVORS

I knew me and my family was going to be alright because of the power of prayer.

I've come this far by faith and trust in God.[86]

This is the source of morale and motivation. I strengthen myself with prayers and supplication.

I think God wanted to test me to absolve me of my sins. I think that is why I was not martyred, I stayed alive; a chemical warfare victim. He doesn't want me to effortlessly join the legion of martyrs.[87]

From a social capital perspective, religion and spirituality create supportive networks that can be effective in the recovery process in spiritual, psychological, emotional, and physical ways. It is not unusual for people to form attachments to spiritual and religious leaders similar to those within family units. One study of older black survivors of Hurricane Katrina, not all of whom attended church, found that "Without exception, the findings indicate that this population coped with Katrina and its aftermath through reliance on a Higher Power."[83] Their coping mechanisms included "(1) regular communication with a supernatural power; (2) miracles of faith through this source of guidance and protection; (3) daily reading of the Bible and various spiritual and devotional materials; and (4) helping others as a consequence of faith and devotion to a supreme being." This group of survivors had a long history of religious beliefs helping them cope with centuries of slavery and discrimination. In another study of Iranians injured by mustard gas during a war with Iraq,[84] religious sentiment was identified as a primary coping mechanism, with subthemes of divine will, illness as a means of absolving sin, saying prayers in the anticipation of divine rewards, and self-sacrifice as a source of pride. Spirituality also appears in First Nations' approaches to disaster planning (Chapter 6.5.9). Aboriginal "networks are embedded in and sustained by value systems that include notions of personhood, ethics, and religion or spirituality"[85] and employ the medicine wheel as a useful tool for holistic thinking and balance. In their culture, ceremony and spirituality is used by people to make sense of their place in the world and to transfer the wisdom of the elders to the larger community.

A secular approach to development or recovery that ignores this important human factor will be less successful than one that embraces it, particularly for highly religious cultures; unfortunately, it is unusual for mainstream disaster managers to use religion

---

BOX 4.2   A PRAYER[96]

O God, our refuge and strength,

Our help in times of trouble.

Have mercy on the lands where the earth has given way.

Have mercy on the lands where the weather has destroyed livelihoods.

Prosper those who rebuild houses,

and strengthen those who rebuild hope

so that entire communities

may face the future without fear.

Amen

and spirituality as a resource or to consider it as a significant factor affecting resilience.[88] Some research has even shown that there is "...a certain wariness between faith-based and secular organizations around the issue of proselytizing and credentials" and that there are noticeable professional and cultural differences between the secular and faith-based approaches to response.[89] There have been documented cases of proselytizing or negative impacts postdisaster as a result of religious interpretations of the event (such as God's punishment for a lack of righteous behavior),[90] which likely accounts for much of the secular bias. In one study following a cyclone disaster in the Cook Islands in the South Pacific, Taylor[91] noted that "To an outsider, the moral obligation imposed on the survivors in the immediate aftermath of the disaster seemed to endanger their already fragile sense of security and self-esteem. Such self-blame was considered a maladaptive method for coping with continuing trauma.... It was also thought to reinforce feelings of helplessness in the survivors, create an extra burden, and impede the recovery of the survivors at a time when encouragement, inspiration, and support from all sources were more likely to help them maintain their desperate existence. It also introduced an element of discord at a time when community needed to affirm its bonds, share its grief, acknowledge the heroic deeds of its members who saved many lives, and ponder its future location." There are practical difficulties as well. Most faith-based community organizations involved in a disaster response have no prior experience and therefore would not be familiar with other disaster management organizations or protocols such as the Incident Command System.

Two lengthy guidebooks on the topic of disaster and spirituality are *Light Our Way, A Guide for Spiritual Care in Times of Disaster*[82] and *NYDIS Manual for New York City Religious Leaders: Spiritual Care and Mental Health for Disaster Response and Recovery*.[93] They address issues for the caregiver, as well as victims.

The first guidebook lists three definitions of spirituality:

- *Spirituality is a personal quest for the transcendent, how one discerns life's meaning in relation to God and other human beings. Healthy spirituality fosters healthy relationships and affirms all of life's experiences as part of the journey.*

  Rabbi Eric Lankin

- *There is no essential demarcation between sacred and mundane, or the secular and spiritual. All of life's activities are infused with a spiritual dimension—echoing as it were, Divine remembrance—so as not to consider the material (including our earthly life) as an end unto itself.*

  Dr. Faiz Khan

- *Spirituality is the essence of life—the beliefs and values that give meaning to existence and that which is held sacred. It is one's understanding of self, God, others, the universe, and the resulting relationships.*

  Rev. Naomi Paget

Spiritual people are noted to have the following characteristics:

- "A sense of awe and wonder
- A sense of community
- A sense of personal mission
- Enthusiasm for continuous discovery and creativity
- A sense of well-being and joy."[94]

Spirituality is part of the way that people and communities understand and give meaning to disasters. However, disasters affect people's spirituality, often by making them reconsider their beliefs and disengaging them from faith-based practices and communities, and by creating a sense of despair and hopelessness. As noted above in the Cook Islands example, religious responses can be maladaptive, although that appears to be much less common than positive adaptive responses.[95] The guidebooks suggest that providers of spiritual care:

- "Offer security, presence and hospitality
- Meet, accept and respect persons exactly as they are
- Listen and provide support
- Focus on the needs of the survivor
- Do No Harm – Never evangelize, proselytize or exploit persons in vulnerable need."

It seems clear that religion and spirituality are important components of understanding human response to disaster, and that their effects are generally positive for people and communities. Unfortunately, disaster planning is handicapped by insufficiently including this aspect of the human condition.

Ethics and Resilience: There seems to be little, if anything, written on the topic of ethics and resilience unless one considers the topic of corruption (which is really about the lack of ethics). Certainly, data and anecdotal stories exist to demonstrate the negative impacts of corruption on society in general and disaster impacts in particular (Chapter 4.5.2). It seems intuitive that an ethical community will be more disaster resilient than one that is not. A community that is guided in its decision making by ethical considerations is more likely to have strong bonds between people and, therefore, a stronger human network and larger social capital. At an individual level, people would help each other because they believe it is the right thing to do, while at a community level decisions would be more likely to evaluate outcomes in a broad way, so as not to be unfair or unnecessarily alienate portions of the community.

What is an ethical community? It is more than simply a community composed of ethical people, although that may be an essential criterion. Kant, in his *Groundwork for the Metaphysics of Morals*[97] published in 1785, suggests that an ethical community would result if all people were to act according to moral law. I believe that an ethical community is one that also has processes and structures in place that enhance ethical thinking and provides avenues through which ethical thought can be translated into policy making, governance (through the use of procedural ethics), and technological design.[98] It would have to be a community that values democratic processes, transparency, a sense of obligation to the greater social good, and the development of trusting relationships. Fundamentally, it would be one with a strong social contract between the state and citizens, rooted in social norms. The topic of ethics and resilience is ripe for research and may provide valuable insights into community resilience.

4.7 Grassy Narrows

It is a place where rage has been turned inward against one's kin and one's Self.
                                                                    Anastasia Maria Shkilnyk

Academics normally make it a point not to react emotionally to what they are studying. I found this particularly hard to do with Grassy Narrows–not because I know any of the people involved or was in any way connected to the disaster but because of the implacable nature of the institutional evil that created it and because it happened in my own province and country.

I just used the word *evil* to describe what happened. I cannot think of a single case study or textbook on disaster that uses that word, but I use it unapologetically. Disasters destroy, and where the vulnerability that led to them is knowingly created, then that destruction happens through deliberate intent. I certainly do not claim that all such processes are evil, for we live in a dangerous world that cannot be made completely safe (as much as we would like it to be), and trade-offs are inevitable. But at times, some of us are callously and unjustly delivered into the chaos of destruction, a frequent reason being economic advantage or power, without any specific ill will directed toward the sacrificial victims. I cannot decide if the impersonal aspect makes it more, or less, dreadful.

Prior to 1970, no suicides were ever recorded at Grassy Narrows. In 1977–1978, there were 28 attempted suicides of people aged 11–49 years of age out of a population of about 500.[99] Three succeeded, two 12-year-old girls and a 17-year-old boy. Shkilnyk[100] estimates that 17% of the population of Grassy Narrows between 11 and 19 wanted to die. As an indicator of community suffering, this is both extraordinary and tragic. The Canadian Broadcasting Corporation produced a very good documentary on this disaster that is well worth viewing, available at: http://archives.cbc.ca/.[1]

The short story of the Grassy Narrows disaster focuses on poisoning of the environment. Between 1962 and 1970, Dryden Chemicals and Reed Paper Limited released about 10 tonnes of mercury into the Wabigoon River from the pulp and paper mill that was upstream of Grassy Narrows. The mercury then bioaccumulated in the food chain, finally reaching toxic levels in the human population, who tragically consumed large numbers of fish from the river. In 1975, a Japanese doctor who specializes in Minimata disease (the result of mercury poisoning) visited Grassy Narrows and found that the residents had been poisoned by eating contaminated fish and that mercury levels were pervasive in the local ecosystem. Levels in residents at Grassy Narrows were up to three times the acceptable limit as defined by Health Canada and seven times that limit at Whitedog. He also commented that the Health Canada limits were, in his opinion, too low to protect people from long-term cumulative effects, especially for pregnant women. Symptoms experienced by the residents included pain in the limbs (45%), numbness (31%), cramps (18%), sensory

---

[1] Search using the terms "disasters" and "grassy" and "narrows".

disturbances (17%), constriction of the visual field (10%), ataxia (9%), and tremor (24%). As a result, he diagnosed them as having mild Minamata disease. This process, where polluting industries preferentially create environmental devastation near First Nations communities, has been referred to as environmental racism.[101]

Though the Ontario and Federal governments did not officially acknowledge the presence of this disease since there were no severe cases, commercial fishing was banned following this evaluation, and a warning was issued against eating contaminated fish. The result of the ban was that Grassy Narrows went from a 95% employment rate to a 5% employment rate within one year; as of 2001, the employment rate was still only 55%. The result of this loss of employment was to create a culture of dependency and a further descent "into social disintegration and self-destruction."[102] As a result of many years of negotiation, a Mercury Disability Board that provided compensation by the government of Canada was established in 1986. The government of Ontario, Reed Incorporated, and Great Lakes Forest Products Limited (the owner of the plant at the time of the agreement) allocated between $250 and $800/month (not indexed to inflation) for the 140 residents who could establish disability through mercury poisoning. In the opinion of Dr Harada, many affected residents are not being compensated. Dr Harada returned to Grassy Narrows in 2002 to follow up his research and found that 65% of the people he examined had Minamata disease (possibly with complications), and another 14% possibly had it. Developmental and cognitive delays have resulted from this chronic exposure. Research shows that mercury levels in the area have been decreasing as it diffuses throughout the environment and ecosystem (Figure 4.16), suggesting that there will be no new cases. The ban on eating fish has been lifted.

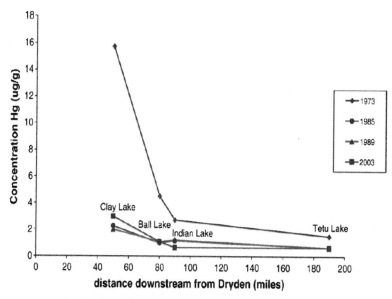

FIGURE 4.16 Mercury contamination in Walleye, in micrograms/gram. *Source: Kinghorn, A., Solomon, P., and Chan, H.M. (2007). Temporal and spatial trends of mercury in fish collected in the English–Wabigoon river system in Ontario, Canada. Science of the Total Environment 372(2), 615–623.*

In the short story, the disaster has been largely resolved. The hazard has been eliminated, compensation has been given to affected people, and the fish are now again safe to eat.

For the long story, one must take an historical perspective. The Ojibwa traditionally relied on hunting and trapping for sustenance, and their settlement and kinship patterns evolved to suit that way of life. With the advance of European colonization, however, they adapted by engaging in the fur trade, and became subject to the vagaries of trade wars, monopolies, and price fluctuations. In the 1820s, the Ojibwa turned to fishing as a major food source due to depletion of the animal population.

In 1873, they signed a treaty with the new government of Canada, in which they gave up 55,000 square miles of land and moved to a reserve where each family of five had only 1 square mile, not nearly enough for their traditional way of life. It was agreed that they would maintain the right to pursue their traditional hunting and fishing activities, subject to governmental needs related to settlement, mining, lumbering, etc. Therefore, they moved to two reserves, Grassy Narrows and Wabauskang. Interaction with "white" society increased with the construction of the railway, the presence of a Hudson Bay Company trading post, logging and mining activities, a pulp mill, and Catholic missionaries and schools.

In 1919, an influenza epidemic killed all of the population in Wabauskang except for those away on their traplines. Overall, 75% of the population of the communities died, and local medicine men were helpless. This event was also important from a cultural and spiritual perspective, since it resulted in a loss of faith in traditional medicine.

Residential schools were the next real assault on Ojibwa culture. It was the federal government's policy to create boarding schools, where children were removed from their families (by force if necessary) and sent to "white" schools for education. The following quote helps to understand the agenda behind this policy: "I want to get rid of the Indian problem. I do not think as a matter of fact, that the country ought to continuously protect a class of people who are able to stand alone… Our objective is to continue until there is not a single Indian in Canada that has not been absorbed into the body politic and there is no Indian question, and no Indian Department, that is the whole object of this Bill."[103] About 200,000 children attended residential schools; if they did not go voluntarily, they were forcibly removed from their families.[104] Physical, emotional, and sexual abuse were common in many of these schools, and children were not allowed to speak their native languages or practice their cultural traditions. Illness was common, particularly tuberculosis, and in some schools death rates reached between 24% and 42%.[105] Another estimate put mortality rates among school age children in a range between 35% and 60%.[106] Paradoxically, the goal of assimilating them into white Christian society was far from attained since they were rejected by the same society that denied them their cultural heritage. Children from the schools lost the old ways and had nothing to replace them. Removed from parents, family, and the wisdom of elders, it is hard to imagine how they could not become deeply damaged human beings. They were a lost generation in the deepest meaning of the word. Residential schools were a form of cultural genocide, and the last one did not close until 1990.[107]

During the 1940s and 1950s, changes to resource management and commercial fishing forced the Ojibwa to pursue a cash economy, a change that ultimately had drastic consequences for a culture based on trade and the family unit. Historically, the entire family participated in hunting and fishing; everybody had their role and contributed their part. With a single wage earner, the tradition of the family as the fundamental economic unit was lost. This led to the fracturing of the family in other ways as well. "The new way of life has rid the community of any need for responsibility, accountability, and desire to improve their own conditions. In the eyes of most parents and children, education no longer has value."[108]

In 1956, an amendment to the Indian Act allowed Ojibwa to purchase liquor. Alcoholism and spree drinking became rampant in the community as a dysfunctional way to cope with adversity and contributed further to the downward spiral of traditional moral standards and family. By the late 1970s, almost half of the reserve income went to the Hudson Bay Company, while 21% was spent at the Ontario Liquor Control Board.

In 1961, the government relocated the Grassy Narrows band to a new site, against their will. The new site had poor soil that was not good for gardening, was far from trapping and hunting grounds, left no room for play areas or good access to water, and grouped families together spatially (in a grid format) in a way that was inconsistent with their historical patterns of community space. They referred to the new place as a cage, a corral, a prison, and a concentration camp.[109] The forced closeness resulted in tensions between clans and families. As well, the Ojibwa believed that the new reserve was sited in a bad place, inhabited by evil spirits. On every level important to the Ojibwa culture, the new location furthered the social disintegration of the community. I doubt this was the intent of the Canadian Government, but by imposing the values and structures of a capitalist European society on an animist First Nations community, the result was catastrophic.

Colonizers do not simply steal the land from Aboriginal people—as Europeans continue to arrive they proceed to steal biography, history, and identity.[110]

The mercury poisoning of the river was more than just a health threat to Grassy Narrows. To them the river, like the land, had spiritual associations, and that the "living waters" had turned against them was a shattering blow to their worldview, made all the worse by the invisible and dreaded nature of the threat. Cultural and social practices related to water include "places of prayer and bathing, stories, dancing and oral histories about water or water bodies, sweat lodges, purification ceremonies, drinking water collection sites, spring water of spiritual significance, medicinal plants nourished by water of spiritual significance, and medicine making."[111] The problem was not simply medical and economic but also psychological and spiritual.

The history and experiences of the Ojibwa at Grassy Narrows resulted in a broken culture, without resilience or capacity to heal itself from the mercury poisoning. It is within this context that the disaster needs to be understood. Vecsey[112] argues convincingly that much of the disaster response addressed symptoms but not root causes, and that what is needed is not subsidy but rather the provision of land and the autonomy needed for the Ojibwa to control their own destiny. As I write this book, much of the ancestral land around Grassy Narrows is under the threat of clear-cut logging, a further risk to their traditional way of life. They have been fighting this since 2002, when they established a blockade on a logging road in their territory. They are pursuing activism as a way to gain more control over their lives, and if justice is served, they will achieve it.

Further Reading

Chan, L., Solomon, P., and Kinghorn, A. (n.d.) Grassy narrows and Wabaseemoong first nations: "*Our waters, our fish, our people*": Mercury contamination in fish resources of two Treaty #3 communities.

Dhillon C, Young MG: Environmental racism and first nations: a call for socially just public policy development, *Canadian Journal of Humanities and Social Sciences* 1(1):25–39, 2010.

Duclos, C. (2010). Band-Aid Solutions to Self-Destruction? Development in Canada and the Case of Grassy Narrows, pp. 105–119. Kanata, Volume 3. *Undergraduate Journal of the Indigenous Studies* Community of McGill, Montreal, Quebec.

Erikson K: The Ojibwa of Grassy Narrows. In *A New Species of Trouble: Explorations in Disaster, Trauma, and Community*, New York, 1984, W.W. Norton & Company.

Ilyniak, Natalia (2012). Colonialism and Relocation: An Exploration of Genocide and the Relocation of Animist Aboriginal Groups in Canada. *Journal of Religion and Culture*: Conference Proceedings. 17th Annual Graduate Interdisciplinary Conference, Concordia University. Montreal, QC (March 2012), pp. 75–83.

Masazumi Harada, Masanori Hanada, Takashi Miyakita, Tadashi Fujino, Kazuhito Tsuruta, Akira Fukuhara, Tadashi Orui, Shigeharu Nakachi, Chihito Araki, Masami Tajiri, Itsuka Nagano: Long-term study on the effects of mercury contamination on two indigenous communities in Canada (1975–2004), *Research on Environmental Disruption* 34, 2005. No. 4 Spring 2005.

Shkilnyk, A.M. (1981). Pathogenesis in a Social Order: A Case Study of Social Breakdown in a Canadian Indian Community. PhD Thesis. MIT.

Vecsey C: Grassy Narrows Reserve: Mercury Pollution, Social Disruption, and Natural Resources: A Question of Autonomy,, *American Indian Quarterly* 11, 1987. No. 4 (Autumn, 1987), pp. 287–314.

Milloy, J.S. (1999). A national crime: The Canadian government and the residential school system, 1879 to 1986 (Vol. 11). Univ. of Manitoba Press.

4.8 Responsibility and Response Ability—Comments on Vulnerability and Community by John (Jack) Lindsay[113]

When David asked me to comment on this chapter, I immediately said yes. As I read it, I reflected on a paper I had prepared for the Public Health Agency of Canada (PHAC) in 2007 explaining vulnerability in a way that would be accessible to a wide range of potential readers but was still academically defendable.[114] To meet that challenge, I decided to ask journalism's classic 5 W's: who, when, where, what, and why? The process of asking these questions led me to ask the final question of *how* can we make a difference to vulnerability? I concluded that we must promote a sense of individual responsibility for community safety and collective responsibility for vulnerability.

How did I get to this conclusion? The first three W's are fairly easy. The disaster studies literature is well stocked with papers discussing who is vulnerable and how that vulnerability varies with time and space. We can identify socioeconomic and physical determinants of vulnerability based on past studies of disaster events and by looking, as David has in this chapter, and I have in my own research, at what other fields of study have to offer. We can group populations by these characteristics or consider vulnerability, as FEMA has done, in terms of what functions may need support rather than at the cause of the functional loss. In the end, we will probably come up with a list or matrix or some other way to distinguish those in the population who are more vulnerable than the rest of the population.

This concerned me when I wrote my paper for PHAC, which only received limited circulation in the purpose for which it was commissioned. Yet I feel my misgivings and conclusions still resonate after reading David's chapter and deserve revisiting here. My experience as an emergency management professional has a number of times involved me in looking at vulnerability grouped around a single "factor" such as working with people with physical disabilities, the elderly, or with new immigrants. This practical involvement with vulnerability gave me one perspective. As the father of a young man with significant physical and social challenges, I also have a personal stake in the matter—I am a member of a "vulnerable" population.

These two angles on the same issue, as a member of the community in need of support and as a professional tasked with providing that support, led me to two large issues regarding how emergency management can address vulnerability. They both are distilled from an understanding of vulnerability as a product of our society and seeing that how we make decisions as a community is integral to creating vulnerability. In my original paper, when answering the easy parts of who, when, and where, I kept highlighting that an individual is only more vulnerable in the context of, and in comparison to, the wider community. To me this is the critically important cornerstone to understanding vulnerability—we create vulnerability through the decisions we make as a community or, sadly and too often, the decisions we fail to make.

One drawback of seeing vulnerability this way is that it seems to take vulnerability out of the traditional jurisdictions and mandates of emergency management. Just as Canada's public health community has recognized that better population health will be achieved by more than just better health care services, emergency managers must accept that decreasing community vulnerability will require more than just better response plans. Dealing with vulnerability requires emergency management to become integrated in community decision making.

The other challenge is to know when to consider vulnerable groups and when to think about vulnerable individuals. It is certainly a step forward to see our communities as heterogeneous and to identify portions of our communities, both geographically and socially, which are at greater risk. We must take the next step to then understand the dynamics and interactions that further differentiate the risk within those populations. We must identify the functional factors that increase vulnerability and the demographic groups where these factors are concentrated.

To do this we must expand our emergency management activities to take an "All Hazard + All People" approach. This is not achieved by suggesting one hazard, such as terrorism, is an adequate representative of all the hazards facing a community or that an individual with the demographic and functional characteristics of the majority is an adequate representative of the whole diverse population. We must look at the full range of hazards that threaten our communities, consider their unique and common characteristics, and then juxtapose these with the functional and demographic determinants of vulnerability.

Engaging the affected communities is the best approach for an undertaking of this complexity. Participatory planning has been recognized in the urban planning literature and in practice as a useful way to work through difficult issues within our communities. Often these urban planning issues revolve around the same set of socioeconomic factors and the varied demographic influences that drive disaster vulnerability. It makes sense to work with the vulnerable people in the community, for instance with seniors groups, in order to identify the factors contributing to their vulnerability and, perhaps even more importantly, to recognize their capacities and resiliency.

The danger with taking a participatory approach is that it could devolve into blaming the victim or pit one vulnerable group against another for limited resources. It is important that in our work with the most vulnerable in our communities we continue to acknowledge that their vulnerability is felt by them as individuals but is the result of larger social processes they have limited, if any, influence on. We must not fall into a rhetoric of "the (insert vulnerable group) know their own needs best so let them solve it themselves."

Another concern is focusing too much on individual preparedness as the only way to meet the demands of a disaster. The current approach of promoting individual responsibility for disaster preparedness may imply that an individual is not entitled to support in a disaster, especially in the initial period, unless their needs are especially urgent. For the most vulnerable in our communities this is the time they may need the greatest support, even though their needs might not appear life threatening, and their inability to prepare themselves is an integral part of their vulnerability. This approach also fails to emphasize that reducing vulnerability and increasing resiliency is mutually beneficial. There is an unlimited supply of preparedness: one individual's preparedness does not come at the expense of another's. In fact, the more prepared one home is, the better off their neighbors are, as they will place less demands on the limited community resources. This logic extends to groups within the community. The rest of the population will gain in a disaster when the needs of the most vulnerable have being addressed through preparedness, as this will allow limited resources to be directed to the areas of greatest need.

For all aspects of emergency management we must consider who is accountable to take action and who has the skills and assets to act. This is the difference between responsibility and the ability to respond. When we are considering vulnerability we are, in part, considering those who do not have a full ability to respond without additional support and who, in turn, is accountable for their situation. This can be a complex relationship that involves a series of decisions and number of different decision makers acting over time and culminating in an individual's or a community's collective vulnerability to specific circumstances.

This is why I feel that the emergency management community must promote a sense of individual responsibility for community safety and collective responsibility for vulnerability. Each individual needs to contribute as much as possible to the overall level of preparedness in the community. This may start with individual preparedness for the immediate impacts of a disaster, but it must be extended to include helping those who require some support. The determinants of vulnerability are collectively generated in our communities, and therefore a community approach is needed to resolve them. We are all responsible for our own preparedness *and* for contributing to our whole community's ability to respond.

John "Jack" Lindsay
Associate Professor, Applied Disaster and Emergency Studies
Brandon University. lindsayj@brandonu.ca

End Notes

1. Van der Leeuw S. E., "Vulnerability and the Integrated Study of Socio-Natural Phenomena," *IHDP Update* 2/01 (2001): 6–7.
2. Burton I., and Kates R. W., "Perception of Natural Hazards in Resource Management," *The Natural Resources Journal* 3 (1963): 412.
3. Hewitt K., ed., *Interpretations of Calamity: From the Viewpoint of Human Ecology*, vol. 1 (Unwin Hyman, 1983). Hewitt K., *Regions of Risk: A Geographical Introduction to Disasters* (London: Longman, 1997).
4. Weiss R., "German, Hungarian Towns Battle Floods; Elbe River Yet to Crest," Bloomberg.com (June 8, 2013), http://www.bloomberg.com/news/2013-06-08/german-hungarian-towns-battle-floods-elbe-river-yet-to-crest.html.
5. "Wealthy US Coastal Enclaves Battle Surging Seas," *The Age: Environment* (August 20, 2013), http://www.theage.com.au/environment/climate-change/wealthy-us-coastal-enclaves-battle-surging-seas-20130820-2s7wf.html.
6. Burton I., Kates R. W., and White G. F., "The Human Ecology of Extreme Geophysical Events," Natural Hazard Research Working Paper (1968). Available from FMHI Publications, http://scholarcommons.usf.edu/cgi/viewcontent.cgi?article=1077&context=fmhi_pub&sei-redir=1&referer=http%3A%2F%2Fscholar.google.ca%2Fscholar%3Fq%3Ddisaster%2Bhuman%2Becology%26btnG%3D%26hl%3Den%26as_sdt%3D0%252C5#search=%22disaster%20human%20ecology%22.
7. UN ISDR definition of Hazard: "*A dangerous phenomenon, substance, human activity or condition that may cause loss of life, injury or other health impacts, property damage, loss of livelihoods and services, social and economic disruption, or environmental damage.*" Comment: The hazards of concern to disaster risk reduction as stated in footnote 3 of the Hyogo Framework are "… *hazards of natural origin and related environmental and technological hazards and risks.*" Such hazards arise from a variety of geological, meteorological, hydrological, oceanic, biological, and technological sources, sometimes acting in combination. In technical settings, hazards are described quantitatively by the likely frequency of occurrence of different intensities for different areas, as determined from historical data or scientific analysis.
8. Ian B., Kates R. W., and White G. F., The Environment as Hazard (New York, NY: Guilford Press, 1993).

9. Slovic P., "Perception of Risk: Reflections on the Psychometric Paradigm," *Social Theories of Risk* (1992), 117–152.
10. McCaffrey S., "Thinking of Wildfire as a Natural Hazard," *Society and Natural Resources* 17, no. 6 (2004): 509–516.
11. Mitchell J. K., Devine N., and Jagger K., "A Contextual Model of Natural Hazard," *Geographical Review* (1989): 391–409.
12. Ibid.
13. 2010 National Building Code of Canada seismic hazard values, http://www.earthquakescanada.nrcan.gc.ca/hazard-alea/interpolat/index_2010-eng.php.
14. Canadian Snow Load Calculations, http://www.hsh.k12.nf.ca/technology/cmhc/english/features/snow/.
15. Arnold M., ed., Natural Disaster Hotspots: Case Studies, no. 6 (World Bank, 2006). http://siteresources.worldbank.org/INTDISMGMT/Resources/0821363328.pdf.
16. Severe Weather Information Centre, World Meteorological Organization, http://severe.worldweather.wmo.int/.
17. United States Geological Survey Earthquake Maps, http://earthquake.usgs.gov/earthquakes/map/.
18. European-Mediterranean Seismological Centre, http://www.emsc-csem.org/#2.
19. USGS Earthquake Hazards Program, http://earthquake.usgs.gov/hazards/.
20. FEMA Mapping Information Platform, https://hazards.fema.gov/femaportal/wps/portal.
21. Hazus: The Federal Emergency Management Agency's (FEMA's) Methodology for Estimating Potential Losses from Disasters, www.fema.gov/hazus.
22. Environment Canada: Flood Damage Reduction Program, http://www.ec.gc.ca/eau-water/default.asp?lang=En&n=0365F5C2-1.
23. Globe and Mail "Alberta Scuttles Flood-Risk Plan for Home Buyers," *Toronto Globe and Mail, News* (Saturday August 17, 2013), A7.
24. Homer-Dixon T., The Ingenuity Gap: Facing the Economic, Environmental, and Other Challenges of an Increasingly Complex and Unpredictable Future (Random House Digital, Inc., 2002).
25. Gunderson L. H., Panarchy: Understanding Transformations in Human and Natural Systems (Island Press, 2001).
26. Zhou H., Wan J., and Jia H., "Resilience to Natural Hazards: A Geographic Perspective," *Natural Hazards* 53, no. 1 (2010): 21–41.
27. Birkmann J., and Wisner B., *Measuring the Un-measurable: The Challenge of Vulnerability*, no. 5 (United Nations University, Institute for Environment and Human Security, 2006).
28. Gallopín G. C., "Linkages Between Vulnerability, Resilience, and Adaptive Capacity," *Global Environmental Change* 16, no. 3 (2006): 293–303.
29. Nadeau R., The Environmental Endgame: Mainstream Economics, Ecological Disaster, and Human Survival (Rutgers University Press, 2006).
30. Birkmann J., ed., Measuring Vulnerability to Natural Hazards. Towards Disaster Resilient Societies (Tokyo, New York, Paris: UNU-Press, 2006).
31. Ibid.
32. Hewitt K., Interpretations of Calamity from the Viewpoint of Human Ecology, no. 1 (Australia: Allen & Unwin, 1983).
33. Ibid.
34. Ambraseys N., and Bilham R., "Corruption Kills," *Nature* 469 (2011): 153–155.

35. Lesson P. T., and Sobel R. S., "Weathering Corruption," *Journal of Law and Economics* 51 (2008): 667–681.
36. Ambraseys N., and Bilham R., "Corruption Kills," *Nature* 469 (2011): 153–155.
37. Oliver-Smith A., "Peru's Five Hundred-Year Earthquake: Vulnerability in Historical Context," The Angry Earth (1999): 74–88.
38. Pelling M., and Uitto J. I., "Small Island Developing States: Natural Disaster Vulnerability and Global Change," *Environmental Hazards* 3 (2001): 49–62.
39. Wisner B., "'Vulnerability' in Disaster Theory and Practice: From Soup to Taxonomy, then to Analysis and Finally Tool," *International Work-Conference* (Disaster Studies of Wageningen University and Research Centre, 29/30 June 2001).
40. Turner B. L. et al., "A Framework for Vulnerability Analysis in Sustainability Science," *Proceedings of the National Academy of Sciences* 100, no. 14 (2003), 8074–79.
41. Cutter S. L., Boruff B. J., and Shirley W. L., "Social Vulnerability to Environmental Hazards," *Social Science Quarterly* 84, no. 2 (2003): 242–61.
42. Benson C., and Twigg J., Tools for Mainstreaming Disaster Risk Reduction: Guidance Notes for Development Organisations (Geneva, Switzerland: Prevention Consortium, 2007), http://www.preventionweb.net/files/1066_toolsformainstreamingDRR.pdf.
43. Hazard, Risk and Vulnerability Analysis Tool Kit, http://embc.gov.bc.ca/em/hrva/toolkit.html.
44. Kuban R., and MacKenzie-Carey H., *Community-Wide Vulnerability and Capacity Assessment* (Office of Critical Infrastructure Protection and Emergency Preparedness (Now Public Safety Canada), 2001), http://www.gripweb.org/gripweb/sites/default/files/CVCA2001_meth.pdf.
45. See note 27 above, Source.
46. I would like to give me thanks to Professor Emdad Haque from the University of Manitoba, for his thoughtful and helpful review of this chapter.
47. Castleden M., McKee M., Murray V., and Leonardi G., "Resilience Thinking in Health Protection," *Journal of Public Health* 33, no. 3 (2011): 369–377.
48. Norris F. H., Stevens S. P., Pfefferbaum B., Wyche K. F., and Pfefferbaum R. L., "Community Resilience as a Metaphor, Theory, Set of Capacities, and Strategy for Disaster Readiness," *American Journal of Community Psychology* 41, no. 1–2 (2008): 127–150.
49. Barton W. H., "Methodological Challenges in the Study of Resilience," in *Handbook for Working with Children and Youth*, ed. Ungar M. (Thousand Oaks, London, New Delhi: Sage Publications, 2005), 135–148. Kirmayer L. J., Sehdev M., Whitley R., Dandeneau S. F., and Issac C., "Community Resilience: Models, Metaphors and Measures," *Journal of Aboriginal Health* (National Aboriginal Health Organization, November, 2009).
50. Kirmayer L. J., Sehdev M., Whitley R., Dandeneau S. F., and Issac C., "Community Resilience: Models, Metaphors and Measures," *International Journal of Indigenous Health* 5, no. 1 (2009): 62–117.
51. Hardin G., "The Tragedy of the Commons," *Journal of Natural Resources Policy Research* 1, no. 3 (2009): 243–53.
52. Tuohy R. and Stephens C., "Older Adults' Narratives about a Flood Disaster: Resilience, Coherence, and Personal Identity," *Journal of Aging Studies* 26, no. 1 (2012): 26–34.
53. Resilience Engineering Association, http://www.resilience-engineering-association.org/.
54. Holling C. S., "Resilience and Stability of Ecological Systems," *Annual Review of Ecology and Systematics* 4 (1973): 1–23.
55. Gunderson L., Holling C. S., Pritchard L., and Peterson G. D., "Resilience," vol. 2, in The Earth System: Biological and Ecological Dimensions of Global Environmental Change, ed. A. Harold Mooney and G. Josep Canadell, 530–531, in *Encyclopedia of Global Environmental Change*, (2002). ISBN: 0-471-97796-9, Editor-in-Chief Ted Munn.

56. Berkes F., "Understanding Uncertainty and Reducing Vulnerability: Lessons from Resilience Thinking," *Natural Hazards* 41, no. 2 (2007): 283–295.
57. Walker B., Holling C. S., Carpenter S. R., and Kinzig A., "Resilience, Adaptability and Transformability in Social–Ecological Systems," *Ecology and Society* 9, no. 2 (2004): 5.
58. They also make the useful observation that trying to find exact and precise definitions can be counterproductive.
59. Young O. R., Berkhout F., Gallopin G. C., Janssen M. A., Ostrom E., and van der Leeuw S., "The Globalization of Socio-Ecological Systems: An Agenda for Scientific Research." *Global Environmental Change* 16, no. 3 (2006): 304–16.
60. Castleden M., McKee M., Murray V., and Leonardi G., "Resilience Thinking in Health Protection," *Journal of Public Health* 33, no. 3 (2011): 369–77.
61. Masten A. S., "Ordinary Magic: Resilience Processes in Development." *American Psychologist* 56, no. 3 (2001): 227–38.
62. Brown and Perkins note that "An examination of disruptions in place attachments demonstrate how fundamental they are to the experience and meaning of everyday life" Brown B. B., and Perkins D. D., "Disruptions in place attachment," in *Place Attachment* (US: Springer, 1992), 279–304.
63. Sheftell J., "Recovery is a 'Shore' thing as New Jersey Rebuilds from Superstorm Sandy." Daily News (2013), http://www.nydailynews.com/life-style/real-estate/new-jersey-shore-winner-sandy-article-1.1278113#ixzz33y5ecs3a.
64. Castleden M., McKee M., Murray V., and Leonardi G., "Resilience Thinking in Health Protection," *Journal of Public Health* 33, no. 3 (2011): 369–77.
65. Masten A. S., and Obradovic J., "Disaster Preparation and Recovery: Lessons from Research on Resilience in Human Development," *Ecology and Society* 13, no. 1 (2008): 9.
66. Murthy R. S., and Isaac M. K., "Mental Health Needs of Bhopal Disaster Victims & Training of Medical Officers in Mental Health Aspects," *Indian Journal of Medical Research* 86, Suppl. (1987): 51–58.
67. Becker S. M., "Psychosocial Care for Adult and Child Survivors of the Tsunami Disaster in India," *Journal of Child and Adolescent Psychiatric Nursing* 20, no. 3 (2007): 148–155.
68. Ibid.
69. Tidball K. G., Krasny M. E., Svendsen E., Campbell L., and Helphand K., "Stewardship, Learning, and Memory in Disaster Resilience," *Environmental Education Research* 16, no. 5–6 (2010): 591–609.
70. Bonanno G. A., Brewin C. R., Kaniasty K., and Greca A. M. L., "Weighing the Costs of Disaster Consequences, Risks, and Resilience in Individuals, Families, and Communities," *Psychological Science in the Public Interest* 11, no. 1 (2010): 1–49.
71. Enarson E., and Morrow B. H., The Gendered Terrain of Disaster (Westport: Praeger, 1998).
72. Cutter S. L., Burton C. G., and Emrich C. T., "Disaster Resilience Indicators for Benchmarking Baseline Conditions," *Journal of Homeland Security and Emergency Management* 7, no. 1 (2010).
73. Mignone J. and O'neil J., "Social Capital and Youth Suicide Risk Factors in First Nations Communities," *Canadian Journal of Public Health* 96, no. S1 (2005): S51–S54.
74. Cutter S. L., Burton C. G., and Emrich C. T., "Disaster Resilience Indicators for Benchmarking Baseline Conditions," *Journal of Homeland Security and Emergency Management* 7, no. 1 (2010).
75. Clauss-Ehlers C. S., and Levi L. L., "Violence and Community, Terms in Conflict: An Ecological Approach to Resilience," *Journal of Social Distress & the Homeless* 11, no. 4 (2002): 265–78.
76. Twigg J., Characteristics of a Disaster-Resilient Community: A Guidance Note (DFID Disaster Risk Reduction Interagency Coordination Group, 2007), https://practicalaction.org/docs/ia1/community-characteristics-en-lowres.pdf.
77. Hyogo Framework for Action, http://www.unisdr.org/we/coordinate/hfa.

78. See Note 76.
79. Public Safety Canada, National Platform on Disaster Risk Reduction: Working Groups. (Public Safety Canada, October 9, 2013), http://www.publicsafety.gc.ca/prg/em/ndms/drr-wg-eng.aspx.
80. Mooney M. A., Faith Makes Us Live: Surviving and Thriving in the Haitian Diaspora (University of California Press, 2009). Shaw A., Joseph S., and Linley A., "Religion, Spirituality, and Posttraumatic Growth: A Systematic Review," *Mental Health, Religion and Culture* 8, no. 1 (2005).
81. Hussain A., Weisaeth L., and Heir T., "Changes in Religious Beliefs and the Relation of Religiosity to Posttraumatic Stress and Life Satisfaction After a Natural Disaster," *Social Psychiatry and Psychiatric Epidemiology* 46, no. 10 (2011): 1027–32.
82. Alawiyah T., Bell H., Pyles L., and Runnels R. C., "Spirituality and Faith-Based Interventions: Pathways to Disaster Resilience for African American Hurricane Katrina Survivors," *Journal of Religion & Spirituality in Social Work: Social Thought* 30, no. 3 (2011): 294–319.
83. Lawson E. J., and Thomas C., "Wading in the Waters: Spirituality and Older Black Katrina Survivors," *Journal of Health Care for the Poor and Underserved* 18, no. 2 (2007): 341–54.
84. Ebadi A., Ahmadi F., Ghanei M., and Kazemnejad A., "Spirituality: A Key Factor in Coping among Iranians Chronically Affected by Mustard Gas in the Disaster of War," *Nursing & Health Sciences* 11, no. 4 (2009): 344–50.
85. Kirmayer L. J., Sehdev M., Whitley R., Dandeneau S. F., and Isaac C., "Community Resilience: Models, Metaphors and Measures," *International Journal of Indigenous Health* 5, no. 1 (2009): 62–117.
86. Lawson E. J. and Thomas C., "Wading in the Waters: Spirituality and Older Black Katrina Survivors," *Journal of Health Care for the Poor and Underserved* 18, no. 2 (2007): 341–54.
87. See note 84 above.
88. Alawiyah T., Bell H., Pyles L., and Runnels R. C., "Spirituality and Faith-Based Interventions: Pathways to Disaster Resilience for African American Hurricane Katrina Survivors," *Journal of Religion & Spirituality in Social Work: Social Thought* 30, no. 3 (2011): 294–319.
89. Ibid.
90. Gillard M. and Paton D., "Disaster Stress Following a Hurricane: The Role of Religious Differences in the Fijian Islands," *The Australasian Journal of Disaster and Trauma Studies* 1999-1(1999), http://www.massey.ac.nz/~trauma/issues/1999-2/gillard.htm Taylor A. J., "Value-Conflict Arising from a Disaster," Australasian Journal of Disaster & Trauma Studies 1999-2: 1999–2002.
91. Taylor A. J., "Value-Conflict Arising from a Disaster," *Australasian Journal of Disaster & Trauma Studies* 2 (1999): 1999–2002.
92. Massey K., *Light Our Way, A Guide for Spiritual Care in Times of Disaster for Disaster Response Volunteers, First Responders and Disaster Planners* (National Voluntary Organizations Active in Disaster, 2006).
93. Harding S., *NYDIS Manual for New York City Religious Leaders: Spiritual Care and Mental Health for Disaster Response and Recovery* (New York Disaster Interfaith Services, 2007), www.NYDIS.org.
94. See note 92 above.
95. Mooney M. A., "Disaster, Religion and Resilience," The Immanent Frame: Secularism, Religion and the Public Sphere (February 24, 2010), http://blogs.ssrc.org/tif/2010/02/24/disaster-religion-and-resilience/.
96. Christian Aid. Prayers about emergencies and disasters. Source: http://www.christianaid.org.uk/resources/churches/prayer/emergencies.aspx.
97. Kant I., in. Religion and Rational Theology, ed. Wood A. W. and Di Giovanni G. (Cambridge: Cambridge University Press, 1996), 154.
98. Devon R. and van de Poel I., "Design Ethics: The Social Ethics Paradigm," *International Journal of Engineering Education* 20, no. 3 (2004): 461–69.

99. Ibid.

100. Shkilnyk A. M., Pathogenesis in a Social Order: A Case Study of Social Breakdown in a Canadian Indian Community, Ph.D. Thesis (MIT, 1981).

101. Dhillon C. and Young M. G., "Environmental Racism and First Nations: A Call for Socially Just Public Policy Development," *Canadian Journal of Humanities and Social Sciences* 1, no. 1 (2010): 25–39.

102. Duclos C., *Band-Aid Solutions to Self-Destruction? Development in Canada and the Case of Grassy Narrows, Kanata,* vol. 3, (Undergraduate Journal of the Indigenous Studies Community of McGill: Montreal, Quebec, 2010), 105–119.

103. Quote by Dr. Duncan Campbell Scott, Head of the Department of Indian Affairs from 1913 to 1932, http://www.wherearethechildren.ca/en/blackboard/page-7.html.

104. Mackenzie D., "Canada Probes TB 'Genocide'," *New Scientist* 194, no. 2602 (2007): 11.

105. Milloy J. S., A National Crime: The Canadian Government and the Residential School System, 1879–1986, vol. 11 (Univ. of Manitoba Press, 1999).

106. Ibid.

107. History of Indian Residential Schools, http://clfns.com/images/people/documents/history_of_indian_residential_schools.pdf.

108. See note 102 above.

109. Vecsey C., "Grassy Narrows Reserve: Mercury Pollution, Social Disruption, and Natural Resources: A Question of Autonomy," American Indian Quarterly (1987): 287–314.

110. Gregory D., Geographical Imaginations (Oxford: Blackwell, 1994), 104.

111. First Nations Traditions and Knowledge, http://www.livingwatersmart.ca/watersmart/firstnations.html.

112. See note 109 above.

113. John "Jack" Lindsay is an associate professor in the Applied Disaster and Emergency Studies (ADES) department at Brandon University where he combines research with his 20 years of experience in emergency management. Jack worked in New Zealand for 6 years, first as a hazard analyst in Wellington and then managing the Auckland City Council emergency management program. He returned to Canada, joining Manitoba Health as a disaster management specialist from 1999 to 2005. He started teaching part time at Brandon University in 2001, joined the ADES department full time in 2005, and received tenure in 2009. In 2011, Jack began his PhD in Emergency Management at Massey University's Joint Centre for Disaster Research in NZ. He previously received the degree of Master of City Planning from the University of Manitoba in 1993 with a research focus on urban planning and emergency management. Jack contributes to the disaster management profession through research and at numerous conferences as both an organizer and speaker. He is a member of the International Association of Emergency Managers, the Canadian Standards Association Technical Committee on Emergency Management, and the American Red Cross Scientific Advisory Council. He was the Manitoba chapter president of the Canadian Emergency Preparedness Association and served as ADES department chair from 2005 to 2010.

114. This commentary contains excerpts from "Vulnerability – Identifying a Collective Responsibility for Individual Safety: An Overview of the Functional and Demographic Determinants of Disaster Vulnerability," a background policy paper prepared by the author for the Public Health Agency of Canada, Centre for Emergency Preparedness and Response, in 2007. The author wishes to acknowledge this initial support.

5

Disasters and Complexity

> **For Want of a Nail**
> For want of a nail the shoe was lost.
> For want of a shoe the horse was lost.
> For want of a horse the rider was lost.
> For want of a rider the message was lost.
> For want of a message the battle was lost.
> For want of a battle the kingdom was lost.
> And all for the want of a horseshoe nail.
>
> Proverb, the origin of which possibly dates back to the 14th century. (Figure 5.1)

FIGURE 5.1 Shoeing a horse. http://www.gutenberg.org/files/22603/22603-h/22603-h.htm.

CHAPTER OUTLINE

5.1 Why This Topic Matters	153
5.2 Recommended Readings	153
5.3 Introduction	154
5.4 Characteristics of Complex Systems	155
5.4.1 Emergence (the Whole Is Greater Than the Sum of the Parts)	159
5.4.2 Self-Organization	160

5.4.3 Adaptation 162
　　　5.4.4 A Comment on Probabilities 163
5.5 Normal Accident Theory 166
　　　5.5.1 High Reliability Theory 168
5.6 Discussion 169
5.7 Close Calls or Near Misses 171
Further Reading 174
5.8 Conclusion 174
5.9 Case Study: Flooding along the Mississippi and Missouri Rivers, Hurricane Katrina and the New Orleans Catastrophe 174
　　　5.9.1 Hurricane Katrina 182
Further Reading 187
End Notes 188

CHAPTER OVERVIEW

Complexity theory changes the way we need to think about how we manage systems, and the degree of control we have over outcomes. Instead of viewing systems as being deterministic, complex systems must be viewed as having properties such as self-organization and emergence that make them somewhat unpredictable. This has important implications in terms of management strategies and risk assessment, as discussed by Charles Perrow in his books on Normal Accident Theory. This is especially true since the world is becoming more complex and tightly coupled because of globalization and technological and scientific advancements. There is still much research that needs to be done to explore how complexity theory must be incorporated into disaster management, particularly detailed empirical case studies that relate theory to practice.

KEYWORDS

- Adaptation
- Chaos
- Close calls
- Complexity
- Coupling
- Emergence
- Feedback
- Governance
- High Reliability Theory
- Hurricane Katrina

- Normal Accident Theory
- Power law
- Self-organization
- Tipping point

5.1 Why This Topic Matters

Brinsmead and Hooker begin their paper on complex systems with the comment, "What complex systems dynamics makes necessary, and possible, is a profound change in the structure of policy making."[1] Several paradigms have been used to understand disasters—why they happen, how they evolve, and how to better manage them (i.e., the hazards paradigm, the vulnerability paradigm, and the resilience paradigm). However, within the past few decades an emerging literature on complexity and disaster, which began in the physical sciences, has been making an increasing impact in fields such as economics, ecology, the social sciences and more recently, disaster studies. A number of articles have applied notions of complexity and chaos to disaster issues, concluding that there are important lessons to be taken from that field; hence the inclusion of complexity and chaos as a topic in this textbook.

The application of chaos or complexity theory to the study of disasters is still in its infancy, but it promises to provide a strong theoretical background to some of the empirical social science research describing what sorts of organizations and disaster management styles are effective in large disasters or catastrophes. There is great potential for future research in this area, both theoretically and in the development of case studies to validate its application and/or to clarify its limitations.

■■ 5.2 Recommended Readings ■

- Duit, A., and Galaz, V. (2008). Governance and complexity—emerging issues for governance theory. Governance: An International Journal of Policy, Administration, and Institutions, 21 (3), 311–335.
- Marais, K., Dulac, N., and Leveson, N. (2004, March). Beyond normal accidents and high reliability organizations: the need for an alternative approach to safety in complex systems. In Engineering Systems Division Symposium. MIT, Cambridge, Mass.
- Miller, R. L. (2001). What disaster response management can learn from chaos theory. Handbook of crisis and emergency management, 93, 293.
- Perrow, C. (2008). Normal accidents: living with high risk technologies. Princeton University Press.
- Drabek, T., and McEntire, D. (2003). Emergent phenomena and the sociology of disaster: lessons, trends and opportunities from the research literature. Disaster Prevention and Management, 12(2), 97–112.

Questions to Ponder

- Under what conditions is predictability of risk low, and under what conditions is it high?
  - What are the implications of this for disaster management?
- What is the difference between power and control (see exercise below)?

STUDENT EXERCISE

Complete the chart below, and compare your solution with those of your fellow students.

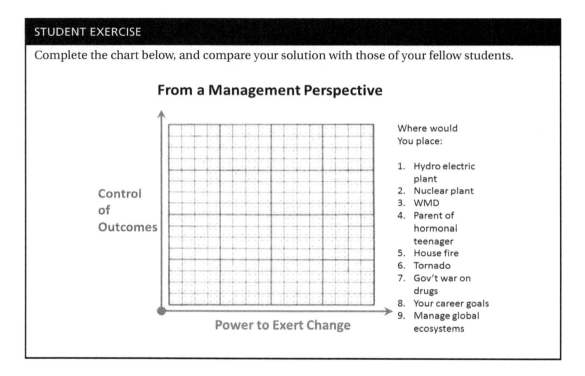

5.3 Introduction

One water molecule is not fluid,

One gold atom is not metallic,

One neuron is not conscious,

One amino acid is not alive,

One sound is not eloquent.

Jochen Fromm, The Emergence of Complexity[2]

My first introduction to chaos theory occurred in the late 1970s when I studied meteorology. The atmosphere is a system that shows all of the characteristics of complexity and chaos and is particularly good at humbling new weather forecasters who are presented with daily reinforcement about the limits of predictability in general, and of their fallibility in particular.

The sensitivity of nonlinear dynamical systems has been known since Henri Poincare's research in the late 19th century into the motions of planets around the sun. However, the first quantitative examples of chaos were discovered in the 1950s when scientists found that computer models solving simple equations would come up with diverging solutions if initial conditions were changed by even tiny amounts, because of the self-amplification of small variations. This is the main reason why there are theoretical limits to our ability to predict weather; much of the initial research in this area was done by atmospheric scientists. In a landmark study in atmospheric physics, Edward Lorenz noted, "All the solutions are found to be unstable."[3] This notion has been made famous by the question asked by Lorenz at a meeting in 1979: "Does the flap of a butterfly's wings in Brazil set off a tornado in Texas?" and is sometimes called the butterfly effect.

This does not mean that nothing can be known about these systems. Outcomes are still bounded even though they are not predictable. For example, as I write this, I have absolutely no ability to predict the weather six months from now; nevertheless, I know that it will not snow in Cuba in July. Chaotic systems exhibit particular patterns of behavior.

Complexity theory has strong implications for our understanding of how disasters should be managed and of what constitutes disaster resilience. It may well be a metaframework that can be used to better understand this field of study.[4]

5.4 Characteristics of Complex Systems

> **THE THREE SETS OF COMPLEXITY SCIENCE CONCEPTS ARE AS FOLLOWS:**
>
> **Complexity and systems**—These first three concepts relate to the features of systems that can be described as complex:
>
> - Systems characterized by interconnected and interdependent elements and dimensions are a key starting point for understanding complexity science.
> - Feedback processes crucially shape how change happens within a complex system.
> - Emergence describes how the behavior of systems emerges—often unpredictably—from the interaction of the parts, such that the whole is different from the sum of the parts.
>
> **Complexity and change**—The next four concepts relate to phenomena through which complexity manifests itself:
>
> - Within complex systems, relationships between dimensions are frequently nonlinear; i.e., when change happens, it is frequently disproportionate and unpredictable.
> - Sensitivity to initial conditions highlights how small differences in the initial state of a system can lead to massive differences later; butterfly effects and bifurcations are two ways in which complex systems can change drastically over time.

Continued

> **THE THREE SETS OF COMPLEXITY SCIENCE CONCEPTS ARE AS FOLLOWS:—Cont'd**
>
> - Phase or state space[6] helps to build a picture of the dimensions of a system and how they change over time. This enables understanding of how systems move and evolve over time.
> - Chaos and "edge of chaos" describe the order underlying seemingly random behaviors exhibited by certain complex systems.
>
> **Complexity and agency**—The final three concepts relate to the notion of adaptive agents, and how their behaviors are manifested in complex systems:
>
> - Adaptive agents react to the system and to each other, leading to a number of phenomena.
> - Self-organization characterizes a particular form of emergent property that can occur in systems of adaptive agents.
> - Co-evolution describes how, within a system of adaptive agents, co-evolution occurs, such that the overall system and the agents within it evolve together, or co-evolve, over time.
>
> *Source:* Overseas Development Institute[5]

First, it is important to differentiate between complicated systems and complex systems. Untangling a chain might be very complicated, but it is not complex in the way that the word is used here. We have learned that some systems, although they display patterns and structures, confound a deterministic approach to prediction. This means they do not unfold in a perfectly predictable manner as a result of universal laws. In a deterministic system, the same initial conditions will always result in an identical outcome. One problem with trying to apply this in the real world is that it is not possible to be infinitely precise in describing initial conditions. Partly this is because of measurement error and incomplete data sets, but there are also theoretical considerations related to Heisenberg's uncertainty principle.[7] As a result these errors sometimes "blow up," which means that they create enormous uncertainties in prediction. This sensitivity to initial conditions is called dynamical instability or chaos, and the systems are called complex or chaotic. Such systems are found at all scales, from the microscopic to galactic.

Chaotic systems are non-linear, which means that cause and effect are not proportional. An example of a linear system is the well-known law published by Sir Isaac Newton in 1687, Force = Mass × Acceleration. According to this law, if you double the force applied to a given mass then acceleration also doubles.[8] Linear relationships show up as a straight line on a linear graph, unlike nonlinear systems, in which the lines are curved and may be discontinuous. Nonlinearity depends on feedback, either negative feedback that damps a response or positive feedback that amplifies an initial perturbation; loudspeaker feedback is an example of the latter. Many systems are a combination of both positive feedback (that creates instabilities) and negative feedback (that stabilizes systems by bringing them back toward equilibrium). Positive feedback can result in sudden changes in a system's direction, character or structure, called bifurcations.[9] An example of this is shown in Figure 5.2.[10] Note how the system moves from a single state to two states (a bifurcation, where it jumps between states), and then to a chaotic state where it appears to constantly move between

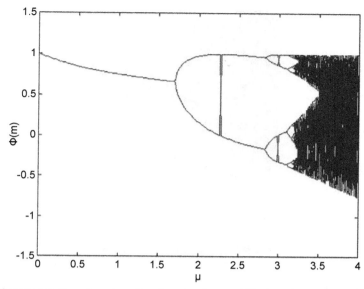

FIGURE 5.2 The bifurcation section of neutron flux changing from stability into chaos, in one-dimension. *Source: Liu, S. T. (2006). Nuclear fission and spatial chaos. Chaos, Solitons & Fractals, 30(2), 453–462.*

the system boundaries. While there may be some predictability in terms of when bifurcations may occur, their outcomes are not very predictable; systems appear to be creative in terms of how they achieve new structure and complexity.

There are visual examples of this in everyday life, such as cigarette smoke (Figure 5.3). Note how the smoke begins as a smoothly flowing stream, which then bifurcates and finally transitions into a random turbulent flow. One lesson to be taken from this is that a particular system may appear to be well behaved, but can transform itself into a badly behaved system as it evolves or when conditions change.

As dynamic nonlinear systems evolve, they have a strong tendency to be attracted to particular locations in phase space called "strange attractors"; these systems might appear to be totally random, but within chaotic systems there is order. Maps of these strange attractors can provide insights into system behavior, as shown in the motion of atmospheric air in Figure 5.4 (known as Lorenz's butterfly). Because of the inherent instabilities in these systems, it is not possible to predict exactly where the system will locate itself in phase space in the future, but studying its history can reveal what attractors it is bounded to and what their range of influence is.

The climate system is known to exhibit these properties. For example, the oceanographer Wally Broecker has suggested that global climate has alternate states depending on the presence, or not, of the great ocean conveyor belt.[11] This belt transports energy and chemicals and has significant impact on regional air temperature and precipitation patterns. One concern related to global climate change is that increased fresh water runoff and melting of the Greenland icecap might abruptly turn the belt off, thereby initiating a domino effect of regional climate changes throughout the world, with potential catastrophic effects. Studies

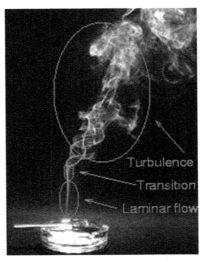

FIGURE 5.3 Smoke coming from a cigarette. Note the smooth flow of the smoke when it first rises and how the pattern becomes chaotic aloft. *Source: Zuccher, S. (2012). Flow stability.* http://profs.sci.univr.it/~zuccher/research/blstability/

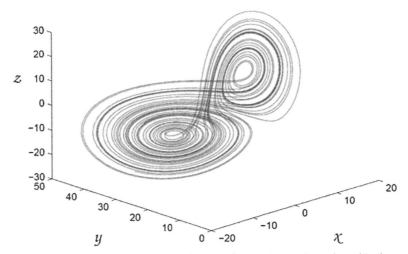

FIGURE 5.4 Lorenz strange attractor. The numerical solutions of Lorenz's equations plotted in three-dimensional space. The curves are functions of time, so imagine them as a roller coaster tracing an invisible object called a strange attractor. *Source: Axelsen, J.B. (nd). Chaos theory and global warming: can climate be predicted?* http://www.skepticalscience.com/print.php?r=134

of Greenland ice cores show that this circulation, also called the thermohaline circulation or meridional overturning circulation, has shut down many times before. Generally, scientists have considered such an event unlikely during this century[12] (known tipping elements include the Atlantic thermohaline circulation, West Antarctic ice sheet, Greenland ice sheet,

Amazon rainforest, boreal forests, West African monsoon, Indian summer monsoon, and El Niño/Southern Oscillation). However, a 2011 review article of climate tipping points[13] note that "recent (re)assessments give an increased likelihood" of passing such a critical threshold. "Even with the most conservative assumptions... results suggest it is more likely than not, that at least one of five tipping points considered will be passed in a >4 °C warmer world." The Intergovernmental Panel on Climate Change suggests that under a high carbon emission scenario, the 4°C threshold will likely be passed before the end of the current century.[14] In my opinion, this conclusion is beyond alarming.

Policy decisions also can result in society evolving along different pathways. One example is the U.S. National Flood Insurance Program, which had the ultimate unintended effect of allowing, if not encouraging, development within flood plains by institutionalizing moral hazard (see the case study on Mississippi–Missouri River flooding and Hurricane Katrina). A more paternalistic approach of preventing such development results in a different pathway, where flood plain development is restricted.

Chaotic systems have several properties that are important to understand. These include emergence, self-organization, and adaptation.

5.4.1 Emergence (the Whole Is Greater Than the Sum of the Parts)

A system displays emergence when parts of it do together what they would not do individually, or when a system does that which it would not do by itself, by virtue of its relationship to its environment. The properties of a system emerge from the interactions between system elements (or agents) and cannot be understood simply by studying individual elements. Even simple sets of rules can create complex patterns such as snowflakes, which can take on a near infinite number of patterns. Computer simulation models have been able to replicate complicated behaviors of schools of fish and bird flocks through the use of just three rules addressing avoidance, alignment, and attraction. Other examples of emergence include ant colonies, neural networks like the brain, and shell shapes and patterns. Emergence is often exhibited during and after disaster as groups of people or organizations form, transform, merge, and dissipate as they attempt to adapt to or manage the situation.[15] Within disaster response, the following categories of emergence have been recognized[16]:

- Supraorganization (a combination of organization types)
- Quasiemergence (minor alterations in structure or function)
- Structural emergence
- Task emergence
- Group emergence
- Emergence based on latent knowledge, and
- Interstitial groups (groups that form between existing organizations).

Research by Drabek[17] found that the following factors affect the stability and longevity of emergent organizations: group size, composition, structure, level of networking, degree of formalization, breadth of issue domain, re-establishment of normal relationships, and goal achievement. What structures emerge and how effective they are depends on cultural, religious, gender, and ethnic variables and can be influenced by management strategies. When values are shared and there is a culture of responsibility, emergent behavior will be more effective. Even roles such as leadership are being considered as emergent phenomena.[18] Successful emergence, from a management perspective, requires planning, nurturing, and resources; it is unlikely to happen in an effective way by chance.

There is an interesting connection between emergence in chaotic organizations and garbage can theory. The original paper on the latter was published in 1972 by Michael Cohen, James March, and Johan Olsen in the *Administrative Science Quarterly*. According to GOOGLE Scholar, it has been cited more than 6,000 times, an impressive achievement most academics can only dream of (I am quite envious). In this theory, organizational choices are a function of "problematic preferences, unclear technology, and fluid participation."[19] The bucket of elements swirling about within the garbage can include choices looking for problems, issues and feelings looking for decision situations, and solutions looking for problems they might answer. The choices that emerge depend on the connections between the different players; particularly important are resources, power structures, and personal agendas. This chaotic mix, described as "organized anarchy" by the authors, is a good partial description of many organizations. Working in the Canadian federal government for many years, I was frustrated by the seemingly irrational process used to make some decisions. After reading about garbage can theory, I understood the process much better and felt much less frustrated (if more jaded), as I shed some of my illusions. It also made me think differently about how to affect positive change; a nicely written, logical memo passed up the chain of command may seem like a rational approach that should be effective, but in reality it addresses only a few of the factors that actually impact organizational change. Networking, as well as addressing resource and power issues important to key decision makers and gate keepers, can be much more effective.

5.4.2 Self-Organization

Clearly related to emergence, self-organization occurs when new kinds of organizations, forms, and/or levels of complexity spontaneously arise out of the relations between individual organisms. These patterns are internal, not imposed by an outside agent, and occur in physics, chemistry, biology, economics, sociology, medicine—in fact, in any field where complex systems exist. Examples include crystallization, thunderstorm development, pedestrian patterns of walking, bird flocking (Figure 5.5), and group-think.

Self-organizing systems generally have critical states. One well-known example is a pile of sand; there exists a critical slope below which the pile is stable, but above which slides and avalanches occur. The criticality results from the balance between friction and gravity. Other examples of this property are earthquakes, lightning, insight, and laughter.[20] The notion of criticality is important in the study of disasters. It can be seen in the fragility curves of our built systems, where little damage occurs below a critical threshold, but

FIGURE 5.5 Bird flocking as an example of self-organized behavior. *Source: U.S. Fish and Wildlife Service.* http://digitalmedia.fws.gov/

beyond which it escalates rapidly. The same is true for ecosystems and social systems; they can absorb damage to a point, but beyond that lack robustness. This issue is one of the major concerns of environmentalists, who fear that because of environmental degradation and climate change the world will pass a critical threshold, after which damage will increase rapidly and systems may shift to different strange attractors. Understanding where critical thresholds exist is fundamental to effective risk analysis.

Self-organization has been linked to resilience, particularly in situations of low predictability in which agents use a variety of strategies to survive or meet some objective. This underlies the adaptive response described in the panarchy model (Chapter 6.5.7). The nature of a self-organization in society depends largely on existing structures, perceptions, and values; for example, emergent groups occur more frequently in the United States and Canada than in Japan, which has a culture with a greater emphasis of reliance on authority. Research has shown that people and groups typically become more unified and cohesive in disaster situations; these groups include those directly affected, such as volunteers, emergency workers, churches, businesses, government agencies, and other interested parties.[21] Tasks undertaken by emergent groups typically include search and rescue, the collection of emergency supplies, the provision of shelter, and emotional support. Emergence can be a great asset; most of the lives saved immediately following disasters are by local citizens and victims, but at the same time the convergence of people and goods at disaster sites can create serious logistical problems. One of the problems with command-and-control models of disaster management is the difficulty they have working with emergent groups, which "minimizes ritual behavior, tolerates decentralization and learning, and fosters effectiveness as it is flexible and innovative."[22] The emergence of multiple actors in disaster situations, many of whom acknowledge different authorities, suggest a coordinating, as opposed to a command-and-control, approach.

5.4.3 Adaptation

Nonlinear complex dynamic systems exhibit adaptive properties and have been named complex adaptive systems (CAS).[23] The important parts of these systems are the relational parts, the interactions, and networks that connect agents. Such systems can be very stable for long periods, but then show fast, irreversible changes that are surprising to observers. CAS consists of agents (e.g., people, species, actors, nations) that can self-organize within networks in which agents communicate with each other. Response is often not smooth, but shows threshold effects, tipping points, or abrupt change as perturbations ripple rapidly through the system. When ripples cross scales or systems, effects can be huge. The global recession of 2009 is an example of this, as is the Arab Spring and the French Revolution.

Duit and Galaz[24] note that "In sum, humans erect institutions and establish norms of cooperation and reciprocity in order to achieve predictability, stability, and low costs for social interactions… But with stability comes rigidity. Institutions are path dependent, sticky, and products of circumstances and power struggles present at the time of construction …the trade-off … in governance systems, is rooted in a much more fundamental tension between the dual needs for institutional stability and change." They describe four governance approaches (Figure 5.6), rigid, robust, fragile and flexible, each of which has its own strengths and weaknesses. In addition they conclude that only the robust strategy, which has a high capacity for both exploration and exploitation, will perform well under

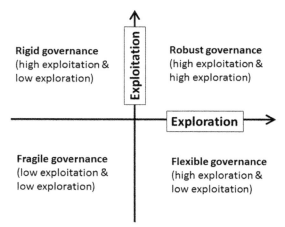

FIGURE 5.6 Four approaches to governance, based on organizational exploitation and exploration capacities. *Adapted from Duit, A., and Galaz, V. (2008). Governance and complexity—emerging issues for governance theory. Governance: An International Journal of Policy, Administration, and Institutions, 21 (3), 311–335.*

uncertainty and rapid levels of change (such as we have in the world today). Exploration refers to risk-taking, experimentation, flexibility, and innovation, while exploitation refers to efficiency, production, implementation, and execution. There is a tension between the two; for example, distributed networks would emphasize exploration, while centralized networks would emphasize exploitation. Many of the social systems in our society work to maximize exploitation, but not exploration, and thus fall into the rigid quadrant. One exception has traditionally been universities, which fall into the flexible quadrant, although now, with research being much more industry driven, it has (sadly, from my perspective) become far more exploitive.

> **STUDENT EXERCISE**
>
> Which of the four quadrants in Figure 5.6 does the disaster management organization that you are most familiar with fall into?

It is very difficult for large institutions or bureaucracies to be flexible, but easy for them to become rigid. This is probably particularly true for governments, which have a tendency to become very rule-bound and are driven by political considerations. This makes it very difficult for them to deal with novel situations. Along these lines, Dunman and Petrescu[25] comment, with respect to terrorism, "The levels of uncertainty that can be created ...are likely to surpass the preparation levels of these agencies," and "bureaucracies are not well-equipped for dealing with unpredictable conditions and ill-structured problems." Louise Comfort, a well-known disaster sociologist at the University of Pittsburgh, discusses the types of organizations needed for effective disaster management,[26] emphasizing the use of networks of organizations characterized by nodes of dense interactions with weaker links to connecting nodes and actors, as opposed to a random network. This structure is typical of the sort of emergent self-organization that happens after large-scale disasters. The challenge of organizations working in disaster management is to shift from a hazard or technocentric approach that is highly structured to one that incorporates aspects of chaotic systems. The need for networking is becoming generally recognized, and many professions and institutions are making it more of a priority; examples include the Canadian Risk and Hazards Network,[27] the mission of Manitoba's Independent Living Resource Center, which includes the statement "To minimize the risk to Manitobans with disabilities as a result of an emergency or disaster through networking...,"[28] and the U.N. Prevention website, which has a section on networks and communities.[29]

5.4.4 A Comment on Probabilities

Disasters occurring in complex systems are represented by different statistical distributions than many other types of systems. Most people are familiar with the "bell curve" or normal distribution (examples include people's height or air temperature distributions), in which the two tail ends rapidly approach zero probability (Figure 5.7). This means

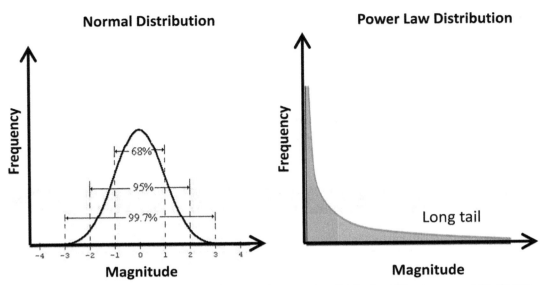

FIGURE 5.7 Illustration of how a normal distribution and a power law distribution differ. In a normal distribution, most events occur near the mean; an example of this is temperature. Very cold or very hot days are rare. In the power law distribution, most events occur near the minimum; for example, most earthquakes or tornadoes are very weak, while very strong ones are rare.

that the probability of extreme events fade rapidly. Disasters are better represented by a power law, which has "fat tails" (Figure 5.7). On a log–log graph, a power law becomes a straight line (Figure 5.8). This means that the probability of extreme events does not fade nearly as quickly as with a normal distribution. This is an important result. We need to be more concerned about hazards represented by power laws, than by those that are represented by normal distributions. Mathematically, a power law is represented by the equation $f(x) = ax^k$. Examples of power laws include: fractals, the number of species present in a habitat, the size of extinction events, the Pareto principle (or 80–20 rule), earthquake frequencies (Figure 5.8), and avalanche frequencies.

Another example of a power law can be seen in the cost of disasters. The National Climatic Data Center in the United States publishes a list of weather-related disasters exceeding $1 billion.[30] Most disasters represent a very small fraction of the total costs, but the highest ranked ones are enormous by comparison. Figure 5.9 shows a plot of their cost versus rank. One interesting feature of this graph is the sharp change in slope that occurs between rank 7 (2.2%) and rank 6 (3.4%). It is unclear why this exists; it may be an artifact of the methodology of cost estimation or reflect a threshold in vulnerability.

Why are power laws important? "One consequence is that events with power laws are scale-free; there is no characteristic size that is typical of the system. What power laws challenge us to do then is give up the view of the world as consisting of typical events with infrequent random variations. Instead, we must accept that there is no "average" event. There are simply many

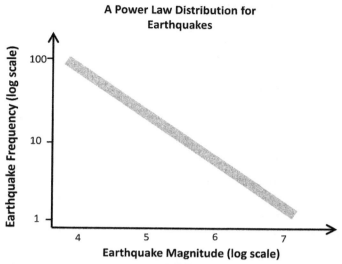

FIGURE 5.8 The power law relationship between earthquake frequency and size. This curve is equivalent to the one on the right hand side of Figure 5.7, but appears different because the axes are logarithmic instead of linear (earthquake magnitude is logarithmic by definition).

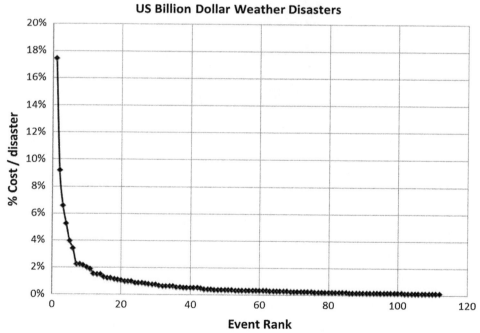

FIGURE 5.9 U.S. billion dollar weather disasters (1983–2011), which show a power law distribution. Note how the top few events dominate the impacts. *Data from NOAA National Climate Data Center,* http://www.ncdc.noaa.gov/billions/.

small ones, a few larger ones, and occasionally extremely large ones."[31] Because of the fat tails, including rare events in risk analyses is much more important for hazards described by power laws than for hazards described by a normal distribution, vulnerability considerations aside.

5.5 Normal Accident Theory

In 1984, Charles Perrow (Figure 5.10) published *Normal Accidents: Living with High Risk Technologies*[32] (another publication with more than 6,000 citations), a work that has had a substantial influence on our understanding of disaster. His main thesis is that some systems, particularly those that are complex and tightly coupled, are subject to **accidents that are fundamental properties of the system** and are therefore unavoidable. This differed substantially from the previous dominant paradigm of High Reliability Theory, which concluded that accidents could be totally prevented through good design and management, an emphasis on a culture of reliability and safety, and effective learning.

FIGURE 5.10 Charles Perrow, the creator of Normal Accident Theory.

By placing various technologies or hazards on a two-dimensional grid of complexity and coupling, Perrow made inferences about their relative risk, using the assumption that risk increases with tighter coupling and greater complexity (Figure 5.11). Examples of disasters that have been analyzed using this theory include Three Mile Island, Chernobyl, analysis of complex system risk:

- **D**esign of the system
- The **E**quipment that makes up the system
- **P**rocedures used to operate the system
- The **O**perators of the system
- **S**upplies and materials that make up the equipment
- The **E**nvironment in which the system operates.

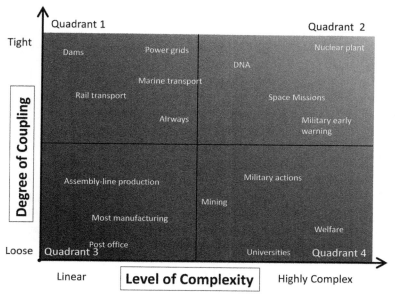

FIGURE 5.11 Coupling–complexity chart. *Adapted from* Normal Accidents: Living with High Risk Technologies *by Charles Perrow.*

He notes that centralized management structures are needed for tightly coupled or linear systems, while decentralized structures work for loosely coupled and complex systems. This leads to an inherent contradiction in tightly coupled but complex systems (quadrant two on Figure 5.11), which require the management system to be both centralized and decentralized simultaneously. Where such systems involve great risk, such as nuclear power plants, he advocates for their abandonment.

Normal Accident Theory has provided valid and useful insights, but has also been critiqued for being too pessimistic, not taking into sufficient account the rarity of some events, and not considering ways to increase safety other than redundancy; examples are the use of nonhazardous materials, reducing complexity, decoupling, designing for controllability, monitoring, interlocks, and reducing the potential for human error.[33]

Given the trends toward increased complexity and coupling in the globalized world of today, the implications of Perrow's theory is that the world is becoming a riskier place. Careful thought should be given to the construction of risk with respect to complex tightly coupled systems.

This theory has interesting implications for the differences between emergencies, disasters, and catastrophes. Quarantelli,[34] among others, has noted that the differences between these categories are both qualitative and quantitative. Overlaying them on Perrow's chart (Figure 5.12) supports that argument and emphasizes the need for different management strategies. Quadrants one and three would seem to be suitable for the more traditional top-down, control-oriented strategies. As systems become

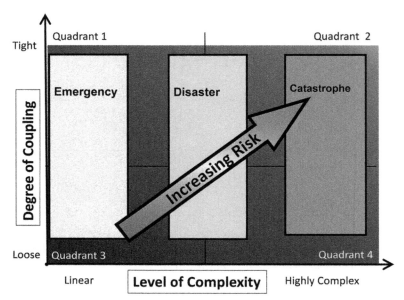

FIGURE 5.12 Implications of Normal Accident Theory for disaster management. For large disasters and catastrophes, both complexity and coupling are likely to be large. This suggests that typical top-down management structures are inadequate, but raises the issue that there may be a fundamental difficulty in created effective strategies that incorporate both top-down and bottom-up simultaneously.

more complex, the effect of emergent properties, uncertainty, and self-organization become important.

5.5.1 High Reliability Theory

This theory[35] was developed by Todd LaPorte and Paula Consolini at the University of California, Berkeley, to better understand how some organizations apparently operate in complex hazardous environments and yet are virtually failure free. The organizations they studied were air traffic control and naval air operations at sea. Their analyses concluded that organizations can achieve outstanding safety records, even when they are complex and tightly coupled, if they create a structure to allow decentralized decision-making based on common rules and understanding, include redundancy in their design, embrace several theoretical perspectives to avoid blind spots, and ensure understanding of the complexities of highly technical production processes.[36] If High Reliability Theory (HRT) is viewed as a set of processes to make organizations more reliable, then it offers a useful perspective. By itself it seems to be too optimistic, and the case studies are not a good counterpoint to Normal Accident Theory (NAT) since they differ from Perrow's definition of complexity. Overall, NAT appears to offer a more substantive view of organizational accidents than HRT, although it makes sense to engage in the processes suggested by HRT to create as safe an organization as possible.

5.6 Discussion

The above discussion has important implications for understanding disasters and catastrophes and for approaches to disaster management. After the terrorist attacks of September 11, 2001, the return "to a renewed emphasis on and a revived authority to the command and control paradigm with its emphasis on hazard control, response activities, hierarchical management, and legislated and defined authority manifested through mandated government agencies"[37] was a backward step that made disaster management less effective and is typical of the knee-jerk type of response that often follows political crises. Theories of complexity suggest that simple scaling up approaches that are effective for small events with limited complexity will probably not be effective for large-scale complex events. Consider the Incident Command System (ICS). The various organizations involved in disaster response need to be well acquainted with it for it to be effective. This is probable when only fire, emergency medical services, and police are involved, but is highly unlikely in situations involving emergent local community groups and nongovernmental organizations (NGOs). Currently considered to be best practice within the field of emergency management, it is likely to have limited applicability to large-scale complex disasters where there are strong arguments for an adaptive management approach.[38]

In a study of the 2004 Asian tsunami, Lassa[39] found that the number of NGOs "increased uncontrollably" to more than 1,000 organizations, creating serious coordination problems and "unnecessary and unhealthy competition of aid players." A network analysis revealed that central nodes were not always within government and that the network that emerged was polycentric and decentralized, as discussed by Comfort.[40] Effective disaster planning needs to be flexible enough to accommodate such developments. Lassa suggests that network theory has the potential to improve postdisaster response, particularly in developing countries where governance is more complex.

A conference in May 1995 called "What Disaster Response Management Can Learn from Chaos Theory"[41] presented a number of interesting conclusions that I believe are just as relevant today as they were then:

- "… the planned emergency response system will probably not be the one that emerges. The one that does emerge will probably have a tendency to be locally self-organizing, somewhat unpredictable in its inter-organizational linkages, and have a tendency to succeed or fail in unpredictable ways." The book *Leave No One Behind: Hurricane Katrina and the Rescue of Tulane Hospital* by Bill Carey notes how "the order of the day was improvisation." Effective local response was often facilitated because of chance, such as when someone happened to have mapping software on his computer because he was a Boy Scout leader, or because an executive at the hospital had just played golf with a representative from a much-needed ambulance service.[42]

- "The best efforts of disaster managers to develop a disaster response system may achieve relatively poor results because:
  - The problems they face at the onset of the disaster are often ambiguous, unclear, and shifting;
  - Information is unavailable, unreliable, problematic and subject to multiple, competing interpretations;
  - Resources are limited, not immediately available, and are being used up at unknown rates;
  - Lacking a clear problem definition as to where the most injured are located and the profile of the injuries and clear measures of success relative to how the entire response is proceeding makes it difficult to systematically prioritize resources during the earliest part of the response; and
  - Attempts to organize may actually be counter-productive due to incomplete knowledge about the nature of the disaster and availability of resources.
- Simple relationships, even deterministic ones, appear to generate indeterminate behavior because of varying response rates between individuals and organizations.
- The application of simple rules (all messages requesting assistance must come to a specific message center and not go directly to response organizations) might generate complex results (message delivery is slowed increasing problems with obtaining resources in a timely manner).
- Small changes such as keeping a shift of workers for two shifts at a hospital appear to have an amplifying effect across the entire response, resulting in large changes later (other hospitals have to take additional casualties which may strain their resources)."

One of their findings is that disasters do not actually enter the region of chaos, but rather exist at its edge, a place where the degree of organization lies somewhere between complete order and complete chaos. This is the region where organisms tend to be most adaptive. Related to this is the window of opportunity emergency managers often refer to—opportunities to make positive adaptive changes. The dark side of this window is what Naomi Klein refers to as disaster capitalism,[43] a perspective much debated by classical economists.

The 1995 conference papers emphasize that conditions of stability versus chaos require different management strategies. Success (or the lack of it) may not always be a function of management skill, but rather "sensitivity to initial conditions and fortunate relationships with other organizations."

STUDENT RESEARCH PROJECT
Obtain a set of disaster plans for your community, state, or province, and analyze them for rigidity versus flexibility. Specifically consider their incorporation of volunteers and emergent organizations.

Not all academics are enamored with complexity theory as a paradigm to understand society; for example, Forthea Hilhorst[44] provides an interesting critique of the complexity approach. She comments, "While complexity thinking looks promising… in practice it is divided by an old schism between structure and agency thinking. Much of complexity theory is based on 'system-thinking'. It denies agency and diversity, and puts unwarranted boundaries around people and phenomena. The study of social domains may be a way out of this problem, since it allows us to focus on the everyday practices and movements of actors negotiating the conditions and effects of vulnerability and disaster." I think her comment about agency and diversity is not correct, but it is true that people do not simply react to their environment; they process their experiences and use that information to (re) create world views that result in new and different relationships. Thus, human systems differ from other systems in a fundamental way. Social domains or "areas of social life that are organized by reference to a series of interlocking practices and values" incorporate contradictions, conflict, and negotiation within the interactions, and thereby provide an alternate and potentially richer paradigm for analysis. Other critics[45]:

- "Dismiss the relevance of complexity science beyond the natural sciences.
- Indicate that more work is needed to demonstrate specific applicability required.
- Criticize fad-driven 'complexologists' selling their services in the realm of organizational management.
- Suggest that complexity sciences adds nothing to existing approaches to understand social phenomena and only offers recommendations already reached elsewhere."

5.7 Close Calls or Near Misses

Normal Accident Theory and Chaos Theory suggest that unexpected accidents and disasters will occur in complex systems irrespective of how much planning is done, although the frequency of such events can be greatly diminished by the incorporation of good preventative measures. In part, a critique of those theories is based on what appears to be the highly reliable record of some industries; students not infrequently question a pessimistic approach (for example, with respect to Perrow's suggestion of abandoning nuclear plants, given the benefits they provide in terms of energy). These are good arguments that, if not well answered, suggest that the theories above may not be of great importance to disaster management.

One approach to gaining insight into this issue is to examine close calls, which are accidents that would have happened had some unexpected variable not been in place to prevent it. Although it appears that close calls or near misses are vastly underreported, there is a new but growing literature on their importance that needs to be taken seriously.[46] At local scales, disasters are rare, but because their occurrence follows a power law, near misses are far more frequent and offer the possibility for serious statistical and case study analysis. The exact ratio of accidents to near misses is bound to vary greatly by industry, but one assessment within the chemical industry puts it at a ratio of 10,000 errors for every near miss and about 100 near misses for every accident.[47]

Consider the threat of nuclear war. There are a number of situations when the world may have fallen into such a situation, the consequences of which would be truly catastrophic.[48] One of these occurred during the Cuban missile crisis of October 1962. A Russian submarine commander, out of communication with Moscow and being hunted by American battleships (that were unaware that the submarines had nuclear weapons), felt desperate enough to seriously consider launching a nuclear missile. He was prevented from doing so by deputy commander Vasili Arkhipov, who persuaded him and the other officers to surface the submarine and await orders from Moscow. "At the conference on the 40th anniversary of the Cuban missile crisis in Havana in October 2002, Vadim Orlov recounted his story in detail but emphasized that the utmost danger came not from an intentional launch of a nuclear torpedo, which even in the tense atmosphere of the last days before the surfacing remained very unlikely, but from malfunctioning equipment or an accident, which could have happened even under less trying conditions."[49] However, "The only instructions concerning nuclear weapons that the captains remember receiving were given in that briefing. As Nikolai Shumkov recalls, he heard Admiral Fokin say 'if they slap you on the left cheek, do not let them slap you on the right one.'" The threat of a nuclear strike should not be underestimated.

There have been other documented close calls of nuclear strikes.[50] Noam Chomsky comments that "actually, nuclear war has come unpleasantly close many times since 1945. There are literally dozens of occasions in which there was a significant threat of nuclear war."[51] What is common is that a number of these could have escalated to include the use of nuclear weapons, either through a deliberate decision or by accident, but that they were prevented by a fortuitous event or by an individual's decision. Clearly, the system is loosely enough coupled so that preventive human interventions are possible. But should we take comfort in the fact that these events were contained or be disturbed by the fact that they even happened? What should we make of the analysis that "the US government, after the Union Carbide Bhopal and West Virginia accidents, calculated that there had been 17 releases in the US with the catastrophic potential of Bhopal in 20 years, but the rest of the conditions that obtained at Bhopal were not present"?[52]

Research by Dillin-Merrill, Tinsley, and Cronin[53] shows that "when near misses are interpreted as disasters that did not occur, people illegitimately underestimate the danger of subsequent hazardous situations and make riskier decisions (e.g., choosing not to engage in mitigation activities for the potential hazard)", while "if near misses can be recognized and interpreted as disasters that almost happened, this will counter the basic "near-miss" effect and encourage more mitigation." The former is common, and is often interpreted as a success, while the latter tends to occur when the near miss has affected others in tangible ways (e.g., our neighbors got flooded, even though we did not). It seems that it is common for organizations to learn the wrong thing from near misses and perceive them as successes instead of harbingers of future disasters, and therefore managers are not held accountable for them, lessons are not recognized let alone learned, and the probability of future disasters is increased. Apparently British Petroleum ignored "many...near misses on the Deepwater Horizon rig prior to the April 2010 accident." What if they had not?

Research shows that better near-miss reporting can result in significant increases in safety, if organizations use the data to implement lessons learned.[54] There are, however, several barriers to this occurring, including potential recriminations for reporting, lack of incentives, lack of management commitment, and confusion regarding what a near miss is and how it should be reported.[55] Data on near misses and research in terms of how they are perceived support the relevance of Normal Accident Theory and Chaos Theory to disaster management and provide opportunities for the creation of much safer systems, if properly used.

A good example of both a disaster and a near miss is Three Mile Island, a nuclear power plant in Pennsylvania that had series of failures in March 1979. Thirteen seconds into the event, there was "a false signal causing the condensate pumps to fail, two valves for emergency cooling out of position and the indicator obscured, a PORV that failed to retreat, and a failed indicator of its position. The operators could have been aware of none of those."[56] Charles Perrow uses this accident as a case study in his book *Normal Accidents: Living With High-Risk Technologies*, as an example of a complex tightly coupled system. Eventually, it was determined that the core had melted down about half way, so it was a near miss in the sense that there was not a complete meltdown. One might view it as a success since human ingenuity prevented the worst case scenario, but only because about 2-½ h into the accident a new shift supervisor discovered the stuck valve. Had that been delayed, a complete meltdown would have occurred. Were the operators smart or just fortunate?

STUDENT EXERCISE

A very good video of the Three Mile Island disaster is available on YouTube. Search for the title "Meltdown at Three Mile Island."

During this video, there are a number of themes that emerge. As the events unroll, analyze each of the themes below.

THEMES:
1. Risk communication
   a. Subtheme: conflicting messages
   b. Subtheme: trust
2. Risk perception
   a. Subtheme: factors that affect perception.
3. Preparedness Plans
   a. Subtheme: evacuation plans
4. Human error versus technological failure
5. Controllability of complex systems
   a. Normal Accident Theory versus HRT
6. Human response, public and government
7. Decision-making under uncertainty

Further Reading

Dillon RL, Tinsley CH, Cronin M: Why Near-Miss Events Can Decrease an Individual's Protective Response to Hurricanes, *Risk Analysis* 31(3):440–449, 2011.

Jones S, Kirchsteiger C, Bjerke W: The Importance of Near Miss Reporting to Further Improve Safety Performance, *Journal of Loss Prevention in the Process Industries* 12(1):59–67, 1999.

Lindberg AK, Hansson SO, Rollenhagen C: Learning from Accidents—What More Do We Need to Know? *Safety Science* 48(6):714–721, 2010.

Phimister JR, Oktem U, Kleindorfer PR, Kunreuther H: Near-Miss Incident Management in the Chemical Process Industry, *Risk Analysis* 23(3):445–459, 2003.

Ritwik U: "Risk-Based Approach to Near Miss. Safety Mangement," *Hydrocarbon Processing* 93–96, 2002.

Savranskaya SV: New Sources on the Role of Soviet Submarines in the Cuban Missile Crisis, *Journal of Strategic Studies* 28(2):233–259, 2005.

5.8 Conclusion

The science of complexity has developed significantly over the past few decades and is beginning to be applied to disaster management. Clearly, disasters are complex events that fit within the chaos paradigm. Thus far complexity theory has not provided specific guidance not already noted elsewhere within the disaster management literature. Nevertheless, it does provide a theoretical basis for many of the empirical studies related to emergence and self-organization and a useful "lens" through which to understand, examine, plan for, and respond to disasters. Much work is needed, both at a theoretical and applied level, to gain a better handle on its applicability and usefulness.

I argue that we are creating a world in which complex tightly coupled systems are increasingly common, and we need to be prudent in terms of how we manage them, probably with greater use of the precautionary principle.

5.9 Case Study: Flooding along the Mississippi and Missouri Rivers, Hurricane Katrina and the New Orleans Catastrophe

As quickly as these crises arise, they tend to fade from the public consciousness. More disturbing is the fact that we have seen few gains in terms of knowledge and commitment to changing land use policies and practices in order to avoid repeat performances. Instead, we remain destined to repeat the past, perhaps with even worse consequences, as ongoing development takes place in flood-prone areas.

Richard Ward[57]

Rivers are composed of two main parts, the stream channel where the river flows in normal times and the floodway and flood fringe, areas adjacent to the floodway that are subject to occasional flooding (Figure 5.13).

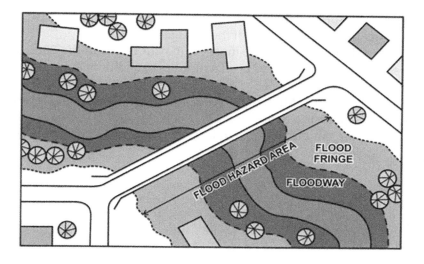

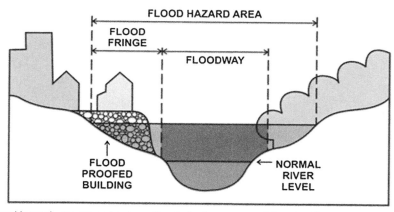

FIGURE 5.13 Flood hazard area around a river. Alberta Environment and Sustainable Resource Development. *Source:* http://environment.alberta.ca/.

Land under the floodway and flood fringe is called the flood plain and is normally fairly flat, fertile, and close to water. For these reasons, it has always been an attractive place for human settlements, with a resultant shift of use from forest to agriculture and urban. These changes have, however, exposed people and communities to flood hazards, which have resulted in the construction of dams and dikes (or levees), the known history of which extend back to the eighth century AD in China. This process has been ongoing along many of the world's rivers, including the Mississippi (the sixth largest river in the world) and Missouri River basins, which drain about 40% of the territory of the United States and parts of Ontario and Manitoba in Canada.

It is normal and expected for rivers to flood. These floods provide important ecological services, recharging groundwater and wetlands, and depositing sediments and nutrients along the floodplain. Some of these functions help reduce flood risk as well; peak stream flow decreases significantly as a result of the presence of wetlands.

Figure 5.14 and Table 5.1 show the most significant floods of the 20th century in the United States. Since 2000, there have been a number of additional severe flooding events, including Hurricane Katrina in 2005, the Midwest flood of 2008, Midsouth flooding and Northeast flooding in 2010, Mississippi River/upper Midwest flooding and Hurricane Irene in 2011, and Superstorm Sandy in the Northeast in 2012[58]. It is clear from these data that flood disasters are not uncommon; in fact, they are cumulatively the most expensive type of disaster in the United States.

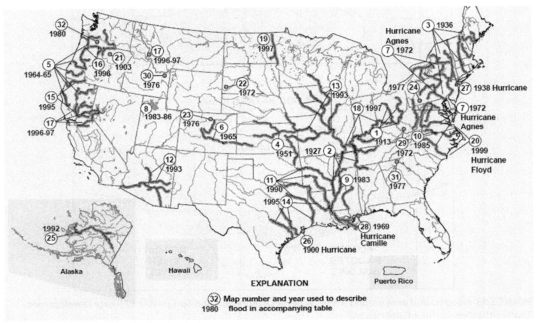

FIGURE 5.14 Most significant U.S. floods of the 20th century. Source: Tarlock, A. D. (2012). United States flood control policy: the incomplete transition from the illusion of total protection to risk management. In Duke Environmental Law & Policy Forum, Vol. 23 (1), pp. 151–183, Duke University School of Law.

There are several approaches to dealing with flood risk including:

- Passive adaptation (such as retreating during flooding events)
- Channel control (dykes, floodwalls, dredging)
- Upstream water retention (dams, reservoirs)
- Floodplain management through the active use of regional dams and dikes, and
- Integrated flood-risk management, which uses all the available strategies (including restoring ecosystems) coordinated by all levels of government on a watershed basis.[61]

Table 5.1 Significant Floods of the 20th Century[60]

Flood Type	Map No.	Date	Area or Stream with Flooding	Reported Deaths	Approximate Cost ($ Uninflated) (M, Million; B, Billion)	Comments
Regional flood	1	March–April 1913	Ohio, statewide	467	143 M	Excessive regional rain
	2	April–May 1927	Mississippi from missouri to Louisiana	Unknown	230 M	Record discharge downstream from Cairo, illinois
	3	March 1936	New England	157	300 M	Excessive rainfall on snow
	4	July 1951	Kansas and Neosho River basins in Kansas	15	800 M	Excessive regional rain
	5	Dec. 1964–Jan. 1965	Pacific Northwest	47	430 M	Excessive rainfall on snow
	6	June 1965	South Platte and Arkansas Rivers in Colorado	24	570 M	14 in of rain in a few hours in eastern Colorado
	7	June 1972	Northeastern United States	117	3.2 B	Extratropical remnants of Hurricane Agnes
	8	April–June 1983; June 1983–1986	Shoreline of Great Salt Lake, Utah	Unknown	621 M	In June 1995 the Great Salt Lake reached its highest elevation and caused $268 M more in property damage
	9	May 1983	Central and Northeast Mississippi	1	500 M	Excessive regional rain
	10	November 1985	Shenandoah, James, and Roanoke Rivers in Virginia and West Virginia	69	1.25 B	Excessive regional rain
	11	Apr. 1990	Trinity, Arkansas, and Red rivers in Texas, Arkansas, and Oklahoma	17	1 B	Recurring intense thunderstorms
	12	Jan. 1993	Gila, Salt, and Santa Cruz Rivers in Arizona	Unknown	400 M	Persistent winter precipitation
	13	May–Sept. 1993	Mississippi River basin in central United States	48	20 B	Long period of excessive rainfall
	14	May 1995	South-central United States	32	5–6 B	Rain from recurring thunderstorms
	15	Jan.–Mar. 1995	California	27	3 B	Frequent winter storms
	16	Feb. 1996	Pacific Northwest and Western Montana	9	1 B	Torrential rains and snowmelt

Continued

Table 5.1 Significant Floods of the 20th Century—cont'd

Flood Type	Map No.	Date	Area or Stream with Flooding	Reported Deaths	Approximate Cost ($ Uninflated) (M, Million; B, Billion)	Comments
	17	Dec. 1996–Jan. 1997	Pacific Northwest and Montana	36	2–3B	Torrential rains and snowmelt
	18	Mar. 1997	Ohio river and tributaries	50+	500M	Slow-moving frontal system
	19	Apr.–May 1997	Red river of the North in North Dakota and Minnesota	8	2B	Very rapid snowmelt
Flash flood	20	Sept. 1999	Eastern North Carolina	42	6B	Slow-moving Hurricane Floyd
	21	June 14, 1903	Willow Creek in Oregon	225	Unknown	City of Heppner, Oregon, destroyed
	22	June 9–10, 1972	Rapid City, South Dakota	237	160M	15 in of rain in 5 h
	23	July 31, 1976	Big Thompson and Cache la Poudre Rivers in Colorado	144	39M	Flash flood in canyon after excessive rainfall
	24	July 19–20, 1977	Conemaugh River in Pennsylvania	78	300M	12 in of rain in 6–8 h
Ice jam flood	25	May 1992	Yukon River in Alaska	0	Unknown	100-year flood on Yukon River
Storm surge	26	Sept. 1900	Galveston, Texas	6000+	Unknown	Hurricane
	27	Sept. 1938	Northeast United States	494	306M	Hurricane
	28	August 1969	Gulf Coast, Mississippi and Louisiana	259	1.4B	Hurricane Camille
Dam failure	29	Feb. 2, 1972	Buffalo Creek in West Virginia	125	60M	Dam failure after excessive rainfall
	30	June 5, 1976	Teton River in Idaho	11	400M	Earthen dam breached
	31	Nov. 8, 1977	Toccoa Creek in Georgia	39	2.8M	Dam failure after excessive rainfall
Mudflow	32	May 18, 1980	Toutle and Lower Cowlitz Rivers in Washington	60	Unknown	Result of eruption of Mount St. Helens

Structural mitigation along the Mississippi and Missouri River basins began in 1717 at New Orleans, and the length of the dikes has grown to more than 3,000 km (Figure 5.15). About 95% of them are agricultural or privately owned and designed to relatively low standards, with the remaining 5% being mostly federal and designed to higher levels of protection. Along with this trend of structural protection has been extensive development in the protected flood-prone areas. This development creates the requirement for continued increased protection because separating rivers from their floodplains results in a vicious cycle that increases flood stages[62]; in many parts of the Mississippi and Missouri River basins, flood stages have increased between 2 and 4 m.[63] Dikes and floodwalls along river sides constrict river channels, which results in higher flood stages for given water flows. Additionally, sedimentation occurs along the river bottom instead of across the flood plains, which also raises river levels. Criss[64] comments on the net effect of structural mitigation along the Missouri River as follows: "high flood stages that once were very rare are commonplace now, and this has occurred in spite of the many large reservoirs that have been constructed to reduce the severity of flooding."

FIGURE 5.15 Mainstream flood control and navigation projects of the Mississippi and Missouri rivers.[59] *Source: Rasmussen, J.L. (1999). Levees and floodplain management. U.S. Fish and Wildlife Service.*

There have been many flooding events in the Mississippi/Missouri basin, but a particularly severe flood occurred in 1993 (Figure 5.16) that:

- Inundated 17,000 square miles
- Destroyed more than 10,000 homes
- Damaged 70,000 buildings
- Put 75 towns completely under flood water
- Caused the evacuation of tens of thousands of people

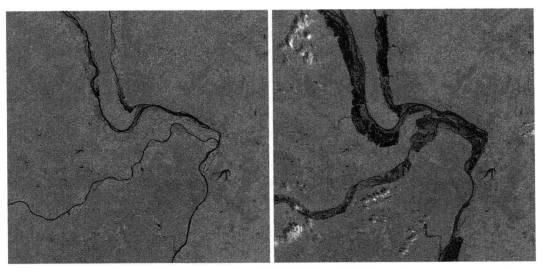

FIGURE 5.16[73] Satellite photos of the Mississippi River in 1991 (left) and 1993 (right). This image pair shows the area around St. Louis, Missouri, in August 1991 and 1993. The 1993 image was captured slightly after the peak water levels in this part of the Mississippi River. Flood waters had started to recede, but remained well above normal. This false-color image was created by combining infrared, near infrared, and green wavelengths. Water appears dark blue, healthy vegetation is green, bare fields and freshly exposed soil are pink, and concrete is gray. The deep pink scars in the 1993 image show where flood waters have drawn back to reveal the scoured land.[65] *Source: NASA Earth Observatory.*

- Killed 50 people and
- Resulted in more than $12 billion in damages.

Almost 150 rivers were affected. In some places peak discharges were estimated to be between a 300 and 500 year return period. More than 85% of the levees along the Missouri and Mississippi Rivers were either damaged or overtopped (none of them were federal urban ones designed to higher standards).[66] Hydrologically, the event was so severe because of very wet antecedent conditions, a long period of frequent rainfall (the Mississippi River near St. Louis was above flood stage for 144 days), and precipitation events that covered large areas.

There are two arguments at play in the debate over the use of structural mitigation; the first is that an over reliance on structural flood protection leads to a vicious cycle of increased flood disasters. The second is that these engineering structures allow for economic development with positive cost–benefit ratios. Most analyses see the 1993 disaster as a failure of flood protection works, but a few view it as a success. For example, Lovelace and Strauser[67] note that "The Upper Mississippi River Flood Control System could not prevent all damages caused by a flood like the one in 1993, because it was not designed to do so. About 15 billion dollars of flood damage actually occurred in 1993, as much as a third of which occurred in upland areas out of the flood plain. The flood control system, however, did an excellent job and prevented 19 billion dollars of additional flood damage and averted an even worse disaster."

> **STUDENT EXERCISE**
>
> Examine the logic of the argument by Lovelace and Strauser.

One of the main policy instruments used to manage flood risk in the United States is the National Flood Insurance Program (NFIP), established in 1968. This program is based on (1) the mapping of 100 year floodplains, (2) limitations on construction within floodplains, and (3) requirements to purchase insurance to qualify for loans and mortgages. The NFIP has run into many problems, including (1) low market penetration, (2) lack of disclosure of flood risk by real estate agents, and (3) lax enforcement by banks. Additionally, many grandfathered structures suffering repeated floods have not been relocated.[68] After Hurricane Katrina in 2005, the debt level of the NFIP increased to almost $27 billion.[69] Some blame Congress for this debt level because of its undermining some of the program's provisions for political reasons—this is part of the risk associated with public–private partnerships. Special interest groups and moral hazard can derail a program from its intended mandate.

Policy changes followed the 1993 flood. The National Flood Insurance Reform Act of 1994 had the intent to improve compliance with the NFIP, particularly with regard to mortgages. It also provided funding for flood mitigation and rebuilding according to local floodplain management ordinances, put more emphasis on floodplain mapping, and incorporated a Community Rating System into the NFIP that encouraged increased flood protection through the use of insurance incentives. It is unclear that the results have been to reduce flood risk overall, as also noted by Kunreuther and Pauly, who suggest "the need for having explicit enforced rules restricting development in these areas, rather than giving people discretion to choose where they would like to live"[70] and making disaster insurance mandatory for all homeowners. Note the following two quotes:

> ...history is repeating itself as some residents who moved into the floodplain since 1993, with assurances from FEMA and other officials that they would be safe, have seen their towns submerged this month, creating personal financial disasters for people who in many cases have no flood insurance.[71]

> The National Flood Insurance Program, now over thirty-five years old, was intended to "encourag[e] sound land use by minimizing exposure of property to flood losses." Its application in Missouri has done just the opposite. Missouri leaves all floodplain management decisions to local jurisdictions, effectively promoting development on floodplains. Missouri needs to adopt statewide floodplain laws, as many other states have already done.[72]

The 1993 flood disaster resulted in recommendations to reduce development on floodplains. FEMA bought 7,700 properties at risk and relocated the town of Valmeyer, Illinois. More recently, however, there is evidence of massive redevelopment on floodplains in many areas.[74] As a result of increasing flood protection to the 500-year level, there has been $2.2 billion worth of development near the city of St. Louis in areas that were flooded in 1993. More development is planned. "(M)any areas on floodplains, such as the Chesterfield Valley along the Missouri River, have seen more development on the floodplain, in areas that were under …1–3 m of water in the 1993 floods, than in the entire previous history of the region."[75] This trend is not universal however, as evidenced by tougher regulations imposed by many other municipalities and states.

The importance of ecosystems in terms of flood risk was also highlighted by this disaster (and many others as well). "Both the 1993 and 2008 floods on the Mississippi River above Grafton, Illinois could have been contained within a small portion of the 100-year floodplain with little flood damage."[76] Because of the weight of structures and lowering of the water table, development behind dikes and flood walls can result in subsidence that increases flood risk. An ecosystem approach provides many benefits to the natural environment, maintaining ecosystem function and biodiversity. These systems contribute to the purification of water and air and the production of food and meet the needs of people who value the natural environment for a variety of reasons.

One interesting twist in flood policy is that the Federal Emergency Management Agency (FEMA) considers areas protected from flood by dikes designed to a 100 year level of protection (or better) as not being at risk of flooding and have removed these areas from flood hazard maps. This is based on the assumption that dikes do not fail, but as we all know this assumption is problematical because history has shown us that dikes do sometimes fail. Additionally, as one looks toward the future, climate change seems likely to result in sea level rise that will put many coastal communities at risk and increase the number of heavy precipitation events such that storms with return periods of 100 years may occur every seven years in the central plains,[77] and storms with return periods of 20 years may occur every five years in the U.S. Midwest.[78] Should this climate trend occur, the risk profile of the United States and many other countries will increase dramatically.

5.9.1 Hurricane Katrina

Perhaps nowhere is this downward spiral of flood risk more evident than in New Orleans (Figure 5.17), where the creation of a flood disaster was made inevitable through development in low lying areas experiencing subsidence—in a city surrounded by high water levels contained by levees along the Mississippi River, Lake Borgne, and Lake Pontchartrain. Protected by improperly maintained levees designed for a category 3 hurricane in an area exposed to category 5 hurricanes, the city was highly vulnerable. The construction of the Mississippi River Gulf Outlet by the US Army Core of Engineers (for which they were later sued) created a conduit for the storm surge to breach some of the city levees. Additionally, the destruction of coastal wetlands and the construction of canals for the purpose of more efficient transportation contributed to increasing the risk of a flood disaster. The impact of Hurricane Katrina on New Orleans and other parts of Louisiana, Mississippi, and Alabama on August 29, 2005, was unparalleled in the history of the United States and

FIGURE 5.17 Elevation profile of New Orleans. Note the area below water level, which is safe only because of levees and seawalls. *Source: Gesch, Dean, "Topography-based Analysis of Hurricane Katrina," US Geological Survey.*

provides a textbook case of how to create a disaster and respond to it badly. This should not have been a surprising event. The first levee failed in 1722, and more recently floods damaged the city in 1812, 1819, 1837, 1856, 1893, 1909, 1915, 1947, 1956, 1965, 1969, and 2004.[79] New Orleans has a long history of damaging floods, and an exercise called Hurricane Pam the previous year had highlighted the risk associated with a strong hurricane making landfall in the area. Understanding how a society can be so maladaptive is at the root of the disaster problem.

From the point of view of New Orleans, things could have been much worse. Though Katrina was a category 5 hurricane at its worst, it was a weak category 4 or strong category 3 hurricane when it made landfall. The storm surge (up to 28 ft high) from Katrina hit the coast just east of New Orleans and traveled up to six miles inland. Had Katrina tracked to the west of the city, it would have been exposed to the full storm surge. As it was, more than 80% of New Orleans was flooded with waters up to 20 ft deep. More than 1,800 people in the regions affected by the storm died (drowning was the major cause of death, followed by trauma and heart conditions; elderly people were most likely to be victims[80]), and the total economic losses may have been around $200 billion. Insured losses were about $46 billion, almost double the previous record set from Hurricane Andrew in 1992.[81] As predicted by disaster vulnerability theory, the disaster was not equally experienced by all; there were strong racial and class differences in suffering, with low-income black homeowners being the most greatly affected. This bias was also present in media reporting, which indicated "a disproportionate media tendency to associate Blacks with crime and violence, a propensity consistent with exaggerated and inaccurate reports regarding criminal activity in Katrina's aftermath."[82]

In 2006, the Senate Committee on Homeland Security and Governmental Affairs published a report, *Hurricane Katrina: A Nation Still Unprepared*.[83] Their observations include that government failed at all levels in a pervasive way. These failures include that "(1) long-term warnings went unheeded and government officials neglected their duties to prepare for a forewarned catastrophe; (2) government officials took insufficient actions or made poor decisions in the days immediately before and after landfall; (3) systems which officials relied on to support their response efforts failed, and (4) government officials at all levels failed to provide effective leadership. These individual failures, moreover, occurred against a backdrop of failure, over time, to develop the capacity for a coordinated, national response to a truly catastrophic event, whether caused by nature or man-made." Saundra Schneider, a professor at Michigan State University notes, "Many of the problems and criticisms that emerged during late August and early September 2005 are identical to the concerns articulated about the governmental response to major disasters during the late 1980s and early 1990s."[84] Although under the leadership of James Lee Witt FEMA became more effective, the movement of FEMA into the Department of Homeland Security (DHS) resulted in a shift in priorities toward terrorism and away from natural disasters and mitigation. In fact, a former FEMA director called federal disaster assistance "an oversized entitlement program," suggesting that faith-based organizations should help victims instead of government.

Individual failures were also noted from leaders at levels from municipal to federal. These included:

- Not arranging for drivers of buses designated to assist evacuation,
- A lack of boats in the fire and police departments and communication failures (including a lack of compatibility between state and federal electronic systems, which resulted in the need for the manual transfer of requests),
- Failures to understand the magnitude of the disaster by decision-makers,
- An inadequate surge capacity,
- The perceived threat to safety due to unsubstantiated rumors of violence and a general breakdown in civil order,[85]
- A lack of familiarity with emergency management manuals and procedures by the city's Office of Emergency Preparedness,
- Flawed levees (design and construction), and
- Conflicts over who had responsibility for responding to a levee breach and how they should be conducted.

The interesting question to be asked is, what are the characteristics of the disaster management system that allowed for these failures? Part of the answer is lack of funding and chronic staffing issues in emergency management departments and systems. There were other factors as well, such as FEMAs lack of power within the Department of Homeland Security (DHS). Recommendations from the Senate report include the abolishment of FEMA to replace it with a stronger organization, and to change how responsibility and communications are delegated within the government. It also must be noted that the report found successes, particularly from the Coast Guard and from some businesses that had conducted both planning and training exercises and had empowered frontline leaders in terms of decision-making, as opposed to trying to maintain a command-and-control model. Bureaucratic barriers were certainly a problem, generating a great deal of red tape. Interestingly, the Department of Defense, although initially experiencing large delays because of red tape, was able to engage in a cultural switch in which they cut through it in the latter part of their response.[86] There is certainly a useful lesson here for other organizations, in terms of how to create organizational cultures that are responsive to crisis.

Governance problems continued after the Katrina disaster. Twenty-five thousand mobile homes purchased by FEMA could not be used because they did not conform to its own regulations. An inspector general's report in 2013 found about $700 million of Katrina relief was not accounted for.[87] After 2005, crime rates in New Orleans generally returned to pre-Katrina levels or exceeded them by the end of 2007.[88] Incidences of partner violence increased significantly as well, with physical victimization doubling from 4% to 8% after Katrina.[89]

There was not just an evacuation from New Orleans, there was a diaspora. In some parts of the city that were most damaged, only about 30% of residents have returned, as opposed to 90% for other parts of the city.[90] Contributing factors to this are high rents and a lack of rebuilding; the major factor slowing resettlement is the level of housing damage combined with income levels.[91] From a vulnerability point of view, this may be a good thing; sometimes disasters can be a time of redirection toward a safer and better development strategy.

In general, there are some important questions that need to be asked:

- Why are we not more effective at reducing flood damages, given our long history with these disasters?
- How long will the upward trend of damages continue?
- What measures need to be taken to reduce the impact of flood disasters?

There are solutions to the problem of flood disasters, and they are well known. The problem is, of course, that the solutions are not palatable to many people and are inconsistent with the economic/political systems within which we live.

Effective floodplain management needs to use a combination of structural mitigation, upstream storage, and land-use planning that incorporates the importance of the natural environment, with the expectation that occasional events will overwhelm flood protection measures. The problem with an overemphasis on structural defenses combined with a national flood insurance policy such as the United States has is that it creates a moral hazard and encourages development in the floodplains. This has the ultimate effect of increasing social vulnerability. An interesting study comparing a set of storms moving through Michigan, in the United States and Ontario, Canada,[92] showed that the impact in Michigan, which was about an order of magnitude larger than in Ontario, resulted because of the "large differences in the amount of flood plain development." These differences exist because Ontario, instead of allowing flood plain development as has happened in Michigan, restricted it through land-use planning and the power of conservation authorities.

> *The problem is that the United States has been unable to translate the consensus among experts into a coherent policy. The most striking aspect of United States flood policy is the growing gap between, on the one hand, the increasing sophistication of flood and hazard mapping and risk assessment and communication, and on the other, the failure to incorporate these developments into law and policy. We continue to rely on uncoordinated, structural defenses, even as our understanding of the limits of that policy increases and we continue to encourage floodplain development.*
> Dan Tarlock[93]

Raymond Burby discusses the Katrina disaster in terms of two paradoxes, the safe development paradox and the local government paradox.[94] The former occurs because the US federal government has actually increased risk because of their efforts to make hazardous areas safer. The latter relates to local officials paying insufficient attention to their own vulnerability issues, in spite of the fact that they bear the brunt of disaster losses. He argues that the paradoxes reinforce each other and create a vicious downward cycle of catastrophic losses—a cycle that can be reversed through the requirement of local governments to assume greater levels of responsibility for unsafe development and to engage in better planning with respect to natural hazards. In particular, he suggests shifting the NFIP from insuring individuals to insuring communities.

Sobel and Leeson[95] in their interesting analysis of Katrina from a public choice perspective note, "Each major US disaster brings yet another tale of FEMA corruption and failure, and yet another Congressional investigation into the problems in FEMA. The failures of FEMA, and of government disaster relief more generally, which occurred in the wake of Hurricane Katrina in 2005 are nothing new; identical problems manifested themselves after virtually every previous major disaster." The basic problem is that self-interest is the main motivator for decision-makers, and the incentives they are exposed to often do not work toward the public good. Specifically:

- Within a centralized system, there are too many layers of bureaucracy that have veto power, which hamstrings decision-making. Those who did not wait for permission to act were more effective.
- Government agencies are prone to type-two error bias, which means that they tend to be overcautious. This is because type-one mistakes (being undercautious) are generally more visible and more prone to punishment than type-two errors.
- Incentives for politicians and public servants are often not well aligned with disaster management priorities. One example is providing support to those engaged in moral

hazard behaviors, such as living in high-risk flood plains. The study by Sobel and Leeson also found that states that were politically important to the president were more likely to have disaster declarations than other states.
- Local governments have incentives to overrepresent the scale of a disaster to maximize federal relief, which leads to distrust of their estimates on the part of FEMA. Additionally, they argue that reliance on top-down decision-making is inefficient in complex disasters and that a decentralized approach is needed. They offer the following evidence of this: "In the first week of relief activities alone, FEMA refused to ship trailers to Mississippi that could be used as temporary housing for disaster victims, turned away critical generators needed by hospitals and victims for power, turned away trucks with water demanded by many, prevented the coastguard from delivering fuel critical to facilitating recovery activities, and refused Amtrak's offer to evacuate victims who desperately needed to get out of the disaster zone. The last Amtrak train left NewOrleans empty. FEMA clearly had no clue what was needed, or by whom."
- Sobel and Leeson also discuss the effect of glory seeking, in which politicians or others seek to present the perception they are doing good, without necessarily doing the deeds that would justify it.
- A bias toward the present over future needs results in shortsighted policy making. This bias contributed toward a levee system insufficient to protect the city.

What is rational for individuals or small communities can become irrational in the aggregate, with the result that our social and political systems often do not implement aggregate rationality. Vested interests can also derail good social policies because individuals and groups who benefit from creating risk often do not suffer negative consequences when disaster occurs. As long as these structural issues dominate decision-making, flood risk will continue to be created as a by-product of social forces.

Further Reading

Britt RR: History Repeats: The Great Flood of 1993, *LiveScience* June 22, 2008. http://www.livescience.com/7508-history-repeats-great-flood-1993.html.

Brown DW, Moin SM, Nicolson ML: A Comparison of Flooding in Michigan and Ontario: "soft" Data to Support "soft" Water Management Approaches, *Canadian Water Resources Journal* 22(2):125–139, 1997.

Criss RE, Shock EL: Flood Enhancement through Flood Control, *Geology* 29(10):875–878, 2001.

Criss RE, (n.d.): *Rising Flood Stages on the Lower Missouri River*. East–West Gateway Blueprint Paper. http://www.ewgateway.org/pdffiles/library/wrc/rising_flood_stages.pdf.

Davidson B: How Quickly We Forget: The National Flood Insurance Program and Floodplain Development in Missouri Wash, *UJL & Policy* 19:365, 2005.

Hey D, Kostel J, Montgomery D: *An Ecological Solution to the Flood Damage Problem. Finding the Balance between Floods, Flood Protection, and River Navigation*, 2009 (Saint Louis University), 72.

Kusky T: The Role of Rivers and Floods in History and the Role of the Mississippi, Missouri and Illinois Rivers in the Development of the United States. In *Finding the Balance between Floods, Flood Protection, and River Navigation*, 2009 (Saint Louis University), 1.

Landen Consulting: Ecosystem Services Case Study: The Mississippi River Floods of 1993 and 2008. Nature's Flood Regulation Services, Ecosystem Degradation, and the Economic Impacts of Large-Scale

Flood Damage, 2009. http://www.landenconsulting.com/downloads/LC-Case-study–large-scale-impacts-of-wetlands-degradation.pdf.

Lehmann E: Risk: National Flood Insurance Program—A Mighty Engine That Couldn't, *E&E Publishing*, January 5, 2013. http://www.eenews.net/stories/1059974778.

Lovelance JT, Strauser CN: *Perception and Reality Concerning the 1993 Mississippi River Flood: An Engineers' Perspective*. Report of the Interagency Flood Plain Management Review Committee to the Administration Flood Plain Management Task Force (1994).

Perry CA: *Significant floods in the United States during the 20th century—USGS measures a century of floods. USGS Fact Sheet 024–00 2000* US Geological Survey.

Pinter N: "One Step Forward, Two Steps Back on US Floodplains," *Science 8*, 308(5719): 207–208, 2005.

Pitlick J: A Regional Perspective of the Hydrology of the 1993 Mississippi River Basin Floods, *Annals of the Association of American Geographers* 87(1):135–151, 1997.

Rasmussen JL: *Levees and floodplain management* US. Fish and Wildlife Service, 1999. http://www.micrarivers.org/phocadownload/River_Management_Issues/leveesandfloodwalls2.pdf.

Tarlock AD: United States Flood Control Policy: The Incomplete Transition from the Illusion of Total Protection to Risk Management. Duke University School of Law, *Duke Environmental Law & Policy Forum* 23(1):151–183, 2012.

End Notes

1. Brinsmead T., and Hooker C., "Complex Systems Dynamics and Sustainability: Conception, Method and Policy," *Handbook of the Philosophy of Science*. (2009) Volume 10: Philosophy of Complex Systems. Elsevier BV.
2. Fromm J., *The Emergence of Complexity* (Kassel: Kassel University Press, Germany, 2004).
3. Lorenz Edward N., "Deterministic Nonperiodic Flow," *Journal of Atmospheric Sciences* 20, no. 2 (1963): 130–141.
4. Pelling M., *Natural Disaster and Development in a Globalizing World* (Routledge, 2003).
5. Ramalingam B., Jones H., Reba T., & Young J., *Exploring the Science of Complexity: Ideas and Implications for Development and Humanitarian Efforts, Second Edition*. Working Paper 285 (London, UK: Overseas Development Institute, 2009).
6. Phase space is used to describe a space where a system exists. For example, a projectile might be described by its position and speed. The coordinates chosen are the ones needed to describe the system and can be multidimensional.
7. The Heisenberg uncertainty principle states that the position and momentum of a particle cannot be precisely known simultaneously. As the precision of one measurement increases, the other decreases.
8. Of course, thanks to Albert Einstein and the special theory of relativity, we now know that this is not true at all speeds.
9. Murphy P., "Chaos Theory as a Model for Managing Issues and Crises," *Public Relations Review* 22, no. 2 (1996): 95–113.
10. Liu S. T., "Nuclear Fission and Spatial Chaos," *Chaos, Solitons & Fractals* 30, no. 2 (2006): 453–462.
11. Broecker W. S., "Thermohaline Circulation, the Achilles Heel of Our Climate System: Will Man-Made CO_2 Upset the Current Balance?," *Science* 278, no. 5343 (1997): 1582–1588.
12. Jungclaus J. H., Haak H., Esch M., Roeckner E., and Marotzke J., "Will Greenland Melting Halt the Thermohaline Circulation?," *Geophysical Research Letters* 33, no. 17 (2006).
13. Lenton T. M., "Early Warning of Climate Tipping Points," *Nature Climate Change* 1, no. 4 (2011): 201–209.

14. Intergovernmental Panel on Climate Change. "Climate Change 2014: Impacts, Adaptation, and Vulnerability," (2014): SUMMARY FOR POLICYMAKERS. http://ipcc-wg2.gov/AR5/images/uploads/IPCC_WG2AR5_SPM_Approved.pdf.
15. Drabek T. E., and McEntire D. A., "Emergent Phenomena and the Sociology of Disaster: Lessons, Trends and Opportunities from the Research Literature," *Disaster Prevention and Management* 12, no. 2 (2003): 97–112.
16. ibid.
17. Drabek T., "Emergent Structures," in *Sociology of Disasters: Contribution of Sociology to Disaster Research*, ed. Dynes R., De Marchi B., and Pelanda C., (Milan: Franco Angeli, 1987), 190–259.
18. Hazy J.K., Goldstein J.A. and Lichtenstein B.B., ed., Complex Systems Leadership Theory, New Perspectives from Complexity Science on Social and Organizational Effectiveness (Mansfield, MA: ISCE Publishing, 2007).
19. Cohen M. D., March J. G., and Olsen J. P., "A Garbage Can Model of Organizational Choice," *Administrative Science Quarterly* (1972): 1–25.
20. Fromm J., "The Emergence of Complexity," (Kassel University Press, 2004).
21. Drabek T. E., and McEntire D. A., "Emergent Phenomena and the Sociology of Disaster: Lessons, Trends and Opportunities from the Research Literature," *Disaster Prevention and Management* 12, no. 2 (2003): 97–112.
22. Britton N., "*Anticipating the Unexpected: Is the Bureaucracy Able to Come to the Dance?*," Working Paper No. 1, Cumberland College of Health Sciences, (Sydney: Disaster Management Studies Centre, 1989).
23. Duit A., and Galaz V., "Governance and Complexity—Emerging Issues for Governance Theory," *Governance* 21, no. 3 (2008): 311–335.
24. Ibid.
25. Duman S., and Petrescu A.S., "When What We Know Does not Apply: Disaster Response, Complexity Theory and Preparing for Bioterrorist Threats," *Sixteenth International Conference of the Public Administration Theory Network*, Anchorage AK, June (2003): 19–22.
26. Comfort L.K., "Risk, Security and Disaster Management," *Annual Review of Political Science* 8, (2005): 335–356.
27. Canadian Risk and Hazards Network www.crhnet.ca.
28. Independent Living Resource Centre http://www.ilrc.mb.ca/projects/demnet/.
29. PreventioWeb http://www.preventionweb.net/english/professional/networks/?pid:6&pih:2.
30. National Climatic Data Center *Billion-Dollar U.S. Weather Climate Disasters 1980–2012* Retrieved from http://www.ncdc.noaa.gov/billions/events.pdf.
31. Farber D. A., "Probabilities Behaving Badly: Complexity Theory and Environmental Uncertainty," UC Davis L. Review 37, (2003): 145. Retrieved from http://scholarship.law.berkeley.edu/facpubs/614.
32. Perrow C., *Normal Accidents* (Princeton University Press, 1984).
33. Marais K., Dulac N., and Leveson N., "Beyond Normal Accidents and High Reliability Organizations: The Need for an Alternative Approach to Safety in Complex Systems," in *Engineering Systems Division Symposium* (Cambridge, Mass: MIT, 2004), 29–31.
34. Quarantelli E. L., "Catastrophes are Different from Disasters: Some Implications for Crisis Planning and Managing Drawn from Katrina," *Understanding Katrina: perspectives from the social sciences* (2005): http://understandingkatrina.ssrc.org/Quarantelli/.
35. LaPorte T. R., and Consolini, P. M., "Working in Practice but Not in Theory: Theoretical Challenges of High-Reliability Organizations," *Journal of Public Administration Research and Theory: J-PART* 1, no. 1 (1991): 19–48.

36. Rijpma J. A., "Complexity, Tight-Coupling and Reliability: Connecting Normal Accidents Theory and High Reliability Theory," *Journal of Contingencies and Crisis Management* 5, no. 1 (1997): 15–23.
37. Buckke P., "Some Contemporary Issues in Disaster Management," *International Journal of Mass Emergencies and Disasters* 21, no. 1 (2003): 109–122.
38. Farber D. A., "Probabilities Behaving Badly: Complexity Theory and Environmental Uncertainty," UC Davis L., *Review* 37, (2003): 145. Retrieved from: http://scholarship.law.berkeley.edu/facpubs/614.
39. Lassa J.A., "Post Disaster Governance, Complexity and Network Theory: Evidence from Aceh, Indonesia after the Indian Ocean Tsunami 2004," IGRSC Working Paper No. 1 (2012) Retrieved from: www.irgsc.org/publication.
40. Comfort L. K., "Risk, Security, and Disaster Management," *Annual Review of Political Science* 8, (2005): 335–356.
41. Koehler G.A., ed., *What Disaster Response Management Can Learn from Chaos Theory, Conference Proceedings May 18–19* (California Research Bureau, 1995) http://www.library.ca.gov/CRB/96/05/index.html.
42. Carey C., *Leave No One Behind: Hurricane Katrina and the Rescue of Tulane Hospital* (Nashville, Tennessee: Clearbrook Press, 2006).
43. Klein N., *The Shock Doctrine: The Rise of Disaster Capitalism* (Metropolitan Books, 2007).
44. Hilhorst D., "Unlocking Disaster Paradigms: An Actor-Oriented Focus on Disaster Response," Abstract submitted for Session 3 of the Disaster Research and Social Crisis Network panels of the 6th European Sociological Conference, (2003): 23–26 September, Murcia, Spain.
45. See note 5 above.
46. Jones S., Kirchsteiger C., and Bjerke W., "The Importance of Near Miss Reporting to Further Improve Safety Performance," *Journal of Loss Prevention in the Process Industries* 12, no. 1 (1999): 59–67.
Phimister J. R., Oktem U., Kleindorfer P. R., and Kunreuther H., "Near-Miss Incident Management in the Chemical Process Industry," *Risk Analysis* 23, no. 3 (2003): 445–459.
Ritwik U., "Risk-Based Approach to Near Miss Safety Mangement," *Hydrocarbon Processing* October (2002): 93–96.
Lindberg A. K., Hansson S. O., and Rollenhagen C., "Learning from Accidents—What More Do We Need to Know?," *Safety Science* 48, no. 6 (2010): 714–721.
47. Bridges W. G., "Gains from Getting Near Misses Reported," in *7th Global Congress on Process Safety* April (2012).
48. Sagan C., "Nuclear War and Climatic Catastrophe: Some Policy Implications," *Foreign Affairs* 62, no. 2 (1983): 257–292.
49. Savranskaya S. V., "New Sources on the Role of Soviet Submarines in the Cuban Missile Crisis," *Journal of Strategic Studies* 28, no. 2 (2005): 233–259.
50. Philips A.F., (n.d.). *20 Mishaps That Might Have Started Accidental Nuclear War* Project of the Nuclear Age Peace Foundation: www.NuclearFiles.org.
51. Chomksy N., and Polk L., *Nuclear War and Environmental Catastrophe* (Seven Stories Press, 2013).
52. Shabecoff P., *Bhopal Disaster Rivals 17 in US* (The New York Times, April 1989), 30, A1.
53. Dillon R. L., Tinsley C. H., and Cronin M., "Why Near-Miss Events Can Decrease an Individual's Protective Response to Hurricanes," *Risk Analysis* 31, no. 3 (2011): 440–449.
54. Jones S., Kirchsteiger C., and Bjerke W., "The Importance of Near Miss Reporting to Further Improve Safety Performance," *Journal of Loss Prevention in the Process Industries* 12, no. 1 (1999): 59–67.
55. Phimister J. R., Oktem U., Kleindorfer P. R., and Kunreuther H., "Near–Miss Incident Management in the Chemical Process Industry," *Risk Analysis* 23, no. 3 (2003): 445–459.
56. See note 32 above.

57. Ward R. C., "Floodplain Development-Learning from the Great Flood of 1993," *Real Estate Issues-American Society of Real Estate Counselors* 31, no. 3 (2006): 17.
58. National Climatic Data Centre "Billion Dollar U.S. Weather Disasters," 1980–2013: http://www.infoplease.com/ipa/A0882823.html.
59. Perry C.A., *Significant Floods in the United States During the 20th Century—USGS Measures a Century of Floods* USGS Fact Sheet (U.S. Geological Survey: 2000), 024–100.
60. Ibid.
61. Tarlock A. D., "United States Flood Control Policy: The Incomplete Transition from the Illusion of Total Protection to Risk Management," *In Duke Environmental Law & Policy Forum* 23, no. 1 (October 2012): 151–183.
62. Flood Stage: The Level of a River at Which Flooding Begins to Occur at a Level That Is Hazardous to People or Communities.
63. Criss R. E., and Shock E. L., "Flood Enhancement through Flood Control," *Geology* 29, no. 10 (2001): 875–878
    Criss R.E., (n.d.). *Rising Flood Stages on the Lower Missouri River* East–West Gateway Blueprint Paper: http://www.ewgateway.org/pdffiles/library/wrc/rising_flood_stages.pdf.
64. Ibid.
65. Rasmussen J.L., *Levees and Floodplain Management. U.S. Fish and Wildlife Service (1999)* http://www.micrarivers.org/phocadownload/River_Management_Issues/leveesandfloodwalls2.pdf.
66. Pitlick J., "A Regional Perspective of the Hydrology of the 1993 Mississippi River Basin Floods," *Annals of the Association of American Geographers* 87, no. 1 (1997): 135–151.
67. Lovelance J.T., and Strauser C.N., "Perception and Reality Concerning the 1993 Mississippi River Flood: An Engineers' Perspective," *Report of the Interagency Flood Plain Management Review Committee to the Administration Flood Plain Management Task Force* (1994).
68. See note 61 above.
69. Lehmann E., *Risk: National Flood Insurance Program—A Mighty Engine That Couldn't* (E&E Publishing, January 5, 2013) http://www.eenews.net/stories/1059974778.
70. Kunreuther H., and Pauly M., "Rules Rather Than Discretion: Lessons from Hurricane Katrina," *Journal of Risk and Uncertainty* 33, no. 1–2 (2006): 101–116.
71. Britt R.R., "History Repeats: The Great Flood of 1993," *LiveScience* (June 22, 2008) http://www.livescience.com/7508-history-repeats-great-flood-1993.html.
72. Davidson B., "How Quickly We Forget: The National Flood Insurance Program and Floodplain Development in Missouri," *Washington UJL & Policy* 19, (2005): 365.
73. Landen Consulting "Ecosystem Services Case Study: The Mississippi River Floods of 1993 and 2008 Nature's Flood Regulation Services, Ecosystem Degradation, and the Economic Impacts of Large-Scale Flood Damage," (2009) http://www.landenconsulting.com/downloads/LC-Case-study–large-scale-impacts-of-wetlands-degradation.pdf.
74. Pinter N., "One Step Forward, Two Steps Back on U.S. Floodplains," *Science* 308, no. 5719 (2005): 207–208.
75. Kusky T., "The Role of Rivers and Floods in History and the Role of the Mississippi, Missouri and Illinois Rivers in the Development of the United States," *Finding the Balance between Floods, Flood Protection, and River Navigation* (published by Saint Louis University, 2009), 1.
76. Hey D., Kostel J., and Montgomery D., "An Ecological Solution to the Flood Damage Problem," *Finding the Balance between Floods, Flood Protection, and River Navigation* (Saint Louis University, 2009), 72.
77. ibid.

78. CCSP "Weather and Climate Extremes in a Changing Climate. Regions of Focus: North America, Hawaii, Caribbean, and U.S. Pacific Islands," *A report by the U.S. Climate Change Science Program and the Subcommittee on Global Change Research*. Thomas R. Karl, et al. ed., Department of Commerce, (Washington, D.C., USA: NOAA's National Climatic Data Center, 2008).

79. See note 75 above.

80. Brunkard J., Namulanda G., and Ratard R., "Hurricane Katrina Deaths, Louisiana, 2005," *Disaster Medicine and Public Health Preparedness* 2, no. 4 (2008): 215–223.

81. Property Casualty Insurers *The Hurricane Katrina Experience—A Property Casualty Insurance Perspective: Five Years Later* (PCI White Paper, 2010), 11.

82. Sommers S. R., Apfelbaum E. P., Dukes K. N., Toosi N., and Wang E. J., "Race and Media Coverage of Hurricane Katrina: Analysis, Implications, and Future Research Questions," *Analyses of Social Issues and Public Policy* 6, no. 1 (2006): 39–55.

83. Senate U. S., "Hurricane Katrina: A Nation Still Unprepared," *Report to the Committee on Homeland Security and Government Affairs* (Washington, DC: ABC News, 2006) http://abcnews.go.com/Politics/700-million-katrina-relief-funds-missing-report-shows/story?id=18870482.

84. Schneider S. K., "Administrative Breakdowns in the Governmental Response to Hurricane Katrina," *Public Administration Review* 65, no. 5 (2005): 515–516.

85. "They Have People... Been in That Frickin' Superdome for Five Days Watching Dead Bodies, Watching Hooligans Killing People, Raping People"—New Orleans Mayor C. Ray Nagin, The Oprah Winfrey Show, (September 6, 2005).

86. Moynihan D. P., "A Theory of Culture–Switching: Leadership and Red–Tape during Hurricane Katrina," *Public Administration* 90, no. 4 (2012): 851–868.

87. Zeleny J., *$700 Million in Katrina Relief Missing, Report Shows*. (Washington: ABC News, April 3, 2013), http://abcnews.go.com/Politics/700-million-katrina-relief-funds-missing-report-shows/story?id=18870482.

88. Leitner M., Barnett M., Kent J., and Barnett T., "The Impact of Hurricane Katrina on Reported Crimes in Louisiana: A Spatial and Temporal Analysis," *The Professional Geographer* 63, no. 2 (2011): 244–261.

89. Schumacher J. A., Coffey S. F., Norris F. H., Tracy M., Clements K., and Galea S., "Intimate Partner Violence and Hurricane Katrina: Predictors and Associated Mental Health Outcomes," *Violence and Victims* 25, no. 5 (2010): 588.

90. Lewis R., "Eight Years after Hurricane Katrina, Many Evacuees Yet to Return," *Al Jazeera America* (August 29, 2013) http://america.aljazeera.com/articles/2013/8/29/eight-years-afterkatrinalowincomeevacueeshaveyettoreturn.html.

91. Fussell E., Sastry N., and VanLandingham M., "Race, Socioeconomic Status, and Return Migration to New Orleans after Hurricane Katrina," *Population and Environment* 31, no. 1–3 (2010): 20–42.

92. Brown D. W., Moin S. M., and Nicolson M. L., "A Comparison of Flooding in Michigan and Ontario: "soft" Data to Support "soft" Water Management Approaches," *Canadian Water Resources Journal* 22, no. 2 (1997): 125–139.

93. Ibid.

94. Burby R. J., "Hurricane Katrina and the Paradoxes of Government Disaster Policy: Bringing about Wise Governmental Decisions for Hazardous Areas," *The Annals of the American Academy of Political and Social Science* 604, no. 1 (2006): 171–191.

95. Sobel R. S., and Leeson P. T., "Government's Response to Hurricane Katrina: A Public Choice Analysis," *Public Choice* 127, no. 1–2 (2006): 55–73.

6

Disaster Models

Essentially, all models are wrong, but some are useful.
George E.P. Box[1] (Figure 6.1)

FIGURE 6.1 George E.P. Box.

CHAPTER OUTLINE

- 6.1 **Why This Topic Matters** 194
- 6.2 **Recommended Readings** 195
- 6.3 **What Is a Model?** 196
- 6.4 **Philosophical Approaches**... 198
  - 6.4.1 Cause and Effect... 198
  - 6.4.2 Ethics and Values 200
- 6.5 **Disaster Models**.. 202
  - 6.5.1 Comprehensive Emergency Management (CEM) 202
  - 6.5.2 Pressure and Release (PAR) Model 205
  - 6.5.3 CARE Model 207
  - 6.5.4 Linear Risk Management Model.. 209
  - 6.5.5 Incident Command System (ICS) 210
  - 6.5.6 Catastrophe (CAT) Models 213
  - 6.5.7 Ecological Models 215
  - 6.5.8 Disaster Risk Reduction 218

194 DISASTER THEORY

    6.5.9 First Nations Wheel 220

6.6 Conclusion 222

6.7 Case Study: Sarno Landslides 222

Further Reading 223

6.8 A Comment by Joe Scanlon 224

End Notes 226

CHAPTER OVERVIEW

There are a number of models that are used to understand and manage disasters. It is important to remember that models are just tools—they are not reality. Some of the most common ones are the Pressure and Release Model, Comprehensive Emergency Management, and the Incident Management System. Each model has its own strengths and weaknesses, and students of disaster management should have a good understanding of all models to choose the most appropriate one for the task at hand.

KEYWORDS

- Cause and effect
- CAT models
- Comprehensive emergency management
- Determinism
- Disaster risk reduction
- Ecology
- Ethics
- First Nations
- ICS
- Models
- Principles
- Sarno
- Type 1 and 2 errors

6.1 Why This Topic Matters

Disaster models and the theories that support them are important tools used by academics and practitioners to gain a better understanding of disasters, and the creation of policies, strategies, and tactics that minimize their harmful impacts. Like any other tool models have limitations, but are useful for various purposes. Students must have knowledge of the main disaster models, their strengths and weaknesses, and a sense for when to use them. The disaster model chosen will strongly influence the success of disaster risk reduction efforts.

6.2 Recommended Readings

- Blaikie P, Cannon T, Davis I, Wisner B (1994). *At risk: natural hazards, peoples' vulnerability, and disasters.* Routledge, London.
- Buck, D. A., Trainor, J. E., & Aguirre, B. E. (2006). A critical evaluation of the incident command system and NIMS. *Journal of Homeland Security and Emergency Management*, 3(3).
- Canton, L. G. (2007). *Emergency management: concepts and strategies for effective programs.* Wiley-Interscience.
- Etkin, D. and Stefanovic, I.L. (2005). Mitigating Natural Disasters: The Role of Eco-Ethics. *Mitigation and Adaptation Strategies for Global Change*, 10, 467–490.
- Gunderson, L. H. (2001). *Panarchy: understanding transformations in human and natural systems.* Island Press.
- IMS 100 Self Study course, Emergency Management Ontario, (http://www.emergencymanagementontario.ca/english/emcommunity/professionaldevelopment/Training/ims100/ims100.html).
- Read about Disaster Risk Reduction on the following websites:
  - http://www.preventionweb.net/english/hyogo/GP/.
  - http://www.unisdr.org/.

Question to Ponder

- Einstein once said, "*As far as the laws of mathematics refer to reality, they are not certain, as far as they are certain they do not refer to reality.*" How is this thought relevant to the work of a disaster manager?

6.3 What Is a Model?

To begin, let's differentiate between a theory and a model. Laurence Moran, Professor of Biochemistry at the University of Toronto, provides a definition that I like: "A theory is a general explanation of particular phenomena that has withstood many attempts to disprove it. Because of the evidence supporting the explanation and because it hasn't been refuted, a theory will be widely accepted as provisionally correct within the science community." In this sense, the word is used differently from that in everyday language, in which it can refer to an unproven hypothesis.

A model is an application of theory—a simplified representation of the real world. We need simplifying assumptions to make problems tractable. For example, it is common in physics to assume motion without friction or gravity without air resistance when those factors are very small compared to other forces. Such assumptions are made not because anybody thinks that they are true, but because they eliminate complicating or confounding factors and because the errors that they generate are small under specified ranges. A model is usually a tool developed for specific purposes—either to aid understanding, or for some application, such as to estimate or manage risk. Whether it is conceptual, physical, mathematical, statistical, or visual, it is a representation of reality; it is not the "real" world but rather a construct to aid us in understanding it. Users of disaster models should be cognizant of their limitations.[2] There have been far too many examples of people mistaking model outputs for reality. New weather forecasters sometimes fall into this trap; I remember a young and inexperienced meteorologist (he had been on shift for less than a week) writing a forecast that bore no relation to what was actually being observed because it better fit the conceptual model of weather patterns that he had been taught. He had not yet learned the lesson that reality trumps models. If you are a student reading this book, you are meshed in academia; do not forget that academics need to be able to think conceptually, but also need to avoid the "ivory tower" syndrome.

There are many different kinds of models, as shown in Figure 6.2. The Pressure and Release Model (Section 6.5.2) is an example of a conceptual model. An example of a physical analog would be wind tunnels, which are used to test the response of structures to

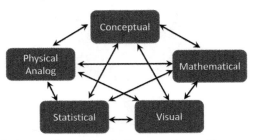

FIGURE 6.2 Types of models. Each type of model has both pros and cons. The process of model development should continuously compare models with each other and with reality, for validation purposes.

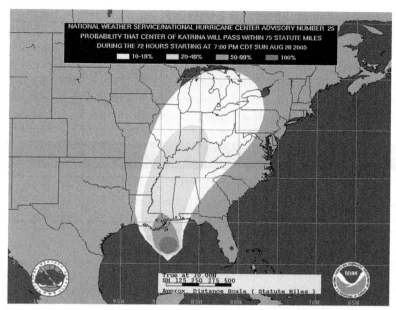

FIGURE 6.3 Example of a mathematical model. Computers, using sets of equations that describe atmospheric processes, predict storm paths such as for Hurricane Katrina in 2005. Since the models are not perfect, predictions are probabilistic; because of this, areas identified as at risk expand with time. As weather data arrive, these models are continuously updated to provide as accurate a prediction as possible. *Source: NOAA.*

strong winds. The output of a mathematical model used to predict hurricane movement is illustrated in Figure 6.3. The insurance industry has devoted significant resources to the development of risk models that estimate damage to communities from various hazards such as earthquakes and floods (see Section 6.5.6). These insurance models combine mathematical, statistical, and visualization methods.

Each of the different types of models can be very useful. However, although patterned after successful approaches used in the physical sciences, given the complexities of human nature, they should be considered more as mechanical metaphors than as reality. It is easy to have too much confidence in model outputs[3]; as discussed above, they need to be evaluated critically. Models can be wrong. Choosing a model exposes us to Type III errors, which relate not to the rejection of a hypothesis when true or its acceptance when false (Table 6.1),

Table 6.1 Type 1 and Type 2 Errors

		Reality	
		Hypothesis = true	Hypothesis = false
Conclusion	Hypothesis = true	Accept hypothesis = success ☺	Accept hypothesis = type 2 error ☹
	Hypothesis = false	Accept hypothesis = type 1 error ☹	Accept hypothesis = success ☺

but rather to the construction of a hypothesis that is poorly related to the issue at stake. When strategies designed to reduce risk actually increase it, they have often fallen into the trap of Type III error. The well-known levee effect[4] is an example of this. It refers to increased development in flood-prone areas after the construction of dykes or dams because people feel protected. Ultimately, however, when a flood occurs, the disaster can be all the greater. One of the most striking examples of this is the development of New Orleans in flood-prone areas, which ultimately led to the disaster associated with Hurricane Katrina in 2005.

> **STUDENT EXERCISE**
>
> Discuss the following two quotes as they relate to disaster management.
>
> - "It is a poor workman who blames his tools." (Unknown)
> - "If all you have is a hammer, everything looks like a nail." (Bernard Baruch).

6.4 Philosophical Approaches

The philosophical approaches that underlie disaster models relate mainly to: (1) our understanding of cause and effect, and (2) ethics and values.

6.4.1 Cause and Effect

Over time there has been a conceptual evolution of how we understand disasters in terms of cause and effect. Historically, there was a strong tendency to view such events as things that happen to us, placing people in a victim role with little or no power. From this worldview, disasters were the result of Fate (Figure 6.4), God, or Nature. In fact, the very word "disaster" comes from Greek roots meaning an ill-favored (dis) star (aster). Faith-based

FIGURE 6.4 The three fates, attributed to Jacob Matham. Print, Engraving. *Source: collection of the Los Angeles county museum of art.* http://commons.wikimedia.org/wiki/Los_Angeles_County_Museum_of_Art.

perspectives on disaster continue to play a large role in how people from some cultures perceive these events.[5] Within these systems, sin, guilt, and punishment often play a predominant role. For example, some people believe that the 2005 Tsunami in Indonesia was sent by God to punish people for their evil ways,[6] or that earthquakes are caused by women's sin.[7] It is difficult to overestimate the importance of this issue, since it largely determines the degree to which many people will engage in mitigation and prevention activities.

Recent models of disasters are based on the notion that we have the ability and capacity to determine our experiences, and view such events as being within our locus of control. Such a positivist, engineering approach to mitigation is embedded in a belief that nature is predictable and controllable by human beings, the roots of which lie in the seventeenth and eighteenth century paradigms of Newton, Descartes, and other rationalist thinkers, and can be traced back to Plato.[8] I do not mean to imply that notions of people managing their environment to be safer do not predate this—they do. Consider, for example the Babylonian Code of Hammurabi, dating back to 1772 BC In part, this approach assumes that science can understand, predict, and perfectly (or almost so) engineer the natural world. It is also based on a belief that it is humankind's natural right to control nature, a perspective that places us above the natural world.[9]

Perspectives on the degree of control that humans have over complex systems have shifted from classical notions as a result of the development of ecological theory (Section 6.5.7) and chaos theory (Chapter 5). Chaotic systems, although bounded, can be highly unpredictable and can exhibit surprising emergent properties. This has led to a shift from management approaches that are deterministic to ones that are adaptive and that recognize limitations to the degree to which humans can control parts of their environment. There is a large and interesting literature on adaptive management[10] that is beyond the scope of this chapter.

John Adams' discussion of the four myths of nature[11] inspired me to think more about the degree of power and control that we have over many of the world's risky systems. This is becoming an increasingly important question as technology increases its potential for destruction. Certainly we have the power to change the surface of the planet, alter genetic codes, and destroy other species, but how much control do we have, in the sense of deterministically creating desired outcomes? The answer varies depending on circumstances; sometimes we have a great deal of control and other times our sense of control is illusory. Depending on how you view these two factors, one would choose very different disaster management strategies. To illustrate the difference between power and control, consider the following: an army may be able to defeat another nation militarily, but may not be able to "win their hearts and minds" and reduce terrorism.

Consider the four quadrants in Figure 6.5. A traditional engineering approach (based on high power and high control) will lie in the upper right hand quadrant. This approach can be very effective for well-defined systems with low complexity that are well understood and for which broad stakeholder agreement exists. This is where traditional top-down risk management strategies are effective. Strategies to increase resilience or ones that use the application of the precautionary principle fit well in the lower right quadrant, where systems are more complex and less well understood, and outcomes are less predictable.

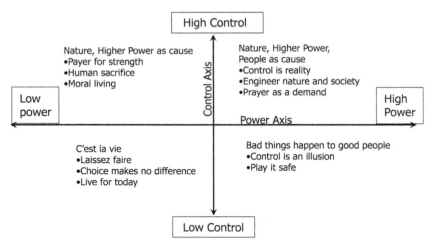

FIGURE 6.5 A conceptual framework for managing disaster, based on degrees.

■ ■ Question to Ponder ■

- *Where would a world view of fatalism lie on Figure 6.5?*
  - *What are your thoughts about fatalism as a disaster management strategy?*

STUDENT EXERCISE

- Place the following on Figure 6.5 (in terms of humans' ability to manage these systems). Explain your choices:
  - [A] nuclear energy plants
  - [B] climate change
  - [C] earthquakes
  - [D] tornadoes
  - [E] broken fire hydrant
  - [F] terrorism
  - [G] hormone-driven teenagers.

6.4.2 Ethics and Values

This topic will be explored in more detail in Chapter 9. The classic ethical tension in the disaster field is between utilitarian and Kantian arguments (Figure 6.6)—that is, should the greater good be emphasized, or should people's rights trump consequentialist thinking? Other ethical approaches that are important include Virtue Ethics and Social Contract Theory. Environmental

FIGURE 6.6 Emmanual Kant. *Source: Wikipedia.*

ethics is beginning to play a larger role in disaster models, largely because of climate change and environmental destruction, and how these trends affect disaster risk. Ethical issues are explored explicitly in the following publications; *Ethical Land Use: Principles of Policy and Planning* by Beatley and *Ethics for Disaster* by Zack, and play a large role in *The Gendered Terrain of Disaster: Through Women's Eyes* by Enarson and Morrow and *A New Species of Trouble* by Erikson.

The International Association of Emergency Managers has addressed this issue by developing a code of ethics and professional conduct based on the three qualities of respect, commitment, and professionalism. Similarly, they developed a set of principles—that emergency management must be comprehensive, progressive, risk driven, integrated, collaborative, coordinated, flexible, and professional. Their definition of collaborative—that "emergency managers create and sustain broad and sincere relationships among individuals and organizations to encourage trust, advocate a team atmosphere, build consensus, and facilitate communication" certainly recognizes the importance of human relationships and implies the need for moral behavior, even if it is not stated explicitly. The Chaplain Network of Nebraska has developed a *Disaster Chaplain Code of Ethics and Guiding Principles* that shows a keen awareness of duties and rights. Some government agencies have ethics guides, although they are often legalistic and generally do not include human relationship issues. For example, FEMA, along with many other U.S. government agencies, has published an ethics guide for their employees that covers ethics prohibitions, travel issues, and when to accept gifts. It is worth noting that some professions have given this issue prominence—see, for example, the *Canadian Code of Ethics for Psychologists.*

It is not unusual for people to be faced with ethical dilemmas, which are situations in which there is a conflict between different moral imperatives. How these are resolved is important. Ethics and values do not appear explicitly in any of the disaster models discussed in Section 6.5, but nevertheless underlie them in important ways. Their inclusion would only serve to make these models more relevant.

> **STUDENT EXERCISE**
>
> - List three ethical dilemmas, related to:
>   - Land use planning
>   - Response
>   - Recovery
> - How would you go about resolving them?

6.5 Disaster Models

In this section, a number of different disaster models will be presented. Some of them, such as comprehensive emergency management and the Incident Command System, are mostly oriented toward management. Others, such as the pressure and release model, are more focused on understanding the causes of disaster.

6.5.1 Comprehensive Emergency Management (CEM)

CEM was developed in the 1980s and rapidly gained widespread acceptance in the field of emergency management. Its great advantage over previous approaches was to explicitly incorporate mitigation and recovery in the emergency management cycle. The four pillars of CEM are as follows:

- **Mitigation** refers to long-term actions that reduce the risk of natural disasters, such as constructing dams and prohibiting people from building homes or businesses in high-risk areas.
- **Preparedness** involves planning for disasters and putting in place the resources needed to cope with them when they happen. Examples include stockpiling essential goods and preparing emergency plans to follow in the event of a disaster.
- **Response** refers to actions taken after a disaster has occurred. The activities of police, firefighters, and medical personnel during and immediately after a disaster fall into this category.
- **Recovery** encompasses longer-term activities to rebuild and restore the community to its pre-disaster state, or a state of functionality. This is also a good time to engage in activities that reduce vulnerability and that mitigate future disasters, such as strengthening building codes or modifying risky land-use policies.

Prevention is sometimes separated out as a fifth pillar. Prior to CEM, the emphasis had been on preparedness and response in a field dominated by persons who had previously worked in the military, fire, police, or EMS. When James Lee Witt became head of FEMA in 1993, he made mitigation the cornerstone of its strategic priorities, an important advance in the practice of emergency management. CEM or something very like it is currently accepted as best practice in many countries, particularly the U.S. and Canada, although a Disaster Risk Reduction (DRR) approach is becoming mainstream in much of the world and has been adopted by the United Nations International Strategy for Disaster Reduction (ISDR).

CEM has been depicted visually in many ways (see Figures 6.7 and 6.8 for two examples). Although this model is very useful, there are some problems with it, in that it suggests that the four pillars are sequential and independent of each other, when in practice they are very interdependent and often coincident. It can sometimes be difficult to decide how to categorize specific actions, which makes the practice of it complicated.

FIGURE 6.7 The four phases of comprehensive emergency management. Prevention is sometimes added as a fifth phase. Though the phases are depicted as separate, they tend to overlap, and some actions can fit into more than one phase.

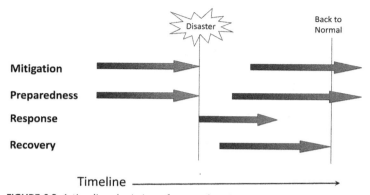

FIGURE 6.8 A timeline depiction of comprehensive emergency management.

■ ■ Question to Ponder ■

- *Consider the following scenario. You live in a community near a forest, and a wildfire has broken out and may move your way. You clear brush away from your yard, increase the limit on your household fire insurance, place valuables in a fire resistant safe, and wet down your roof and trees.*
  - *Are your actions mitigation, preparedness, or response?*

Another critique of the depictions of Figures 6.7 and 6.8 is that they do not explicitly emphasize the importance of capacity, resilience, informal networks, and formal arrangements, although all emergency managers are cognizant of their importance. Figure 6.9 shows the CEM cycle resting on these three platforms.

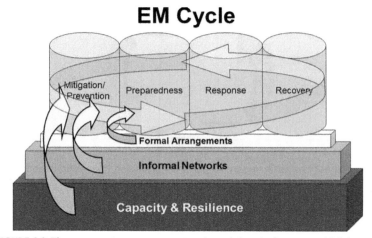

FIGURE 6.9 The emergency management cycle and the platforms on which it rests.

Various standards such as the U.S. NFPA 1600 and the Canadian CSA Z1600 use CEM as a basis for their structure. It is also embedded in some legislation and policy, such as the *Emergency Management Doctrine for Ontario*[12] and FEMA's national emergency management strategy,[13] and therefore this model has a very large impact on how EM is carried out in practice. There are a number of very good texts on Emergency Management that discuss CEM in detail.[14]

The United Kingdom uses a model similar to CEM called Integrated Emergency Management (IEM). This model is mandated for use by the Civil Contingencies Act 2004. IEM covers the six areas of anticipation or horizon scanning, assessment, prevention, preparedness, response, and recovery (Figure 6.10), and is built on the underlying concept of resilience. As in Canada, primary responsibility for response is at the local level; higher levels of government get involved in a combined and coordinated response as situations escalate.

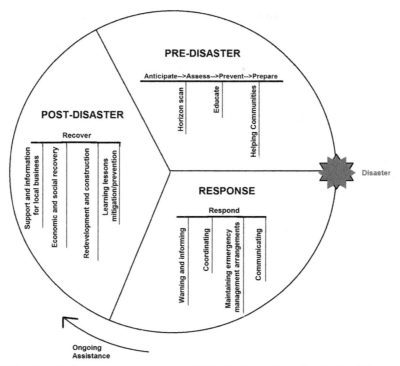

FIGURE 6.10 Integrated emergency management. *Adapted From: Essex County Fire & Rescue Service.*

In the assessment stage, organizations are required to conduct risk assessments of potential threats and hazards, and to identify preventative measures. Such planning needs to be flexible and to consider worst-case scenarios. Prevention is mandated by legislation, regulations, codes, and guidance documents that set standards for the management of various kinds of threats. Preparation includes planning, training, and exercising, and clearly defining roles and responsibilities. These plans need to be integrated within organizations as part of an overall management strategy. Response must be collaborative and coordinated; organizations must have clearly identified trigger points that activate their emergency management plans. Recovery planning should begin as soon as possible and should fully involve the affected community. IEM emphasizes using common consequences of incidents, as opposed to considering various causes; in this sense, it is like an all-hazards approach. A complete description of the model can be found in the Web archives of the UK government.[15]

6.5.2 Pressure and Release (PAR) Model

The PAR model was first published in the book *At Risk*[16] in 1994, and subsequently in the second edition in 2004[17]. It rapidly became widely accepted, and is a very useful model for understanding the way in which social–economic and cultural forces create vulnerable conditions. The main notion behind the model is that disasters occur when a hazard interacts with vulnerability.

After having taught the PAR model for a number of years to students at York University and the University of Toronto, I developed a slightly modified version[18] that students, in my experience, and I have an easier time working with and understanding (Figure 6.11).

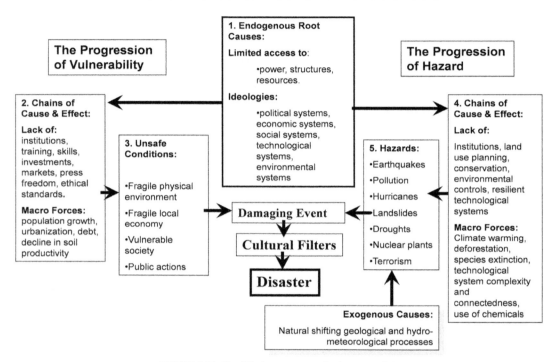

FIGURE 6.11 Modified pressure and release model.

The essence of this modified model is that *chains of cause and effect,* largely rooted in social processes, create a progression of vulnerability (left side of the figure) that results in unsafe conditions in human settlements. This model views disaster primarily as an external event that happens to people, is rooted in the tradition of western rationality, and empowers people and communities to change their disaster experience by altering their external, social, and constructed world.

This model also includes a progression of hazard (right side of the Figure 6.11) which has inputs that are both exogenous (factors over which we have no control) and endogenous (factors that we can affect). The PAR model has been widely adopted within the disaster and emergency management and is a powerful tool to deepen understanding of why disasters happen from a social science perspective. This adaptation of PAR has the advantage of providing a more dynamic view of hazard and of linking hazard and vulnerability in a more formal way. Figure 6.11 can also be used as a schematic representation to illustrate a common definition of risk used within the natural hazards and disaster management community (i.e., Risk=Vulnerability×Hazard).[19]

Many of the root causes in the cause-and-effect model are presented as being common to both vulnerability and hazard, thus creating an important common nexus. An example is an economic system that can result in environmental degradation that makes many hazards worse, and that also creates a critical infrastructure system that lacks resilience because of an emphasis on short-term profit making.

One aspect of the disaster cycle that is not obvious from Figure 6.11 is a return loop, where the social processes that affect hazard and vulnerability are altered by the impact that disasters have on people and communities. For example, many of the policies related to land use and building codes, or the development of the insurance and reinsurance industries, occurred in the aftermath of disaster as society strived for ways to mitigate or prevent these events. This would be portrayed in Figure 6.11 as arrows progressing from disaster to the cause-and-effect chains. Mitigation or prevention can occur in any one of boxes 1–5. Within the original PAR model, this is considered as the release phase.

PAR is a very useful model that students often use to analyze case studies. One trap that students sometimes fall into when using this model is limiting their analyses to the factors listed in the boxes. They should not be considered comprehensive. This model provides a macro perspective on the social and natural forces that create disasters, but is less useful for micro scale analyses. If a household is the center of an analysis, then the CARE Household Model[20] is a useful alternative.

6.5.3 CARE Model

CARE is a nongovernmental organization involved in humanitarian relief and developmental issues. In contrast to the PAR model that emphasizes a macro/meso approach to disaster management, the CARE disaster model (Figure 6.12) focuses on the household level (as embedded within larger scales), and views disasters as one of many external factors (shocks and stresses) within the context box on the left side of the figure. It was developed "to identify constraints to family and community livelihood security and design grassroots programs to overcome them."[21]

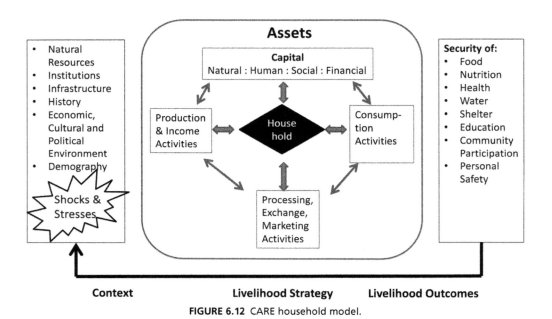

FIGURE 6.12 CARE household model.

Although its focus on the household level differentiates it from the PAR model, it is structurally and topologically similar in that it uses the notion of cause and effect to understand system constraints and how people use resources to cope with hazard. It is less explicit in terms of dynamic processes, which must be inferred. For studies focusing on a micro level where a community-based assessment is required, it is a good alternative to the PAR model.

Capacity and vulnerability are explicit in this model, particularly in Assets and the Livelihood outcomes box that examines such issues as food security. For more information about this model, readers are referred to the toolkit paper published by CARE.[22]

6.5.4 Linear Risk Management Model

A typical risk management process, such as the Canadian Standards Association (CSA-Q634-91), can be adapted for disaster management (Figure 6.13). An example of this is discussed by Tarrant,[23] who notes the following:

- Support from senior management is critical
- Developing a risk management process across the whole organization is vital
- Risk management is a core element of good management practice
- Risk management needs to have a holistic approach
- Risk management helps break down silos and divisions in organisations and results in better understanding of objectives
- Risk management integrates a systematic way to make informed decisions.

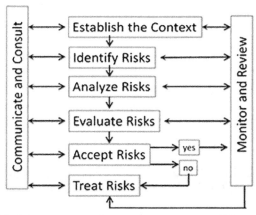

FIGURE 6.13 A traditional model of the emergency risk management process. This model works well when risks are known and well understood, and when stakeholder agreement exists in terms of how to address them.

A typical risk management strategy[24] that might be used within the Treat Risks box on the bottom of Figure 6.13 is shown in Figure 6.14. This type of linear decision-making model works well when uncertainties are small, hazards and vulnerabilities are well understood and subject to known and available controls, and stakeholder buy-in exists in terms of risk management strategies.

Bounded, linear models of this sort have been increasingly critiqued over the years because of their top-down approach, deterministic nature, and reliance on expert judgment. As the complex, chaotic, and unpredictable nature of disasters has become better understood, there has been a shift toward models that incorporate alternative approaches such as the precautionary principle, social discourse, bottom-up strategies,

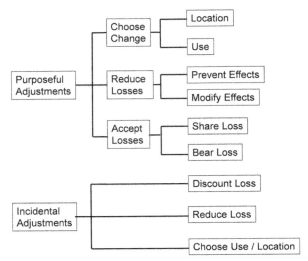

FIGURE 6.14 Choices in treat risks box of Figure 6.13. *Source: Burton, Kates and White: The Environment as Hazard.*

and the engagement of emergent organizations. Readers interested in a critique of this type of risk management process may also wish to explore the Garbage Can Model.[25] Reading about the Garbage Can Model was a "eureka" moment for me, and helped me understand the way in which the Government of Canada Civil Service really works.

I do not mean to suggest that linear decision making models are not useful; they are, depending on the characteristics of the problem being addressed.

STUDENT EXERCISE

Give an example of an emergency or disaster for which:

- A linear decision making model is useful for analysis.
  - Explain why.
- A linear model is a poor choice.
  - Explain why.

6.5.5 Incident Command System (ICS)

The Incident Command System (ICS) or a larger version of it, such as the U.S. National Incident Management System (NIMS), has become the gold standard for many emergency management organizations, including the Department of Homeland Security in the United States. There are many resources available to students who want to learn more, including free distance education courses, such as the IMS-100, IMS-200 (and so on), offered by Emergency Management Ontario or by FEMA.

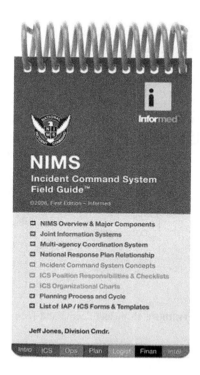

The concepts underlying ICS were first used in military operations, but were officially developed as ICS as a result of massive wildfires in California in the 1970s. Studies of the responses highlighted problems related to communication and management, specifically: (1) non-standard terminology among responding agencies, (2) lack of capability to expand and contract as required by the situation, (3) nonstandard and nonintegrated communications, (4) lack of consolidated action plans, and (5) lack of designated facilities. ICS was developed to address these failures. It is a modular system based on a command-and-control approach that also devolves decision making to local levels as much as possible, and activates only those elements that are needed to deal with crisis. It uses the concepts of unity of command, common terminology, management by objective, a flexible and modular organization, and defined span of control.

At the top of the organizational structure (Figure 6.15) at the strategic level are the Incident Commander and the General Staff, composed of an Information Officer, a Safety Officer and a Liaison Officer. Section Chiefs may be assigned for the areas of Planning, Logistics, Finance, and Operations. Below that is the tactical level, composed of Branch Directors and Supervisors, which is responsible for meeting operational objectives. Finally,

Incident Command System

```
                        Incident Commander
                              ├── Safety
                              ├── Information
                              └── Liaison
        ┌──────────────┬──────────────┬──────────────┐
   Operations      Planning        Logistics      Finance/
   Section         Section         Section        Administration
                                                  Section
```

- **Operations Section**
  - Staging Areas
  - Branches
    - Divisions & Groups
      - Strike Teams & Task Forces
      - Single Resources
  - Air Operations Branch
    - Air Support Group
    - Air Tactical Group

- **Planning Section**
  - Resources Unit
  - Situation Unit
  - Demobilization Unit
  - Documentation Unit
  - Technical Specialist

- **Logistics Section**
  - Services Branch
    - Communication Unit
    - Medical Unit
    - Food Unit
  - Support Branch
    - Supply Unit
    - Facilities Unit
    - Ground Support Unit

- **Finance/Administration Section**
  - Time Unit
  - Procurement Unit
  - Compensation/Claims Unit
  - Cost Unit

FIGURE 6.15 Incident command system.

there is the task level supervised by unit leaders, which has to accomplish specific tasks required to meet tactical objectives.

Although ICS is considered best practice within the EM industry, there is still some controversy regarding circumstances under which it may not be effective. Buck,[26] in a study designed to critically evaluate ICS and NIMS, found that "Results suggest the applicability of ICS in a range of emergency response activities, but point to the importance of context as a largely un-examined precondition to effective ICS. Our findings indicate that ICS is a partial solution to the question of how to organize the societal response in the aftermath of disasters; the system is more or less effective depending on specific characteristics of the incident and the organizations in which it is used. It works best when those utilizing it are part of a community, when the demands being responded to are routine to them, and when social and cultural emergence is at a minimum. ICS does not create a universally applicable bureaucratic organization among responders but rather is a mechanism for inter-organizational coordination designed to impose order on certain dimensions of the chaotic organizational environments of disasters. … Our final conclusions suggest that the present-day efforts in the National Incident Management System (NIMS) to use ICS as a comprehensive principle of disaster management probably will not succeed as intended."

Students of emergency management should become familiar with ICS, given its acceptance within the community, but should also be aware of its limitations. I strongly recommend taking one or more of the ICS/IMS courses that are available online in order to become certified.

6.5.6 Catastrophe (CAT) Models

A number of insurance and private consulting companies began to develop quantitative catastrophe (CAT) models in order to better assess their risks, partly in response to growing payouts by insurance and reinsurance companies during the 1980 and 1990s, and particularly due to such events as 11 insurance companies going bankrupt after Hurricane Andrew in 1992. The main purpose of these models is to estimate the amount of insured damage that would occur due to a catastrophic event such as a severe hurricane, flood, or earthquake. These models are based on meteorology, seismology, engineering, and actuarial science.

The main proprietary models are AIR Worldwide, EQECAT, and Risk Management Solutions (RMS), Inc. There is an International Society of Catastrophe Managers that promotes this profession within the insurance industry. The U.S. government has developed an open source model similar to the CAT models called Hazards United States (HAZUS) that estimates physical, economic and social impacts from hurricanes, floods, and earthquakes using GIS technology. As of 2012, a Canadian version of HAZUS is being developed by Natural Resources Canada.

The basic structure of CAT models is shown in Figure 6.16.[27] Final outputs typically include an Exceedence Probability (EP) curve of financial loss (an example of which is shown in Figure 6.17) and Average Annual Loss (AAL), which is also an estimate of the annual premium required to cover estimated losses. Many of these models are also capable of outputting spatial maps of modeled loss. The main sources of uncertainty (which can be very large) relate to incomplete scientific understanding and a lack of empirical data on hazard and vulnerability. Gaps in information involving business interruption costs and repair costs affect the accuracy of the loss component of the model.

To illustrate the difficulty in estimating damage, consider the sensitivity of fragility curves to different factors. Fragility curves show the amount of damage or the probability

Elements of a Catastrophe Model

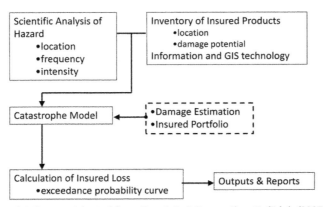

FIGURE 6.16 CAT model. *Source: Adapted from Grossi, P., & Kunreuther, H. (Eds.). (2005). Catastrophe modeling: A new approach to managing risk, Vol. 25, Springer.*

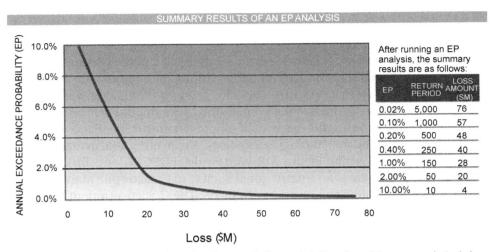

FIGURE 6.17 Exceedence probability (EP) analysis from RMS. The probability of small losses are relatively large, but decrease for large losses, eventually approaching zero. *Source: The Review (2008), A Guide to Catastrophe Modeling.*

of structural failure as a function of hazard intensity. Figure 6.18[28] is an example; note the large difference between construction using a 6d nail and an 8d nail. At critical levels, relatively small changes in wind speed can result in very large changes in damage. For example, in Figure 6.18 a change in wind speed from 80 mph to 100 mph results in a change in the probability of failure from about 5% to 50% for a 6d nail (enclosed). As a result, poor quality of construction and/or incomplete inventories of building type and method of construction can lead to very large errors in model outputs. This also shows how small

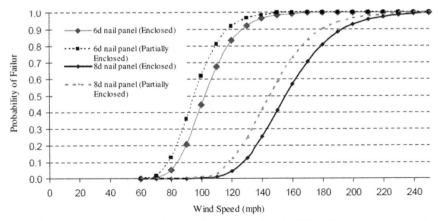

FIGURE 6.18 Fragility curves. Note the high degree of sensitivity of probability of failure to changes in nail type. *Source: Li, Y. (2005). Fragility methodology for performance-based engineering of wood-frame residential construction. A Thesis Presented to The Academic Faculty, School of Civil and Environmental Engineering, Georgia.*

changes in climate can result in large changes in risk, and the importance of having very accurate hazard probability profiles.

Other examples of how small differences in construction can have significant outcomes come from a forensic engineering analyses of houses destroyed after the Joplin tornado of 2011 and the Barrie, Ontario tornado of 1985.[29] Mark Levitan noted that in the Joplin disaster "Fairly modest changes in building design and construction, and code changes and practices can lead to dramatic improvements in wind hazard resistance of structures"[30] (the use of hurricane clips is one example). In the Barrie tornado, many of the houses that were destroyed had been improperly anchored to their foundations, in one case because washers were not put on the anchoring bolts.

CAT models are very complex, and it can be very expensive to buy them or to run scenarios. Larger insurance companies may have staff specifically devoted to running models for them or may contract it out to the supplier. Without the use of such tools, however, insurance companies cannot generate good estimates of their financial risk.

6.5.7 Ecological Models

We have met the enemy and he is us.

Source—Pogo.

In a paper I published in 2005 with Ingrid Stefanovic from the University of Toronto, we considered natural disasters from an eco-ethical perspective[31] (Figure 6.19), emphasizing the interactions and reliance of society on the natural world. This model is, to a

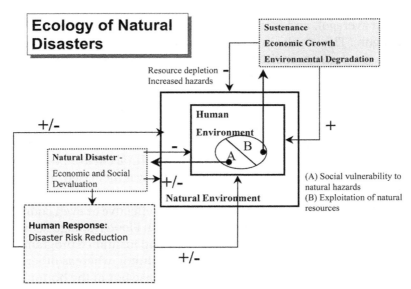

FIGURE 6.19 An ecological model of disaster.

large degree, an extension of the work of Ian Burton as described in *The Environment as Hazard*,[32] which considers the natural environment as both a resource and a hazard, and how mankind's relationship with nature can increase vulnerability.

In the center of Figure 6.19 are two boxes with solid lines, which represent human and natural environments. The human environment box is placed within the natural environment one, emphasizing an ecological perspective. Within the human environment box is a circle representing interaction with those parts of nature that can potentially be resources for society, or can be hazards. Component A represents that part of society vulnerable to natural hazards, and those hazards. An example would be a city built near a fault line and therefore subject to earthquake risk. This is essentially a simple representation of the PAR model. Component B represents that part of nature that is a resource and exploited by humans for sustenance and economic growth (such as harvesting forests for lumber, urban development, paving over land for urban development, or converting the natural landscape into agricultural land). The idea underlying this is that nature is both a resource when it functions within our coping range, and a hazard when it exhibits extremes beyond that range.

From these boxes, there are various arrows pointing in and out, with plus and minus signs beside them. Those signs are meant to represent the average direction of feedback, either positive (constructive to the system) or negative (destructive to the system). Clearly, there are value judgements inherent in these terms, and what one person may consider constructive another may consider destructive. The terms can be interpreted within the context of total resources within the system and complexity; greater resources and increased complexity would be reflected by a plus sign. Therefore, a flux of resources from the natural environment to the social environment would be a plus for the social but a minus for the natural system.

"B" (exploitation of resources) leads to economic growth but also to environmental degradation (on average), and is represented by the dashed box in the upper right-hand corner of the figure. This results in feedbacks into the human and natural environments. One leading to the human environment is positive, reflecting how the use of natural resources enhances our society. However, the feedback into the natural environment is negative, since environmental degradation generally results from resource exploitation. This feedback has the net result of increasing risk by altering the hazards themselves.

Environmental values and the nature of the relationship between humans and nature play a crucial role in the nature of the feedback loops involving "A" and "B." When nature is not valued, or when the links between human and natural environments are discounted, then hazards are ultimately made worse or vulnerability is increased, although short-term benefits may accrue to social systems.

Some mitigation programs appear to have been ineffective or even counter-productive in the long term. Examples of this include the Canadian Flood Damage Reduction program in parts of Quebec[33] and some aspects of the U.S. flood insurance program.[34] The reasons for this are complex; some are political, some are cultural, whereas others are technical. For this reason, the feedback from the Human Response box at the bottom of Figure 6.18, to the Social and Natural Environments box have a plus-or-minus sign.

In order to be effective, mitigation activities need to reduce vulnerability. There are many different ways in which we can be vulnerable, including physical, personal, geographical, structural, environmental, psychological, cultural, spiritual, social, economic, and institutional. These vulnerabilities are often linked in complex ways; for example, a poor economy can lead to a lack of institutional capacity and a greater use/misuse of environmental resources, with consequent environmental degradation. These linkages lead to the notion that any strategy designed to mitigate risk needs to be very broad based. In particular, these strategies should encourage a use of the natural environment that does not degrade it in ways that make hazards worse.

Ecological paradigms tend to fall into two categories—equilibrium and nonequilibrium. Equilibrium models of the world are too simplistic to accurately capture the complexity of environmental interactions because the environment is in a continual state of flux. An equilibrium model views disasters as harmful aberrations to be avoided, whereas a non-equilibrium model incorporates large disturbances as part of the dynamic of change.[35] This duality can have important consequences in terms of how hazards are managed. For example, Smokey the Bear, whose outlook was rooted in an equilibrium model, advised us all to put out all forest fires. Yet doing so had the ultimate effect of creating larger forest fires. Modern fire management strategies incorporate the notion that small fires are part of the natural ecosystem and are needed to prevent massive fires. From this point of view, it is the absence of disturbance, not the disturbance itself, which is the real threat. What are the implications of transferring this paradigm from natural ecosystems to human ecosystems?

Panarchy: Work by Lance Gunderson and C.S. Holling[36] view disaster as part of the evolution of complex systems. Originating from the field of ecology, their work provides a metaphor for any complex system, including society. Figure 6.20 illustrates their basic model of the pathways that systems move through as they evolve.

An example is the ecology of forest systems. New forests exploit the earth's resources and over time become old forests, which are stable. However, as they evolve, the potential for fire increases due to increasing amounts of litter on the forest floor and the development of dense canopies. Eventually a trigger (such as lightening or arson) will create a fire that begins the destructive release phase (this is where a disaster may occur), after which the system reorganizes itself and the cycle repeats. During the release phase, if a critical threshold is passed, a system may flip into a different state such as from a clear lake dominated by fish to a murky one dominated by plankton. In these cases, the process can be irreversible.

A complete description of this model is beyond the scope of this book, and thus far Panarchy has not often been referred to in the disaster literature, although there is a huge potential for its use at the conceptual if not the applied level. The authors of Panarchy frankly acknowledge that it should be considered as an incomplete metaphor; the greatest advantage of this model may be by broadening perspectives to include the following issues; the instability of complex systems, the significance of cross-scale interactions, and the importance of adaptive change and learning.

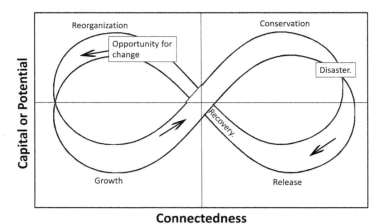

FIGURE 6.20 Panarchy: the adaptive cycle. Panarchy identifies four basic stages of complex systems: exploitation, conservation, release, and reorganization. Exploitation is associated with rapid expansion. During conservation the slow accumulation and storage of energy and material is emphasized. Release occurs rapidly, as during a disaster. Reorganization can also, but not always, occur rapidly, like disaster recovery. Potential sets the limits to what is possible. Connectedness determines the degree to which a system can control its own destiny through internal controls, as distinct from being influenced by external variables. *Source: After Gunderson, L. H. (2001). Panarchy: understanding transformations in human and natural systems. Island Press.*

6.5.8 Disaster Risk Reduction

It might be a stretch to call Disaster Risk Reduction (DRR) a model; it might be better described as a philosophical approach. The UN International Strategy in Disaster Reduction has embraced a DRR philosophy as it works to reduce the harmful impacts of disasters. Its definition of DRR is "The conceptual framework of elements considered with the possibilities to minimize vulnerabilities and disaster risks throughout a society, to avoid (prevention) or to limit (mitigation and preparedness) the adverse impacts of hazards, within the broad context of sustainable development".[37] It is a "concept and practice of reducing disaster risks through systematic efforts to analyse and reduce the causal factors of disasters. Reducing exposure to hazards, lessening vulnerability of people and property, wise management of land and the environment, and improving preparedness for adverse events are all examples of disaster risk reduction."[38] Because of its emphasis on causal factors, it fits well into the PAR model approach to understanding disaster.

At the FEMA Higher Education Conference in June 2012, I talked with several American emergency managers about the CEM model that they embrace, versus the DRR model that much of the rest of the world is moving toward. Aside from the antipathy that some of them showed toward the UN in general, they suggested that there was nothing in DRR that was not already included in CEM—that it was just a different presentation of similar concepts. Proponents of DRR would say that it is more holistic in terms of how it relates to society. I suspect that the success of any project depends less on the choice of CEM versus DRR than on culture, resources, political commitment, and the knowledge and expertise of the persons working on it. Wikipedia has a short but useful summary of DRR at http://en.wikipedia.org/wiki/Disaster_risk_reduction#cite_note-1. I would also recommend going to the ISDR[39] Web site to read about it.

As part of the DRR implementation, countries are developing National Platforms for DRR strategies in order to build more resilient communities. The nature of the platforms differs by country, but broadly speaking they are nationally owned and led, multi-stakeholder forums or committees working on disaster risk reduction. To see which countries have national platforms and to learn more about them visit the ISDR Web site at http://www.unisdr.org/we/coordinate/national-platforms. Particularly in developing countries, DRR helps to bridge gaps between development and humanitarian programs, and works to improve livelihood security. It particularly emphasizes increasing community resilience, reducing vulnerability, and bottom-up community-driven programs. These efforts are important during nonemergency times, but should also be integrated into response and recovery. The Code of Conduct for the International Red Cross and Red Crescent and NGOs in Disaster Response Programmes recognizes this and states in Article 8 that "Relief aid must strive to reduce future vulnerabilities to disaster as well as meeting basic needs".

In addition, there is a Global Platform for DRR based on the Hyogo Framework for Action,[40] and for those with a more academic bent a journal called the *International Journal of Disaster Risk Reduction*.[41] DRR terminology has become mainstream and is in wide use, except in the United States.

One area that is receiving an increasing amount of attention is the overlap between DRR and climate change adaptation. Many of the actions taken to reduce disaster risk, such as increasing community resilience and reducing vulnerability to floods, droughts, or heat waves, are also good adaptations to future increases in climate-related hazards. There is a growing recognition that the strong linkages between these two issues require a coordinated response.[42]

6.5.9 First Nations Wheel

In June 2007 the Assembly of First Nations of Canada published the results of a workshop on pandemic planning.[43] Their approach, which is described as being applicable to emergency planning in general, is based on a holistic policy and planning model (Figure 6.21). Though not well known within the disaster and emergency

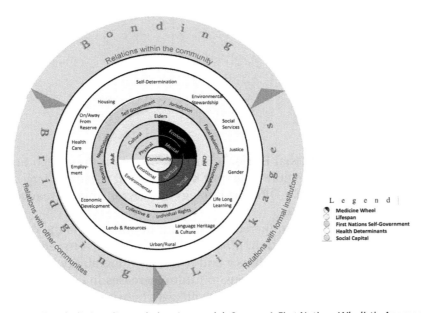

FIGURE 6.21 First nations holistic policy and planning model. *Source: A First Nations Wholistic Approach to Pandemic Planning: A Lesson for Emergency Planning. Assembly of First Nations, June 2007, Ottawa, Ontario.* **Source:** http://64.26.129.156/cmslib/general/pan-planning20078310219.pdf.

management community, there are aspects of it that I find very appealing, including its explicit reference to spirituality, its emphasis on community, references to collective and individual rights, and the respect given to elders. The report describes this approach as being unique in the way it includes stakeholders, and notes that "Each of the major stakeholders can have their own wheel specific to their roles and responsibilities."

The following description of medicine wheels, which date back thousands of years, is taken from an article written by Sandra Laframboise and Karen Sherbina on the Web site *Dancing to Eagle Spirit Society*[44]:

The teachings found on the Medicine Wheel create a bio-psychosocial and spiritual foundation for human behaviour and interaction. The medicine wheel teachings are about walking the earth in a peaceful and good way, they assist in helping to seek; healthy minds (East), strong inner spirits (South), inner peace(West) and strong healthy bodies (North).

As mentioned earlier, a Medicine Wheel can best be described as a mirror within, which everything about the human condition is reflected back. It requires courage to look into the mirror and really see what is being reflected back about an individual's life. It helps us with our creative "Vision", to see exactly where we are in life and which areas we need to work on and develop in order to realize our full potential. It is a tool to be used for the upliftment and betterment of humankind, healing and connecting to the Infinite.

Today, the Medicine Wheel has become a major symbol of peaceful interaction among all living beings on Mother Earth... representing harmonious connections.

The term "Medicine" as it is used by First Nations people does not refer to drugs or herbal remedies. It is used within the context of inner spiritual energy and healing or an enlightened experience often referred to as "spiritual energy's"... The Medicine Wheel and its sacred teachings assist individuals along the path towards mental, spiritual, emotional and physical enlightenment.

In another discussion of medicine wheels, *The Governance Model of the Bent Arrow Society*[45] includes words such as humanness, love, honesty, passion, reflection, and sharing. These are words entirely missing in mainstream planning documents in emergency management. I suspect that it would be both interesting and productive to explore how First Nation approaches to policy and planning model could be adapted and used to enhance mainstream disaster planning.

STUDENT EXERCISE

Pick a disaster with which you are familiar.
- For any three of the above models, list the pros and cons of using them as an analysis tool.

6.6 Conclusion

There are many disaster models available to academics and practitioners who wish to study disaster theory and management or engage in risk reduction measures. The choice of model can have important consequences for the kinds of strategies that are undertaken, and students should not only have a working knowledge of the main models but also a sense for when to use them. This chapter has briefly introduced you to nine of these models. As you progress through your studies, keep them in mind and consider how different models might be useful depending on the kind of disaster being studied and the purpose of the model user. But—most importantly—always remember that *models are not reality*, and their use inevitably introduces simplifications that should not be forgotten.

6.7 Case Study: Sarno Landslides

The City of Sarno is located near the Vesuvius Volcano in south-central Italy, lying in the foothills of the Sarno ridge (Figure 6.22). Eruptions over geological time have created layers of soil a few meters thick that are loose and unstable; the result is a well-known history of landslides in the region. Landslides have, in fact, been occurring annually since the 1960s.

On May 5/6, 1998, after 9 days of rain, over 400 slides were triggered in the region by an intense rainfall event with a 24-h return period of 33 years, and a 4-day return period of 10–15 years. The result of an excessive infiltration of water into pyroclastic material is the reduction of the shear strength of the soil, changing the physical state of the material from rigid to fluid. This is what happened in Sarno. The initial slides developed at higher elevations on steep slopes, but as they traveled they expanded and widened into fluid debris avalanches called slurry flows, incorporating water and material into their paths. The slides reached speeds of 50 km/h, were up to 6 m in height, and traveled 2–3 km in distance. A video of a slide can be seen at http://www.youtube.com/watch?v=7iZzNL1VUWU. The geological

FIGURE 6.22 Sarno, Italy. Note how development occurs in places that are obviously vulnerable to landslides.
Source: Google Earth.

event was exacerbated by loss of vegetation on the slopes due to the burning of trees and brush in order to grow grass for pasture, for criminal activities, or for the purpose of obtaining government relief and reforestation jobs and funds.

The slides, which happened at night, resulted in 160 deaths and 115 injuries, and more than 1200 persons became homeless. The damage was due in large part to uncontrolled development in hazardous regions. In Sarno, zoning laws and environmental regulations were routinely ignored, and much of the construction was poorly built. Additionally, many traditional flood and landslide control systems built in the nineteenth century by the Bourbon rulers of Naples had been paved over or developed for housing. Existing drainage canals had been filled with mud and debris prior to the time of the disaster. Local communities had done everything possible to exacerbate the hazard and to increase vulnerability to a known and well-documented hazard. "It's been on everybody's lips for years. We've discussed it at the City Council. We knew this could happen. We just never understood how bad it could be."[46]

There is an interesting contrast between the City of Sarno, which suffered greatly with 126 deaths, and Quindici, where only 11 people died. Sarno is well known for being controlled by the mafia, whereas the mayor of Quindici had been successful in ridding the town of organized crime. Just before the disaster, the mayor of Sarno had reassured citizens and authorities that the town was safe, which greatly contributed to the death toll. The mayor of Quindici, on the other hand, called for an evacuation on noon the day before the slides.

Caporale[47] notes "the failure of the Italian government to develop a sound policy of land use and control. Over the past 20 years there have been numerous laws to answer this particularly strong need in Italy, but they have been largely ineffective or simply ignored." He critiques the largely ineffective response on the part of the Italian authorities due to overlapping responsibilities from various government bureaus, which resulted in a lack of coordination and communication. He also comments on cultural problems related to transferring a FEMA model of disaster management, based on well-defined lines of accountability and authority, to the Italian government.

This disaster was made inevitable due to failures at levels from national to individual. The systems in place and the cultural context within which they exist simply failed to function as they should have. Corruption, lack of accountability, denial, and apathy appear to have prevented the implementation of policies and actions that could have mitigated landslide risk. It is hard to imagine a scenario that more effectively created the preconditions necessary for a disaster to occur, both in terms of aggravating the hazard and exposing the population.

Further Reading

Caporale R. *The May 1998 Landslides in the Sarno Area in Southern Italy: Rethinking Disaster* Quick Response Report #131. Boulder, CO, 2000. Natural Hazards Center, University of Colorado. http://www.colorado.edu/hazards/research/qr/qr131/qr131.html.

Crosta GB, Negro PD: Observations and modeling of soil slip-debris flow initiation processes in pyroclastic deposits: the Sarno 1998 event, *Natural Hazards and Earth System Sciences* 3:53–69, 2003.

Hervas J. *Lessons Learnt from Landslide Disasters in Europe.* 2000, European Commission Joint Research Center, NEDIES Project.

Stanley A. *Italian Town Buries 90 After Mudslide,* May 11, 2000, New York Times.

Zanchetta G, Sulpizio R, Pareschi MT, Leoni FM, Santacroce R: Characteristics of May 5–6, 1998 volcaniclastic debris flows in the Sarno area (Campania, southern Italy): relationships to structural damage and hazard zonation, *Journal of Volcanology and Geothermal Research* 133:377–393, 2004.

6.8 A Comment by Joe Scanlon[48]

The way communities visualize a disaster and what happens when one occurs affects the way communities plan and respond. To put it another way, faulty modeling can lead to bad planning and ineffective response. That's why it is of concern that there is a perception that all untoward incidents are the same and therefore that the appropriate response for all will be the same, and therefore that all plans are based on the same model. In fact, as Dr. E. L. Quarantelli of the Disaster Research Center, University of Delaware, has argued, there are different types of emergency incidents and—he also argues—these lead to very different problems. That means the models that communities and emergency agencies are using are wrong, and this distorts planning and response.

The first kind of emergency is what I have chosen to call an incident. (Dr. Quarantelli called this an accident, but I use incident because the term accident can suggest cause.) An incident is a site-specific event that occurs, then ends, creating no continuing threat. It could, for example, be a train wreck, a building collapse, or an air crash. It may well occur away from a community and thus pose no threat to anyone other than those directly involved. A train wreck in Hinton, Alberta, for example, occurred on an isolated section of the rail system. The 1985 air crash in Gander, Newfoundland, occurred in a wooded area near the airport, but one where no one lived. Even a building collapse such as the Save-On store collapse in Burnaby, British Columbia posed no continuing threat and did not affect nearby buildings.

Because an incident occurs at a specific and identifiable location, that location can often be fairly quickly controlled and the response can usually be managed for the most part, by cooperating emergency agencies. Police, for example, will control the site and arrange to facilitate movement of emergency personnel to the site and the movement of victims, including the injured, away from the site. Firefighters will deal with fires, spills and heavy rescue. Ambulance personnel will sort out and care for the injured and dictate who should be moved to what medical center and in what order that should happen. The entire response, in other words, may be under the control of emergency personnel; but even then, part of the initial response will be made by those close to the scene. For example, when an Air Ontario plane crashed in the snow near Dryden airport, the first responders were a pilot and surveyor who had seen the plane take off and go down, and went immediately to the crash site.

The second category is a disaster. It will probably not be at a specific site, but may involve a widespread impact. It could be a set of tornadoes that touch down at a number of locations within a short space of time, an earthquake that causes significant damage, or a flood. In complete contrast to an incident, the initial response is by those who are there—uninjured and injured survivors including passers-by. They do the initial search and rescue. They transport the injured to medical centers. By the time emergency agencies get involved, the initial response is over and those involved—the victims and their rescuers—have moved on. One result of this kind of response is that most of the injured go to one or two hospitals (usually the nearest with a trauma center), and the least injured will arrive at the hospital first. Furthermore, the hospitals will have no idea who is coming, and the emergency responders, when they do get involved, have no idea what has happened before they arrived.

The third category is a catastrophe. This is an event with such widespread impact that it overwhelms community resources. Hurricane Katrina was a catastrophe, as was the Haitian earthquake. The only Canadian example was the December 6, 1917, Halifax explosion when a munitions ship exploded in the inner harbor of Halifax, Nova Scotia, with one-seventh the power of the first atomic bomb. One-fifth of the population of Halifax was killed or injured. Thousands of fires were started when stoves tilted over in wooden (frame) houses. All the hospitals were damaged. All senior personnel in the fire department were killed in the blast. A catastrophe is very, very different from an incident and very different from a disaster.

These three types—incident, disaster, and catastrophe—are not variations on a theme but different kinds of events, and this is significant because most emergency plans are based on the assumption that all emergencies will be incidents. Thus emergency plans assume there will be a site, that it will be controlled, and that the response will be by emergency personnel. Such plans may work for such events as the 1985 air crash in Gander, Newfoundland when there was just one road to the crash site and it was quickly controlled by police. But such plans prove less than useful in a disaster and irrelevant to a catastrophe. The problem is that because of their perceptions, emergency agencies fail to develop response plans that would be useful in a disaster or catastrophe. For example, if victims are taken to the hospital by uninjured and injured survivors, it is important that someone at the hospital takes the names, not just of the victims, but also of those who have helped them. That information can then be transmitted to those in the impact areas so they will know who is not missing and who need not be searched for.

One of the simplest and most useful models for one aspect of emergency response, the problem of dealing with the dead, was developed by Tom Brondolo, formerly of the New York Medical Examiner's Office. He identified four criteria that can be used to identify the level of difficulty in dealing with and identifying the dead, in the wake of a mass death incident. The criteria are: the number of dead; the availability of a list; the speed of recovery of the bodies; and the condition of the bodies.

Obviously the number of dead is significant. Ten dead are easier to deal with than 100, and 100 dead are easier to deal with than 1000. It is also easier to deal with them when the identity of those who died is known, such as after an air crash when an accurate passenger list is usually available. That is in contrast to a widespread destructive incident, when it takes time to establish who is missing and likely dead. The speed of recovery of the bodies also affects the response. After the terrorist incident in Oklahoma City, bodies were slowly and carefully retrieved from the wreckage of the Murrah Building, and they were usually found where they were expected to be. Just a few bodies reached the morgue each day, and the identities usually had only to be confirmed. This is in sharp contrast to the Indian Ocean tsunami, where thousands of bodies were piled up on the ground at temples (Thailand) and mosques (Sri Lanka). No one had any idea who they were or where they had come from. Finally, bodies recovered intact, when fingerprints, dental records, clothing and other personal effects are all available, are easier to identify than body parts, as was true after the Swissair crash when only one body was recovered intact.

End Notes

1. Box, G. E. P., and Draper, Norman. 2007. Response Surfaces, Mixtures, and Ridge Analyses, Second Edition [of Empirical Model-Building and Response Surfaces, 1987], Wiley.
2. Whitehead A.F., *Science and the Modern World* (Free Press (Simon & Schuster), 1925).
3. Kahneman D., "Maps of Bounded Rationality: A Perspective on Intuitive Judgment and Choice," *Nobel Prize Lecture* 8, (2002): 351–401.
4. Burton I., Kates R. W., and White G. F., *The Human Ecology of Extreme Geophysical Events*. FMHI Publications. Paper 78. (1968) http://scholarcommons.usf.edu/fmhi_pub/78.
5. Oliver-Smith A., and Hoffman S. M., *The Angry Earth: Disaster in Anthropological Perspective* (Routledge). (1999).
6. Cornell V.J., "Evil, Virtue and Islamic Moral Theology: Rethinking the Good in a Globalized World" in: Gort J.D., Jansen H. and Vroom H.M., eds., *Probing the Depths of Evil and Good: Multireligious Views and Case Studies* (281–304), (Amsterdam: NL Rodopi, 2007).
7. Women to Blame for Earthquakes, Says Iran Cleric (April 19, 2010). *The Guardian*. Retrieved from, http://www.guardian.co.uk/world/2010/apr/19/women-blame-earthquakes-iran-cleric.
8. Stefanovic I.L., *Safeguarding Our Common Future: Rethinking Sustainable Development* (Albany, NY: State University of New York Press, 2000).
9. Devall B. and Sessions G., *Deep Ecology: Living as if Nature Mattered* (Salt Lake City: Peregrine Smith Books, 1985).
10. Berkes F., "Understanding Uncertainty and Reducing Vulnerability: Lessons from Resilience Thinking," *Natural Hazards* 41, no. 2 (2007): 283–95.
11. Adams J., *Risk*. Adams describes four different views or "myths" of nature that people tend to hold. These are (1) nature benign (predictable, bountiful, robust and stable), (2) nature ephemeral (fragile, precarious, unstable, and unforgiving), (3) nature perverse/tolerant (a combination of the previous 2 states, depending on circumstances), and (4) nature capricious (unpredictable). Examples of each of these states exist in nature, and thus supporting arguments can be made for any of the above. A person's risk-taking preference and approach to management will largely depend on which of the above myths he or she adheres to. For example, nature benign or capricious would support a laissez-fair approach to management, nature perverse/tolerant an interventionist approach, while nature ephemeral would suggest use of the precautionary principle (UCL Press, 1995).
12. Emergency Management Doctrine for Ontario (2010), http://www.emergencymanagementontario.ca/stellent/groups/public/@mcscs/@www/@emo/documents/abstract/ec081624.pdf.
13. FEMA. http://www.fema.gov/national-preparedness.

14. Ferrier N., *Fundamentals of Emergency Management: Preparedness* (Toronto, Canada: Emond Montgomery, 2009). Haddow G.D., and Bullock J.A., (2003) Introduction to Emergency Management (pp. 275). New York, USA: Butterworth Heinemann, (Elsevier). Lindell M.K., Prater C., and Perry R.W., (2006). Fundamentals of Emergency Management. Emmitsburg MD: FEMA Emergency Management Hi-Ed Project Lindell M.K., Prater C. and Perry R.W., (2007). Introduction to Emergency Management (pp. 684). USA: John Wiley and Sons.
15. UK Resilence (2004). *Dealing with Disaster (revised 3d edition)* UK Cabinet Office. http://webarchive.nationalarchives.gov.uk/20050523205851/ukresilience.info/contingencies/dwd/index.htm.
16. Blaikie et al., *At risk: Natural hazards, peoples' vulnerability, and disasters*, (London, UK: Routledge, 1994).
17. Wisner et al., *At Risk: Natural hazards, people's Vulnerability and Disasters* (2nd ed.), (London, New York: Routledge).
18. Etkin D, "Patterns of Risk: Spatial Planning as a Strategy for the Mitigation of Risk from Natural Hazards," *Building Safer Communities: Risk Governance, Spatial Planning and Responses to Natural Hazards* 58, (2009): 44.
19. This equation is sometimes incorrectly written as Risk = Hazard + Vulnerability. The relationship is multiplicative, not additive. Another similar expression used by Smith (Environmental Hazards book) is Risk = [Hazard (probability) x Loss (expected)]/Preparedness (loss mitigation). This equation suffers from the problem of being unstable, that is Risk → ∞ when Preparedness → 0, which is physically unrealistic. I would recommend that students not use this formulation.
20. CARE (2002). Household Livelihood Security Assessments: A Toolkit for Practitioners. Prepared by TANGO International Inc., Tucson, Arizona for CARE USA. http://www.proventionconsortium.org/themes/default/pdfs/CRA/HLSA2002_meth.pdf.
21. Lindenburg M., "Measuring Household Livelihood Security at the Family and Community Level in the Developing World," *World Development* 30, no. 2 (2002): 301–318.
22. CARE (2002). Household Livelihood Security Assessments: A Toolkit for Practitioners. Prepared by TANGO International Inc., Tucson, Arizona for CARE USA. http://www.proventionconsortium.org/themes/default/pdfs/CRA/HLSA2002_meth.pdf.
23. Tarrant M., "Regional Workshop on Total Disaster Risk Management," *Emergency Management Australia (EMA)* (2002), Australia. http://www.adrc.asia/publications/TDRM/17.pdf.
24. Burton I., The Environment as Hazard (The Guilford Press, 1993).
25. Cohen M. D., March J.G., and Olsen J.P., "A Garbage can Model of Organizational Choice," *Administrative Science Quarterly* 17, no. 1 (1972): 1–25.
26. Buck D. A., Trainor J. E., and Aguirre B. E, "A critical evaluation of the incident command system and NIMS," *Journal of Homeland Security and Emergency Management* 3, no. 3, (2006), Article 1.
27. Adapted from: Grossi P., and Kunreuther H, eds., (2005). *Catastrophe Modeling: A New Approach to Managing Risk* (Vol. 25). Springer.
28. Li Y., (2005). Fragility Methodology for Performance-Based Engineering of Wood-Frame Residential Construction. (Unpublished thesis). Georgia Institute of Technology, Atlanta, GA, USA.
29. Allen D.E., (1986), Tornado Damage in the Barrie/Orangeville area, Ontario, May 1985. Building Research Note No. 240, National Research Council Canada.
30. Gass, H., (2014) "Tornado Survival Could Improve with Better Building Codes," April 30, 2014. *Scientific American, Climatewire*.http://www.scientificamerican.com/article/tornado-survival-could-improve-with-better-building-codes/.
31. Etkin D., and Stefanovic I. L., (2005). Mitigating Natural Disasters: The Role of Eco-Ethics. In Mitigation of Natural Hazards and Disasters: International Perspectives (pp. 135–158). Netherlands: Springer.
32. Burton I., *The Environment as Hazard*, (The Guilford Press, 1993).

33. Robert B., Forget S., and Rousselle J., "The Effectiveness of Flood Damage Reduction Measures in the Montreal Region," *Natural Hazards* 28, no. 2–3 (2003): 367–385.
34. Larson L., and Plasencia D., "No adverse impact: New direction in floodplain management policy," *Natural Hazards Review* 2, no. 4 (2001): 167–181.
35. Reice S. R., *The Silver Lining: The Benefits of Natural Disasters*, (Princeton University Press, 2003), 217.
36. Gunderson L. H, *Panarchy: Understanding Transformations in Human and Natural Systems*, (Washington, D.C: Island Press, 2001).
37. UNISDR. Living with Risk: "A Global Review of Disaster Reduction Initiatives, 17, New York and Geneva," (2004).
38. What is Disaster Risk Reduction? UNISDR, http://www.unisdr.org/who-we-are/what-is-drr.
39. UNISDR. www.isdr.org.
40. PreventionWeb. Serving the Information Needs of the Disaster Reduction Community. http://www.preventionweb.net/english/hyogo/GP/.
41. International Journal of Disaster Risk Reduction. Elsevier. http://www.journals.elsevier.com/international-journal-of-disaster-risk-reduction/.
42. Gero A., M´eheux K., and Dominey-Howes D., "Integrating Community Based Disaster Risk Reduction and Climate Change Adaptation: Examples from the Pacific," *Natural Hazards Earth System Science* 11, no. 1 (2011): 101–113.
43. The Assembly of First Nations. *A First Nations Wholistic Approach to Pandemic Planning: A Lesson for Emergency Planning*, (Ottawa, Ontario, 2007). Retrieved from 64.26.129.156/cmslib/general/pan-planning20078310219.pdf.
44. Dancing to Eagle Spirit Society. http://dancingtoeaglespiritsociety.org/index.php.
45. Jobin S., *Guiding philosophy and governance model of bent arrow traditional healing society (unpublished Master degree thesis)*, (British Columbia, Canada: University of Victoria, 2005). http://fngovernance.org/resources_docs/Guiding_Philosophy__Governance_Model__Bent_Arrow1.pdf.
46. Stanley A., (1998). Italian Town Buries 90 After Mudslide. New York Times, May 11 http://www.nytimes.com/1998/05/11/world/italian-town-buries-90-after-mudslide.html.
47. Caporale R., (2000). The May 1998 Landslides in the Sarno Area in Southern Italy: Rethinking Disaster Theory. Quick Response Report #131. Natural Hazards Centre, University of Colorado, Boulder, CO. http://www.colorado.edu/hazards/research/qr/qr131/qr131.html.
48. Joe Scanlon is Professor Emeritus and Director of the Emergency Communications Research at Carleton University He has been doing disaster research since 1970. He has published about 200 book chapters, monographs and articles in peer-reviewed and professional journals. In 2002, He Received the Charles Fritz Award for a Lifetime Contribution to Sociology of Disaster.
49. Etkin D., Higuchi K., and Medayle J., "Climate Change and Natural Disasters: An Exploration of the Issues," *Journal of Climate Change* 112, no.3–4(2001): 585–599.

7

Myths and Fallacies

The great enemy of truth is very often not the lie—deliberate, contrived, and dishonest—but the myth—persistent, persuasive and unrealistic.
John F. Kennedy

FIGURE 7.1 Cartoon depicting perception versus reality.

CHAPTER OUTLINE

- 7.1 Why This Topic Matters 230
- 7.2 Recommended Readings 231
- 7.3 Myths of Fact 231
  - 7.3.1 Tornadoes 231
  - 7.3.2 Earthquakes 233
  - 7.3.3 Other Myths 236
- 7.4 Myths of Human Behavior 237
- 7.5 Fundamental Myths of Our Relationship to the World 240
- 7.6 Conclusion 244
- 7.7 Fables: of Little Pigs and Ants 244

7.8 Case Study: the Great Flood 247

7.9 A Comment by Joe Scanlon 248

End Notes 250

CHAPTER OVERVIEW

Research has shown that myths about disaster are commonly believed not just by the general public, but also by many professionals working in the field. Sometimes these myths can have very harmful outcomes, by encouraging people and decision makers to engage in risk reduction or management strategies that are not optimal or even harmful. Some myths are relatively straightforward and easy to dispel, such as the belief that one should open windows when threatened by an oncoming tornado. Others are much more fundamental to belief systems and extremely hard to alter. It is important for the field of disaster management to move further toward a more evidence-based approach, and this can only be done through research and education.

KEYWORDS

- Earthquake
- Fable
- Fallacy
- Fantasy documents
- Flood myths
- Myth
- Panarchy
- Shattered assumptions
- Three little pigs
- Tornado

7.1 Why This Topic Matters

Myths and fallacies about disasters play out in important ways, both in disaster planning and response. There has been a good deal of research about disaster mythology that is often overlooked by planners and responders, resulting in some very unfortunate outcomes. All students of disaster studies should be aware of this research and contribute toward moving the profession of disaster management toward an evidence-based approach.

This chapter will look at several different kinds of myths, beginning with myths about tornado and earthquake facts (other hazards are not included due to space limitations), then considering myths of human behavior that have emerged from several decades of social science research (Figure 7.1), and finally looking at some deeper myths related to worldviews. It is usually fairly easy to convince people about factual myths, more difficult to change belief in myths of human behavior, and almost impossible for people to shift worldviews except under extraordinary conditions.

■ ■ 7.2 Recommended Readings ■

- de Goyet C. D. V., "Stop propagating disaster myths," *Disaster Prevention and Management* 9, no. 5 (2000).
- Fischer H. W., *Response to disaster: Fact versus fiction & its perpetuation: The sociology of disaster*, third edition, University Press of America (2008).
- Quarantelli E. L., "Conventional beliefs and counterintuitive realities," *Social Research: An International Quarterly* 75, no. 3 (2008): 873–904.
- Scanlon T. J., and Unit D. E. P., *Lessons Learned or Lessons Forgotten: The Canadian Disaster Experience*. Institute for Catastrophic Loss Reduction (2001).
- Tierney K., Bevc C., and Kuligowski E., "Metaphors matter: Disaster myths, media frames, and their consequences in Hurricane Katrina," *The Annals of the American Academy of Political and Social Science* 604, no. 1 (2006), 57–81.

■ ■ Questions to Ponder ■

- Why are disaster myths so persistent?
- How important are movies in propagating disaster mythology?

> **STUDENT PROJECT**
>
> *Pick a disaster movie and analyze it for disaster myths and realistic portrayals. Wikipedia has a good list at* http://en.wikipedia.org/wiki/List_of_disaster_films.

7.3 Myths of Fact

7.3.1 Tornadoes[1]

- Myth: The southwest corner of a basement is the safest location during passage of a tornado.
  - Reality: Winds rotate around a tornado, and whether a tornado tracks north or south of a house, and where it is relative to a suction vortex (Figure 7.3), will determine wind direction. There is no way to predict which corner of a basement will be the safest.
- Myth: Some towns are protected, and tornadoes never strike big cities.
  - Reality: There are differences in climatology from place to place, but simply because a place has never been hit by a tornado does not mean that it will never be; it just reflects the low probability associated with a tornado hitting any particular location. If you consider the geographical area covered by cities and compare it to rural areas, it is clear that the probability of an urban tornado is very small compared to a rural one.

FIGURE 7.2 Drawing of a tornado from *Orbis Sensualium Pictus* (Comenius, 1658—1685). This reproduction comes from a 1685 reprint from Levoča (Slovakia) four-language version (Latin, German, Hungarian, and Czech). http://www.sciencedirect.com/science/article/pii/S0169809503000759.

FIGURE 7.3 Tornado structure with multiple suction vortices is seen in this picture of the Tushka, Oklahoma tornado, an EF3 which struck the town on April 14, 2011, during the mid-April 2011 tornado outbreak. The multiple funnels result in a somewhat chaotic and unpredictable surface wind damage pattern. *Source: NOAA,* http://www.srh.noaa.gov/images/oun/wxevents/20110414/stormphotos/austin-garfield/20110414_tushka2.jpg

- Myth: Opening windows to equalize air pressure will save a roof, or even a home, from destruction by a tornado.
    - Reality: There was an old school of thought that houses exploded because of the relatively high pressure inside them compared to the low pressure near a tornado. It is now known that this is not true. Rather, houses usually fail because strong winds blow off roofs, after which the walls collapse from wind pressure (Figure 7.4). In fact, many of the injuries from tornadoes occur due to broken glass being blown around,

so not only does taking time to open windows fail to prevent house collapse (and can even increase the chances of failure if the winds are blowing into the open window), it exposes people to greater risk. In any case, if a tornado is close to a house, windows will be broken because of flying debris.

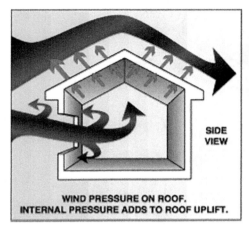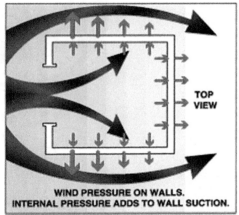

FIGURE 7.4 Wind loads on a house. *Source: FEMA,* http://www.fema.gov/safe-rooms/building-damage

- Myth: Highway overpasses are a safe place to shelter.
  - Reality: In 1991, a video of a tornado in Kansas[2], filmed by a TV crew that took shelter under an overpass, was widely viewed and led many to believe that overpasses would be safe shelters in a tornado. Winds and debris caught up in the tornado blow horizontally, so a shelter with only overhead cover is probably not very helpful in this type of situation.

7.3.2 Earthquakes[3]

- Myth: Some animals can sense an earthquake before it happens.
  - Reality: There is anecdotal evidence that animal behavior has predicted earthquakes, but the behavior is inconsistent and is not often observed prior to an earthquake. One example of when animal behavior may have predicted an event was during the December 2004 tsunami in the Indian Ocean, when elephants are reported to have run for higher ground, presumably because they felt the ground vibrations from the P wave prior to the major event.[4] In general, however, such predictors are unproven.
- Myth: Earthquakes happen during hot and dry spells.
  - Reality: This misconception dates to the ancient Greeks. Earthquakes result from forces occurring deep in the earth and are not related to weather.
- Myth: Big earthquakes always occur early in the morning.
  - Reality: Many large earthquakes have occurred at other times of the day.

FIGURE 7.5 In Japanese mythology, earthquakes were caused by a giant catfish. In this figure, the deity Ebisu falls asleep while guarding the foundation stone (kaname ishi) for Kashima, allowing the Namazu to cause the quake that destroys Edo. Kashima (aka Takemikazuchi, enshrined at Kashima) is seen returning on horseback too late. The thunder deity is seen engaging in "thunder farting." Coins are seen falling from the destruction, indicating redistribution of wealth. *Source: International Research Center for Japanese Studies, Namazu-e collection,* http://shinku.nichibun.ac.jp/namazu/ichiran.php

- Myth: California could fall into the sea because of an earthquake.
  - Reality: "The San Andreas Fault System is the dividing line between two tectonic plates. The Pacific Plate is moving in a northwesterly direction relative to the North American plate. The movement is horizontal, so while Los Angeles is moving towards San Francisco, California won't sink. However, earthquakes can cause landslides, slightly changing the shape of the coastline."
- Myth: The ground can open up and swallow people.
  - Reality: This image makes for good fiction, but an earthquake causes ground shaking, not open fissures. However, at times earthquakes can cause deformation of the ground that possibly includes open fissures that people or cars could fall into, so there is an element of truth here. This is more likely to happen with sink holes.
- Myth: The safest place to be in an earthquake is under a doorway.
  - Reality: "That's true only if you live in an unreinforced adobe home. In a modern structure the doorway is no stronger than the rest of the building. Actually, you're more likely to be hurt (by the door swinging wildly) in a doorway. And in a public

building, you could be in danger from people trying to hurry outside. If you're inside, get under a table or desk and hang on to it."[5]
- Myth: Small earthquakes keep big ones from happening.
  - Reality: Small quakes do release some stress, but generally not enough energy is dissipated to prevent large quakes.
- Myth: We have good building codes, so we must have good buildings.
  - Reality: This is true if there is good enforcement (which in many countries is not the case[6]), but older buildings are not safe unless they have been retrofitted.
- Myth: Earthquakes are becoming more frequent (Figure 7.6).
  - Reality: Although there is some evidence of a cyclical pattern, the average frequency of large earthquakes has remained fairly constant in the past century.[7] The perception that they are more frequent is not factual. It may seem that quakes are more frequent because technological advancements have made scientists better at observing them, and because earthquakes are reported more often by the media than in the past.
- Myth: There's nothing I can do about earthquakes, so why worry about them?
  - Reality: Nobody can stop an earthquake from happening, but we can construct buildings so that they will not collapse, and save lives by knowing how to prepare and respond in the aftermath of an earthquake disaster.

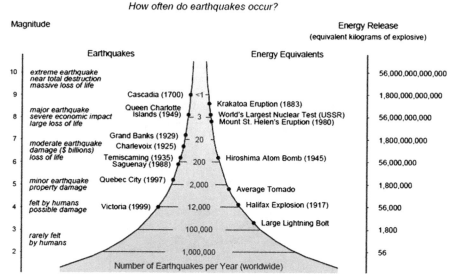

FIGURE 7.6 Frequency of earthquakes. Note the rapid change in frequency as earthquake intensity increases. The change follows a power law, as opposed to being linear (see section 5.4.4). *Source:* http://www.earthquakescanada.nrcan.gc.ca/info-gen/magfreq-eng.php

7.3.3 Other Myths

- Myth: Dead bodies cause epidemics.
    - Reality: One source of this myth may be from beliefs regarding the Black Death in Europe in the Middle Ages, where dead bodies were a reservoir of contagion (Figure 7.7). "Associating epidemics with the presence of human remains after an accident, conflict, or disaster is a deeply ingrained myth in Western culture."[8] But this is not true today; in terms of disease transmission, dead bodies are generally less of a risk than the living.
    - Reality: "On rare occasions when victims of a disaster are carriers of communicable diseases, they are, in fact, a far lesser threat to the public than they were while alive. The result of this mistaken belief is the overlooked and unintended social effect of the precipitous and unceremonious disposal of corpses. This action is just one more severe blow to the affected population, because it deprives them of their human right to honour the dead with a proper identification and burial. The legal and financial consequences of the lack of a death certificate will add to the suffering of the survivors for years to come. Moreover, the summary disposal of bodies, superficial 'disinfection' with lime, mass burial, or cremation of corpses require important human and material resources that should instead be allocated to the people who have survived and remain in critical condition."[9]
- Myth: International response saves a lot of lives.

FIGURE 7.7 The Great Plague of London in 1665. The last major outbreak of the bubonic plague in England.
Source: Love of History, http://theloveforhistory.com/wp-content/uploads/2011/06/plague_380x529_712060a.jpg

- Reality: The vast majority of lives are saved by local response. For example, during Iran's 2003 Bam earthquake, which destroyed 85% of the city, local Iranian Red Crescent rescue teams were deployed within minutes despite losing four team members and their headquarters in the earthquake. They saved 157 lives with just 10 dogs. In contrast, international search and rescue teams from 27 countries took up to two days to arrive. Although they were armed with sniffer dogs and remote sensing equipment, they saved just 22 lives. The massive cost of international response, if put toward mitigation and sustainable development, would end up saving far more lives. This presents some interesting ethical issues.
- Myth: Any kind of aid and relief is useful after disaster, providing it is supplied quickly enough.
    - Reality: Sending aid that is not needed diverts relief resources from other more needed tasks, and can even have a negative effect on the economy of local communities by undercutting livelihoods. Of course delaying sending aid can create problems; a needs assessment can ensure a more appropriate response.

> **STUDENT EXERCISE**
> With a partner, debate the ethics of allocating resources to international response as opposed to sustainable development.

7.4 Myths of Human Behavior

Civilization is hideously fragile... there's not much between us and the Horrors underneath, just about a coat of varnish.

C.P. Snow

...the old instinct of the wild beast broke through the thin veneer of civilization

The Return of Tarzan, Edgar Rice Burroughs (**Figure 7.8**)

There is an important thread of literature on public beliefs about disaster, particularly with respect to human behavior, which has shown that many of these beliefs are myths.[10] A book by Henry Fischer III called *Response to Disaster: Fact Versus Fiction and its Perpetuation* has a very good discussion on this topic. *It must be noted that virtually all of this research is based in the United States, and since human behavior has strong cultural norms, it should be extrapolated with caution to different cultures.* These myths tend to be perpetuated in media and film and are particularly resistant to change. In fact, surveys have shown that even professional emergency managers often believe these myths.[11]

FIGURE 7.8 Tarzan. *Source: Book by Edgar Rice Burroughs published in 1914.*

The basis of some of these myths is reflected in the quotes above—that civilized and moral behavior disappears during disaster, allowing the savage nature within to emerge. But is this so? On the contrary, research has shown that in the vast majority of cases, dominant behaviors during disaster are prosocial and that heroism and selflessness are far more common than in normal times. The research that supports these conclusions does not include long-lasting catastrophes that envelop whole societies, and it may be that antisocial behaviors emerge in those cases. For example, in one famine affecting Egypt from 1199 to 1202 there were reports of cannibalism; those convicted of this offense were put to death and in some cases were then turned over to the masses as legal food.[12] Below is a list of common disaster myths about human behavior. Sadly, these myths are perpetuated by the media and are very persistent.

- Myth: Victims are subject to psychological dependency and shock, are incapable of functioning, and require direction.
    - Reality: First response is normally by victims and people in the community, who tend to respond well to crisis. Paternalistic attitudes toward victims can cause resentment and derail local input to response and recovery.
- Myth: During normal times, people act in a predictable, rational, orderly fashion, but during disaster the norms that govern our behavior collapse and there is a shift to deviant behavior that is irrational, unpredictable, and disorderly (Figure 7.1). This includes looting, mob behavior, and panic flight, when potential victims take any actions necessary to flee due to extreme fear.

- Reality: People flee when there is a perceived avenue of escape, but almost always in a rational way. Looting is rare, and only becomes significant in situations where the social climate is predisposed to such behavior, and the disaster acts as a trigger.
- Myth: People will refuse to evacuate.
  - Reality: Though some people choose not to evacuate because of denial or fear of looting, it often happens because of a lack of choice. For example, residents of New Orleans were criticized for not evacuating before Hurricane Katrina hit, but many of them did not own a car, had no money, or lacked the means to evacuate. In addition, people will often refuse to evacuate to a shelter if pets are not allowed.
  - Reality: Warnings are believed when there is clarity, consistency, frequency, and a history of accurate warnings by a trusted authority.
  - Reality: Most shelters are underused.
- Myth: Price gouging occurs after disaster.
  - Reality: Research shows that this is rare, and when it does occur, it is mainly practiced by outsiders.
- Myth: Martial law is commonly declared after disaster.
  - Reality: In Canada, martial law has never been declared after a disaster. In the United States, it was declared after the Great Chicago Fire of 1871, but was lifted after a few days. Declaring martial law because of a disaster is quite rare.
- Myth: Survivors feel lucky to be alive.
  - Reality: Survivors often feel guilt for having survived when others have not. They may also experience self blame, especially when other family members are victims.

Unfortunately, the beliefs described above can lead to flawed decision making.[13] Examples include delayed evacuation orders and an overemphasis on security as opposed to search and rescue. The latter has been critiqued with respect to the level of military response to Hurricane Katrina[14] in New Orleans, and to the Haiti earthquake. An article by CNN[15] had the following to say about the situation in Haiti: "And in the days that followed, we did not see a population breaking into buildings or fighting over scant resources. The Haiti we saw is not a land of fires and violence and looting. This is a land where desperation has been supplanted by despair." I recall seeing a video clip of the Haitian Minister of Defense at the time of the earthquake, saying in frustration that what they needed was help, not more guns. A high-profile example of delayed evacuation occurred in 2012 when the Italian cruise ship *Costa Concordia* ran aground. The captain of the ship, when interviewed, stated that he delayed issuing an abandon ship order to avoid panic and mob behavior from the passengers; this decision almost certainly cost some passengers their lives. While watching the news about the sinking on TV and reading about it online and in the newspapers, I frequently saw the media refer to panic among the passengers. Yet all the video clips I saw of the event revealed no panic—mainly I saw calm, organized behavior with some examples of fear, but no panic.

Disaster myths are not just false beliefs; they are false beliefs that cause harm.

> **STUDENT EXERCISE**
>
> Survey five friends and family members about their perceptions of looting, panic, shock, and the role of the military during disaster. Ask them to explain the reasons for their answers. Compare your results with other students in the class.

7.5 Fundamental Myths of Our Relationship to the World

There are other myths present in the academic literature that are not explicitly discussed in the more common lists of disaster mythology, but are important for understanding the construction of vulnerability and how people view and respond to risk. The myths of human nature described in Chapter 3.5.3 fit into this category. It is not possible to do justice to these complex topics in the short space available in this chapter; the purpose of this section is simply to introduce these ideas so that you may pursue them in the future if it becomes relevant to your interests or work.

The myth of technological and economic growth: This myth is based on the notion that progress in the human condition can be achieved through technological and economic growth at the expense of environmental and social wellbeing,[16] and relates to the discussion on environmental ethics in Chapter 9.4.4. This belief requires an artificial separation between those endeavors that create wealth and progress from their negative consequences, where progress is guided by performance values that can be quite different from human and social values, and is subject to a narrow accounting construct. According to Willem Vanderberg, professor of engineering at the University of Toronto, "The 'system' is lost in the labyrinth of social and environmental implications of technology, which it constantly expands by rarely getting to the root of any problem."[17] According to Vanderberg, the fundamental contradiction is that we achieve rationality at microscales which, when aggregated at the macro level, become irrational. It is a case where scaling up results in a fundamentally flawed system. There are many examples of this; for example, it lies at the root of society's failure to mitigate climate change. Like the tragedy of the commons,[18] it makes sense for each country to continue to pollute the global oceans and atmosphere, even though as all countries do so, tragic results are likely. This contradiction also applies to nuclear weapons (while a particular country may benefit from having them, proliferation is akin to insanity). This theme echoes many of the sentiments expressed by Ulrich Beck in *The Risk Society*.

The myth of technological and economic growth is also closely related to one described by Denis Mileti—that of people seeking to "make themselves totally safe through the fantasy of using technology to control nature." Since much of our risk is constructed through technology, it is, in part, an attempt to pull ourselves up by our bootstraps. According to Mileti, the result is that "much of what we think we know is wrong, and many of the actions we take to fix problems just make them worse." I think it is easy to get carried away with this perspective and label all technology as harmful, but this is not the case. Many aspects of technology and economics are part of a holistic solution—they are just not the whole solution.

No problem can be solved from the same level of consciousness that created it.
Albert Einstein

There are many positive changes and actions being taken in the world to reduce disaster risk, and yet the number and magnitude of disasters appears to be growing, not diminishing. My sense is that this occurs because many of the causes are structural in nature and strongly rooted in social processes that are obliquely addressed by Disaster Risk Redution (DRR) strategies. Notwithstanding their many and obvious benefits, technological and economic growth increase vulnerability in many ways; examples are the creation of more complex and tightly connected infrastructure, and processes for the short-term maximization of economic efficiency and wealth in a global economy (such as "just-in-time" delivery instead of stockpiling goods, and minimal standards of safety). I believe that the myth of technology as savior is not as strong as it once was. Many institutions now explicitly recognize the need for interdisciplinary solutions. For example, in 2011 I was approached by a professor of engineering to participate in a proposal for a Center of Excellence in Disaster Studies, and he commented on how engineering is only a partial solution to the problem. This recognition is, I find, typical from many physical scientists. Politically, however, that is not always the case; for example, under the Harper conservative government, Canada has shifted deeper into the pathology of sacrificing environmental concerns for economic growth, wiping out decades of progress in this area. I fear that this will haunt us in the future.

> **STUDENT EXERCISE**
>
> List several ways in which an overdependence on technology or economic growth has increased risk beyond what many in society might deem acceptable.

Myths at the core of our internal world: In her book *Shattered Assumptions: Toward a New Psychology of Trauma*,[19] Ronnie Janoff-Bulman talks about the assumptive world, a "strongly held set of assumptions about the world and the self which is confidently maintained and used as a means of recognizing, planning and acting." She proposes that the three most fundamental assumptions held by most people are that: (1) the world is benevolent, (2) the world is meaningful, and (3) the self is worthy. Of course not everyone believes in these assumptions, but they are very common in spite of a great deal of contrary evidence.

These assumptions, though flawed, allow people to trust themselves and their environment, and therefore enable functional relationships. However, they can also lead to maladaptive behavior by encouraging undue risk taking or making false associations between cause and effect. Specifically, they can result in: (1) unrealistic optimism, (2) people believing that they can control more than they can, and (3) situations in which negative events are viewed as punishments and positive ones as rewards, even when it is not true. These assumptions allow us to feel safer and more in control. One particularly insidious outcome is blaming victims for their hardships; the most well-known example of this is blaming rape victims ("if she hadn't been walking down that street, or dressing provocatively…"), but it also explains the general tendency to blame disaster victims for their fate. Janoff-Bulman emphasizes that these myths are not "narrow beliefs, but broad abstract conceptions that are also emotionally potent." Her model explains many aspects of human risk perception and behavior.

Janoff-Bulman's research suggests that victims of disaster become more likely to hold negative assumptions about themselves and the benevolence of the world, particularly if the disaster was caused by people. While on the road to recovery, survivors primarily attempt to rebuild their shattered assumptive world through three major strategies. The first is through comparison with others, especially other victims. The second is interpreting one's own role in the cause of the disaster, often through self blame. The third is reevaluating the disaster through the lens of benefits and purpose, in an attempt to give meaning to the event.

When reading about the destruction of the assumptive world, it has struck me that there may be a similarity between experiencing a disaster and being asked a koan.[20] (A famous koan is "Two hands clap and there is a sound. What is the sound of one hand clapping?" Hakuin Ekaku.) Both create cognitive dissonance, the disaster because it violates our inner assumptive world of safety and meaning and requires us to reestablish a new world view, and the koan because it is not subject to typical rationalistic western interpretation, but is rather meant to create doubt, replace reason with intuitive enlightenment, and explore themes of nonduality. Disasters cause people to doubt, create strong emotional and intuitive reactions, and are often interpreted in terms of dualities. For example, one common metaphor of disaster is that of war, where people fight against a dangerous enemy. The notion of war is rooted in dualities: us versus them, good versus evil, and individual versus state. This metaphor, however, has serious limitations. For example, if winning the fight against flood disasters means using technology to defeat nature so that one might continue to develop in flood plains, then it may well be a losing strategy. An ecosystem approach to living with floods by accepting their occurrence can be much more effective; a good example of this is the Nederland leeft met Water (the Dutch Live Together with Water) program in The Netherlands,[21] launched in 2003, which involves a paradigm shift from relying on levees and river normalization to more inclusion of water retention and floodplain rehabilitation strategies.

Panarchy: The Myth of Management: The book *Panarchy: Understanding Transformations in Human and Natural Systems*[22] was published by Lance Gunderson and C.S. Holling in 2002. It is a brilliant and innovative study that provides great insights into how our world works. A website on panarchy that has some good resources is found at www.panarchy.org.

The traditional approach to managing systems depends on their being well behaved from a mathematical perspective; in other words, systems do not have discontinuous responses and are predictable in the sense that a given set of inputs results in a predictable set of outputs. This, of course, does apply to many of the systems people work with (after all, the systems we depend on tend to be quite reliable most of the time). However, the world is changing, and nature and people interact with each other in increasingly complex ways. Traditional approaches do not address multistable states, and ignore the possibility that "the slow erosion of key controlling processes can abruptly flip an ecosystem or economy into a different state that might be effectively irreversible." Attempts to control complex systems using traditional resource exploitation and management approaches—although often initially successful—eventually create less resilient ecosystems, more rigid institutions, and deeper social dependencies. One example of this is flood control and the "levee effect," where unintended long-term consequences include more damaging floods.

Panarchy discusses the problem of partial truths: the failings of disciplinary boundaries and models based on ecological, economic, or social perspectives to address interdisciplinary complex problems. Each approach alone is insufficient and ignores important interactions. Ecological approaches ignore the need for economic development, economic models make the bad assumption that nature can be replaced by human engineering and management, and social models often assume that there are no limits to what can be achieved by bottom-up and local initiative approaches. Integrated models and solutions are needed, but are extraordinarily hard to create.

The panarchy model proposes that systems tend to follow this cyclic pattern: (1) exploitation, (2) conservation, (3) release, and (4) reorganization. Using the example of a forest, this would be the phases of (1) forest growth, when the nutrients in the soil allow for the creation of the forest, (2) an old-growth forest that is stable over time, (3) a forest fire, and (4) a regrowth of the forest, similar to its old state or perhaps a new one if environmental conditions have changed. This model suggests that disasters (i.e., the forest fire) are not events that can always be avoided, are a part of the cyclical dynamics of complex systems, and serve useful purposes.

The implications of panarchy in terms of disaster management are important, and emphasize the use of adaptive systems and ecosystem resilience models as opposed to more traditional management approaches. I believe that the panarchy model has great potential in terms of its application to disaster theory. Aspects of it are now incorporated into the DRR approach, which emphasizes community resilience as a fundamental aspect of reducing disaster risk, but much work remains to be done at both theoretical and applied levels. A case study applying this model to a complex disaster could make for a very interesting PhD thesis.

STUDENT RESEARCH PROJECT

From a system resilience perspective, describe the long-term effect of the following:

- Suppressing all forest fires[23]
- Monoagricultural practice and the loss of biodiversity of wheat and banana species[24]
- Famine relief in Ethiopia[25]

Fantasy documents: Sociologist Lee Clarke wrote a very interesting book in 1999 called *Mission Improbable: Using Fantasy Documents to Tame Disaster.*[26] He cites the book's theme as being "that organizations and experts use plans as forms of rhetoric, tools designed to convince audiences that they ought to believe what an organization says. In particular, some plans have so little instrumental utility in them that they warrant the label 'fantasy document.'" Sometimes this might be deliberate, but the more frightening truth is that some organizations are simply unaware of the limits of their knowledge and control, and therefore attempt the unachievable. Particularly under conditions of high uncertainty, these plans become unlikely to succeed in practice; what they do achieve is to become "rationality badges" that provide legitimacy. "It is organizations that socially construct risk, danger, and safety… most of us, most of the time, are at the mercy of organizational action."

Clarke uses three case studies to develop his thesis: evacuation plans for nuclear power plants, nuclear war civil defense plans, and contingency plans for major oil spills. Some problems do not have solutions, and yet society may demand them. This conundrum forces institutions to create plans based on assumptions rooted in fantasy, which they may be all too likely to believe themselves. As a weather forecaster in the 1980s, I remember having to provide seven-day precipitation forecasts. There was no skill in our ability to do so, but society demanded it. Politicians wanted votes, senior public servants needed to please those who control budgets and promotions, and therefore as a government agency we were mandated to provide them. I expect the public had the good sense not to believe what we were saying, but it was an example of planning based upon a fantasy.

7.6 Conclusion

We view the world through a series of veils of ignorance and filters of prejudice. Thus, we hold up partial truths as symbols of reality; it is the human condition, to a greater or lesser degree. And although all of us sometimes fall prey to the fallacy of misplaced concreteness and naïve assignments of cause and effect, many of these partial truths and myths can be dispelled through the use of an evidence-based approach. It is important to do so, for many of these false beliefs can result in dreadful outcomes.

7.7 Fables: of Little Pigs and Ants

So the wolf huffed and he puffed and he huffed and he puffed. The house of straw fell down and the wolf ate up the first little pig (Figure 7.9).

Just after the terrible Asian tsunami of 2004, there was a story of a village where everybody ran for higher ground when the water retreated from the coastline, because of a

FIGURE 7.9 The wolf blows down the straw house in a 1904 adaptation of the fairy tale "Three Little Pigs." *Source: Library of Congress.*

nursery rhyme children learned when they were young that told the story of a tsunami disaster. By contrast, in many other places people went down to the beach to look at the dry sea bed, and were then made so much more vulnerable to the wave when it swept in. An educational program based on nursery rhymes has been developed by an Australian NGO, under which "children are taught nursery rhymes with accompanying actions that contain memorable and simple emergency response information."[27]

Risk communication strategies about hazards are normally created by experts who gravitate toward information-driven approaches, which focus on data and a calculative approach to risk estimation. Of course these are very useful, but they can be augmented by other approaches that emphasize stories, narratives, and fables.

Fables and fairy tales have been part of our culture for a very long time, and their longevity speaks to their ability to resonate with parents and children. They are not simply stories, but also provide important lessons for children about their place in the world, duties and responsibilities, and how to live one's life.[28] Most modern children's literature is designed to inform, entertain, or develop skills. Fables, on the other hand, serve a broader goal: to help add meaning to a child's life. According to Bruno Bettelheim, "nothing can be as enriching and satisfying to a child and adult alike as the folk fairytale. True, on an overt level fairy tales teach little about the specific conditions of life in modern mass society… But more can be learned from them about the inner problems of human beings, and of the right solutions to their predicaments in any society, than from any other type of story within a child's comprehension."[29]

Bettelheim uses a deeply Freudian interpretation of fables. But some fables also provide important lessons on how to manage disaster risk; two of these are the "Three Little Pigs" and the "Ant and the Grasshopper."

In the "Three Little Pigs," the pigs are told that they are at risk from a big bad wolf (sometimes a fox), who wants to devour them. After they leave their home, each pig builds a house, but from different materials. The first pig who is lazy uses straw, the second who is selfish uses wood, and the third little pig (who in some versions of the story is the only one to have seen the big bad wolf before building his home), who is neither lazy nor selfish, plans ahead, and builds his house from brick. Soon, along comes the wolf who huffs and puffs and blows down houses numbers one and two. In some versions of the fable the first two pigs get eaten, while in others they escape to their brother's house of brick. The wolf is unable to blow down the brick house and tries to get inside by coming down the chimney where he falls into a pot of boiling water placed there by the resourceful third little pig, who then cooks and eats the wolf in a dramatic shift of events. It is the third little pig who is the hero of the story, the one who listens to his mother, who is virtuous and works hard now in order to reap benefits later. There is a strong juxtaposition of not only good and evil, but also between virtue and vice.

There are several moral lessons in this story. Once is that evil as represented by the wolf or fox is temporary in its ascendancy. A second is that lack of virtue (in the form of laziness and doing a poor job) leads to a bad end, while virtuous behavior is rewarded. A third is that planning and structural mitigation can save your life; it is this third lesson that is most obviously relevant from a disaster planning perspective. There is also the theme of retributive justice; the bad wolf suffers for his crimes, and by becoming dinner for the third little pig does so in an ironically appropriate way.

The story of the "Ant and the Grasshopper" (sometimes a cricket) is similar in some ways to the "Three Little Pigs" in that it provides a moral lesson. There are many versions of this story, which is referred to in the biblical Book of Proverbs (6.6–9): "Go to the ant, you sluggard! Consider her ways and be wise, which having no captain, overseer or ruler, provides her supplies in the summer, and gathers her food in the harvest." In this story, the ant works hard to store food up for winter while the grasshopper sings and plays, and teases the ant for working so hard (Figure 7.10). Then, when winter arrives the grasshopper begs the ant for food. The ant refuses, saying that since the grasshopper had sung all summer it can dance all winter. In some more recent versions of the tale, the ant invites the grasshopper into his home and shares his food. The lesson is about delayed gratification; the virtues of hard work and preparedness are evident. The grasshopper, like two of the three pigs, has little concern for the future and emphasizes the pleasure principle over the importance of dealing with reality and planning for the future. This tale differs from the "Three Little Pigs" in that there is no evil character or hero present, although in some versions of the tale the ant is presented in an unfavorable light.

Narratives, fables, and fairytales offer a largely untapped opportunity for public education about disaster risk reduction.

FIGURE 7.10 "The Ant and the Grasshopper" from Aesop's Fables. *Source: Project Gutenberg etext 19994.*

STUDENT EXERCISE

Create an outline of a program that would use narratives, fables, and fairytales to promote disaster risk reduction to the public, particularly children.

7.8 Case Study: the Great Flood

Many civilizations and cultures around the world, dating back to ancient Sumeria in the seventeenth century BCE (the Epic of Ziusudra) and the Babylonian Gilgamesh Epic, have stories of a great flood that destroyed the world. There may be over 500 such flood myths.[30] A study by James Perloff found that "In 95 percent of the more than two hundred flood legends, the flood was worldwide; in 88 percent, a certain family was favored; in 70 percent, survival was by means of a boat; in 67 percent, animals were also saved; in 66 percent, the flood was due to the wickedness of man; in 66 percent, the survivors had been forewarned; in 57 percent, they ended up on a mountain; in 35 percent, birds were sent out from the boat; and in 9 percent, exactly eight people were spared."[31]

In many of these stories, humanity was saved through the actions of a man who built a boat because of a divine command. Examples of these are the stories of Noah in the Old Testament Book of Genesis (Figure 7.11), and Manu in the mythology of India. The degree to which these stories come from a common origin has been much debated and is unclear.[32]

These flood myths (those who interpret religious documents literally may not view these stories as myths), whether they refer to a global, regional, or local floods, represent

FIGURE 7.11 "Noah's Ark" (oil on canvas) by Edward Hicks, 1846, Philadelphia Museum of Art. *Source:* http://www.cs.berkeley.edu/~aaronson/zoo.html

some of the earliest surviving narratives of catastrophe. The story of Noah[33] recounts how one righteous man and his family, by following the commandments of the Lord, escaped the flood sent by God to punish humankind for his evil ways.[34] In this story, God is both a destroyer and a nurturer. Considered as a metaphor as opposed to a religious tenet, the message of preparedness is obvious. There are other lessons as well, such as subservience and obedience to God's will, punishment of sin, and reward of virtue. This story illustrates the importance of fatalistic worldviews in some societies (see Chapter 6.4.1). Where would you place this story on the conceptual framework illustrated in Figure 6.5?

7.9 A Comment by Joe Scanlon

The persistence of myths about individual and organizational behavior in emergencies raises the question: Where do these myths come from? Some of them appear to come from scholars such as Gustav LeBon who, in his book *The Crowd*, argued that in times of crisis man reverts to what he called the *lower orders*, which he said include animals, women and children. He was basing his theories on observation—probably inaccurate—of the French revolution. Other myths may have survived because of a gender bias: Until recently, emergency agencies were staffed largely by men who saw their role as taking over and assisting what they perceived as the less able elements in society, namely women and children. They incorrectly assumed that individuals, especially women and children, could not cope in a crisis.[35]

But the peculiar aspect of these beliefs is that although they are widespread, they appear to affect organizational behavior only during emergencies. Individuals may believe the myths but do not act in line with them. They do not panic. They do not normally loot. Instead, they look around them, see what needs to be done, and do it. That is not always true for organizations. They sometimes hold back warnings because of a fear of panic. They focus on security because of a belief that looting is an issue. They order evacuations even when there is no continuing threat, in the belief that people will not be able to cope.

Despite decades of research showing that these beliefs are myths, the media also persist in acting as if they were real. That is partly because the media expect emergency agencies to know what has happened—unaware that initial search and rescue has been done by survivors—and partly because they look for and interview persons who match their stereotypes. The belief that individuals cannot cope leads to another phenomenon: determination on the part of emergency personnel that civilians must not take part in organized response. This has been challenged recently by the Amsterdam fire department, which has started teaching its personnel to evaluate the initial response by civilians and where appropriate, incorporate it into their response. This innovative approach was shared in recent articles[36] in a police and a fire journal in Canada, although it has not yet (as far as is known) been incorporated into Canadian emergency response planning.

Perhaps the best example of an informal response occurred after the 9/11 terrorist attacks on the World Trade Center in New York City. With normal transportation systems shut down, several hundred thousand people were evacuated by water. This was accomplished without any plan and without any attempt by the Coast Guard or other agencies to enforce any rules. And it was accomplished without a single mishap.

None of these findings is written in stone, and it is true that Dr. E.L. Quarantelli recently modified his position on looting. He now says it may occur on some occasions and that when it does, it is usually targeted—as is true when riots occur. That means the looters do not usually attack everyone and everything but rather, for example, they target business establishments against which they have a grudge. Sometimes reports of what happens, even official reports, can be misleading. In the aftermath of Hurricane Katrina, there were several hundred charges alleging that people had cheated when they sought and received disaster assistance. Almost all were convicted. An analysis of the charges showed that at the most a handful—perhaps as few as three of four victims—had actually defrauded the system. The rest were not victims, but saw a chance to exploit the system: some of those who received compensation were actually in jail when the hurricane struck.

It is true that most disaster research has been done by Americans and that most of it covers what has happened in U.S. incidents. It is also true that most other research has been done in developed countries like the United Kingdom, Sweden, and Australia, to mention just three where there are very active research communities. However, the data available suggest that behavior in other world regions matches behavior in the western world. For example, an earthquake in Tangshan, China, led to massive over-response—a phenomenon identified by two U.S. researchers, Fritz and Mathewson, as convergence.

In another incident, a forest fire in Mongolia led to many deaths when officials held back warnings for fear of causing panic.

End Notes

1. The Tornado Project, http://www.tornadoproject.com/myths/myths.htm.
2. Overpass Tornado, http://www.youtube.com/watch?v=lHBZylcxIvw.
3. Department of Conservation, State of California, http://www.consrv.ca.gov/index/Earthquakes/Pages/qh_earthquakes_myths.aspx.
4. O'Connell-Rodwell C. E., "Keeping an 'ear' to the ground: seismic communication in elephants," *Physiology* 22, no. 4 (2007): 287–94.
5. See note 3 above.
6. Saatcioglu M., Mitchell D., Tinawi R., Gardner N. J., Gillies A. G., Ghobarah A., Lau D., "The August 17, 1999, Kocaeli (Turkey) earthquake damage to structures," *Canadian Journal of Civil Engineering* 28, no. 4 (2001): 715–37.
7. Shearer P. M., and Stark P. B., "Global risk of big earthquakes has not recently increased," *Proceedings of the National Academy of Sciences* 109, no. 3 (2012): 717–21.
8. De Ville de Goyet C., "Epidemics after Natural Disasters: A Highly Contagious Myth," *Natural Hazards Observer*, (2007): 3–5.
9. *The Lancet* 356, no. 9231 (August 26, 2000): 762–64.
10. Add myth references – see pg 49 in Fischer.
11. Nirupama N., and Etkin D., "Emergency managers in Ontario: An exploratory study of their perspectives," *Journal of Homeland Security and Emergency Management* 6, no. 1 (2009); Nirupama N., and Etkin D., "Institutional Perception and Support in Emergency Management in Ontario, Canada," *Disaster Prevention and Management* 21, no. 5 (2012), 6.
12. Nash J. R., Darkest Hours. Wallaby, Pocket Books (1976).
13. Tierney Etkin. Others E. L., Quarantelli and Russell Dynes. That article, titled "When Disaster Strikes (It Isn't Much Like What You've Heard and Read About)," was published in Psychology Today in February 1972.
14. Tierney K., Bevc C., and Kuligowski E, "Metaphors matter: Disaster myths, media frames, and their consequences in Hurricane Katrina," *The Annals of the American Academy of Political and Social Science* 604, no. 1 (2006): 57–81.
15. O'Brien S., and Arce R., Desperation, but no violence seen in Haiti. CNN, January 30, 2012. http://www.cnn.com/2010/WORLD/americas/01/28/haiti.notebook/index.html?hpt=T2.
16. Vanderburg W. H., The Labyrinth of Technology (University of Toronto Press, 2000).
17. Ibid.
18. Hardin G., "The Tragedy of the Commons," *Journal of Natural Resources Policy Research* 1, no. 3 (2009): 243–53.
19. Janoff-Bulman R., Shattered assumptions (Simon and Schuster, 2002).
20. From Britannica online: "in Zen Buddhism of Japan, a succinct paradoxical statement or question used as a meditation discipline for novices, particularly in the Rinzai sect. The effort to "solve" a koan is intended to exhaust the analytic intellect and the egoistic will, readying the mind to entertain an appropriate response on the intuitive level. Each such exercise constitutes both a communication of some aspect of Zen experience and a test of the novice's competence."
21. New Water Management in the Netherlands (n.d.). Deltawerken online, http://www.deltawerken.com/New-water-management-in-the-Netherlands/353.html.

22. Gunderson L. H., and Holling C. S. eds, *Panarchy: understanding transformations in human and natural systems* (Island Press, 2001) Chapter 1, available at, http://www.resalliance.org/index.php/panarchy.
23. "A century of wildfire suppression in the United States has led to increased fuel loading and large-scale ecological change..." Donovan G. H., and Brown T. C., "Be careful what you wish for: the legacy of Smokey Bear," *Frontiers in Ecology and the Environment* 5, no. 2 (2007): 73–9.
24. Check out the following;a. UN Food and Agriculture Organization, http://www.fao.org/es/esa/pesal/AgRole10.html.b. Lichtfouse E., Navarrete M., Debaeke P., Souchère V., Alberola C., and Ménassieu J., "Agronomy for Sustainable Agriculture. A review," *Agronomy for Sustainable Development* 29, no. 1 (2009): 1–6.
25. The following book is a must read for anybody wanting to go into this topic in depth – De Waal, *Famine Crimes: Politics and the Disaster Relief Industry in Africa* (London, UK: Villiers Publications, 1997). For a shorter and less dense discussion go to: de Waal A., *Democratic Political Process and the Fight Against Famine. IDS Working Paper 107* (Institute of Development Studies, 2000), http://www.ids.ac.uk/files/Wp107.pdf.
26. Clarke, Lee, *Mission Improbable: Using Fantasy Documents to Tame Disaster* (Chicago, IL: University of Chicago Press, 1999).
27. Australia Network News, Pacific kids learn survival through nursery rhymes (2012), http://www.abc.net.au/news/2012-11-27/an-surviving-disasters-through-nursery-rhymes/4393850.
28. Bettelheim B, *The uses of enchantment: The meaning and importance of fairy tales* (New York: Knopf, 1975) 332.
29. Bettelheim B., *The uses of enchantment: The meaning and importance of fairy tales* (Vintage, 2010).
30. Schoch Robert M., *Voyages of the Pyramid Builders* (New York: Jeremy P. Parcher/Putnam, 2003).
31. Perloff J., *Tornado in a Junkyard: The Relentless Myth of Darwinism* (Arlington, MA: Refuge Books, 1999).
32. Van D. N, "The flood myth and the origin of ethnic groups in Southeast Asia," *Journal of American folklore* (1993): 304–337.
33. *Genesis* 5, 32–10:1.
34. "The Lord saw how great the wickedness of the human race had become on the earth, and that every inclination of the thoughts of the human heart was only evil all the time. The Lord regretted that he had made human beings on the earth, and his heart was deeply troubled. So the Lord said, "I will wipe from the face of the earth the human race I have created—and with them the animals, the birds and the creatures that move along the ground—for I regret that I have made them." But Noah found favor in the eyes of the Lord." ... "The Lord smelled the pleasing aroma and said in his heart: "Never again will I curse the ground because of humans, even though[i] every inclination of the human heart is evil from childhood. And never again will I destroy all living creatures, as I have done." Source: http://www.biblegateway.com/passage/?search=Genesis+5:32-10:1.
35. Scanlon J., "Human Behaviour in Disaster: The Relevance of Gender" *The Australian Journal of Emergency Management* 11 no. 4 (Summer 1996–1997): 2–7.
36. Scanlon J., and Jelle G., "When Disaster Strikes, Ordinary Citizens Respond It's Time To Make Them Part Of The Plan," *Royal Canadian Mounted Police Gazette* 75, no. 1 (2013): 30–1.

The Poetry of Disaster

Come, ye philosophers, who cry, "All's well,"
And contemplate this ruin of a world.
Voltaire (Figure 8.1)

FIGURE 8.1 Voltaire.

CHAPTER OUTLINE

8.1 Why This Topic is Important	254
8.2 An Essay by Nicole Cooley	254
8.3 Some Thoughts	258
8.4 Case Study: Burning of the Library at Alexandria	262
Further Reading	266
8.5 A Comment by Joe Scanlon	266
8.6 A Comment by Robin Cox	269
End Notes	270

CHAPTER OVERVIEW

The traditional way scholars study disasters is by developing theories, models, narratives, and empirical studies. Art, however, provides a way of perceiving and experiencing disasters that is not present in these traditional forms, and the study of this field through art creates a broader and deeper understanding of our experience of horrific events.

KEYWORDS

- Cultural narratives
- Disaster poetry
- Ethics
- Library of Alexandria

8.1 Why This Topic is Important

People come to understand disaster through different pathways. There is the language of mathematics used by scientists and statisticians, the language of logic used by social scientists and philosophers, and the language of narrative used by reporters and victims. There is also the art and poetry of disaster, which convey meaning through image and feeling. Ultimately, disaster is a human experience, which is portrayed evocatively through art and poetry.

8.2 An Essay by Nicole Cooley[1]

It is not you who will speak: let the disaster speak in you.

Maurice Blanchot, The Writing of the Disaster

Disaster shuts down language. Disaster cannot be fathomed. Disaster stops all speech because the suffering it causes is so total and complete.

This is a common way we speak about disaster. Yet, as a poet, as a reader, and as a teacher of poetry, I do not and cannot believe this. For the past eight years, which are marked, in our by two large-scale catastrophes, the attacks of September 11, 2001, and Hurricane Katrina, I have been thinking about, reading, and writing the poetry of disaster. I lived in New York City on 9/11. I grew up in New Orleans, and my parents are survivors of the hurricane. These two disasters altered the way I think about poetry and its relation to disaster.

Does disaster render language inadequate? What is the relationship of disaster to the language of poetry in particular? In 1965, cultural critic Theodor Adorno asserted that "to write poetry after Auschwitz is barbaric." Disaster is so often discussed in terms of silence and the inability to speak, but I want to think about how disaster produces speech, writing, and testimony, and how disaster is reproduced through language. I am not talking about disaster as metaphor in poetry, but about a poetry that arises in direct response to a disaster, a *poetry of disaster*.

As I write this, Haiti is still coping with the aftermath of the terrible earthquake of January 12, 2010, and the BP oil spill has devastated the U.S. Gulf Coast. Edwidge Danticat, in a piece titled *Suffering* in the January 25 issue of *The New Yorker*, observes that President Obama "vowed that America would not forsake Haiti, because its tragedy reminds us of 'our common humanity.'" I worry that these communities in Haiti and on the Gulf Coast will be forsaken. I worry that it is too easy to relegate disaster as happening to the "other," that we do not see disaster as part of all of us. Yet I believe language—the language of poetry—can bring us back to that "common humanity" in crucial ways.

In her introduction to *Against Forgetting: Twentieth Century Poetry of Witness*, Carolyn Forché writes, "Poetry of witness presents the reader with an interesting interpretive problem. We are accustomed to rather easy categories: we distinguish between 'personal' and 'political' poems…We need a third term, one that can describe the space between the state and the supposedly safe havens of the personal. Let us call this space 'the social.'" Here, at this juncture, I would situate the poetry of disaster in the "social," the space of community where we might find new understandings of what poetry can do in the world.

Recently, there has been renewed media interest in disaster. In addition to newspapers, websites, and blogs offering accounts of disaster, studies by historians, public intellectuals, and sociologists from the emergent field of disaster studies have been important. Books such as Naomi Klein's *The Shock Doctrine: The Rise of Disaster Capitalism* and Amanda Ripley's *The Unthinkable: Who Survives When Disaster Strikes and Why* attest to that interest. Rebecca Solnit's *A Paradise Built in Hell: The Extraordinary Communities That Arise in Disaster* (2009) is particularly suggestive for thinking about poetry, although she does not mention poetry at all, because she continually takes us back to language. She writes, "The word emergency comes from emerge, to rise out of…An emergency is a separation from the familiar, a sudden emergence into a new atmosphere, one that often demands we ourselves rise to the occasion. Catastrophe comes from the Greek kata, or down, and streiphen, or turning over. It means an upset of what is expected…The word disaster comes from the Latin compound of dis-, or away, without, and astro, star or planet; literally without a star." With Solnit's work in mind, I want to pose two central questions about the poetry of disaster. First, what kind of representation of disaster is possible—and necessary—in poetry? Second, what are the common elements of a poetry of disaster?

My own thinking about the poetry of disaster began on a day of national emergency. Two days after 9/11, in my poetry seminar at Queens College–City University of New York, we were scheduled to discuss Muriel Rukeyser's *The Book of the Dead*, her 1938 sequence about the Hawk's Nest Tunnel disaster in Gauley Bridge, West Virginia, in 1929. The poem relates the experience of the miners, mainly African American men, who contracted silicosis, a deadly lung disease, while building the Hawk's Nest Tunnel for the Union Carbide Company. Poems in the sequence explore the incident from multiple perspectives, including those of the miners, the doctors who made the diagnoses, and a mother who had lost all her sons. The text includes a range of kinds of language, from personal testimony to a Union Carbide stock report.

That afternoon, I stood at the door to my classroom, afraid to enter. I did not want to teach at all that day, and I particularly did not want to teach a poem about disaster. My students might be dead or missing. They might have lost family members. From the campus library, we could see smoke and ash still rising along the horizon. It felt wrong to discuss poetry—but that's exactly what my students wanted to do. We began with Rukeyser's "The Disease," spoken by doctors:

This is a lung disease. Silicate dust makes it.
The dust causing the growth of

This is the X-ray picture taken last April.

A student asked, what was missing? What could not be said, and how could the poem show this? Our discussion started with the omission in Rukeyser's second line, a silence that now spoke to all of us.

We saw a space that poetry could open for us. In the face of disaster, poetry was what we needed. What could the speaker in the poem not say? Why would the doctors not speak the word silicosis? "The Disease" offers an unnamed doctor's description of the diagnosis of the deadly illness set in opposition to a committee's questions. In the poem, there is no room, literally, for the suffering bodies, the dead and dying miners. And so our discussion moved between 9/11 and *The Book of the Dead*. What was the answer or the "official" truth of 9/11 as it was told to us in those early days after the attacks, and in the weeks after, when the United States began to bomb Afghanistan? We talked about the poem "Praise of the Committee," which offers a description of the "facts" of the case, the voice of a senator, and at the poem's end, the perspective of the poet who asks:

Who runs through electric wires?
Who speaks down every road?
Their hands touched mastery; now they
demand an answer.

Our discussions about Rukeyser sparked our own demand for answers throughout that terrible September. For the rest of the semester, my students and I engaged with Rukeyser's text as a poem that might allow us to re-see the disaster around us, in a city that was still burning a few miles from our classroom. A number of us went down to 14th Street to a makeshift poetry memorial in Union Square. Poems had been taped to the iron fence that surrounds the park, and people walked silently around the fence, reading. In class, we returned to the question over and over—what work can poetry do in the world, in the face of catastrophe?

Four years later, Hurricane Katrina left much of the Gulf Coast and my native city, New Orleans, in ruins. In late August 2005, I found myself on the phone with my parents in New Orleans, begging them to leave the city as Katrina approached the Gulf Coast. "We are not leaving," my mother said. "This is our home." For several days, my sister, my brother, and I did not know if they were dead or alive. We called FEMA, the Red Cross, the Louisiana State Police, the local hospital. All the phones were out, circuits busy. In that first week after Katrina, none of us knew the scope of the damage or the number of people who were missing or dead. All I could do was watch the news—the roads in and out of New Orleans shut down, my childhood city filling like a bowl.

Remembering that other autumn, I turned back to Rukeyser, to the final poem in *The Book of the Dead*: "What three things can never be done?/Forget. Keep Silent. Stand alone." As I watched familiar landmarks disappear on my computer screen—the city literally drowning, with my parents still in it—I began to write everything down, as Rukeyser's speaker does, to document e-mails sent by my parents' friends, forms from the Coast Guard website, phone calls with my sister and brother as we tried to imagine how we could get our parents out. After I knew they were safe, after I could go down to New Orleans, three months later, I began to write poems responding to these two disasters, which would become two books, *Breach*, and *Milk Dress*. I began to explore the kinds of representations of disaster that are possible in poetry—and I began to teach and try to write these texts.

Teaching, reading, and writing—as well as personal experience—have led me here. They have led me to speculate on the elements of a poetry of disaster, and to search for its commonalities to think more deeply about the larger question about the work that poetry can do in the world.

A poetry of disaster relies on fragments. From Theresa Hak Kyung Cha's "Dictee," focused on the Japanese occupation of Korea, to Peter Balakian's poems about the Armenian genocide, to Claudia Rankine's recent documentary works that conjoin video and poetry, the poetry of disaster refuses chronology and teleology. When we think of writing about disaster, we envision telling coherent stories of events—but the poetry of disaster has a different relation to narrative. The only way these events can be spoken of or voiced is through broken forms. Chris Llewellyn's *Fragments from the Fire: The Triangle Shirtwaist Company Fire of March 25, 1911* represents the devastating effects of the fire that killed 144 immigrant workers at the Triangle garment factory in New York City. Doors had been locked to keep out union organizers, and workers could not escape when the fire started. From its title on, the book privileges the fragment as a poetic mode, using forms such as the Italian cento, as in the poem "Survivor's Cento," composed of pieces or "patches" of text, and juxtaposing photographs of the dead women with the poems.

A poetry of disaster invokes the collective alongside the individual, often in tension with each other. Notably, this tension is different from the opposition of a split versus a coherent voice, found in much current poetry. A large-scale catastrophe involves the deaths of many, maybe even mass casualties, and much of the work I would consider the poetry of disaster sets the voice of many speakers alongside the voice of the individual poetic subject. In the sonnets that comprise "Love in the Time of War" in Yusef Komunyakaa's *Warhorses*, for example, the poems shift between the collective and individual voices. Within a single poem, the language moves from "When our hands caress bullets & grenades,/or linger on the turrets & luminous wings/of reconnaissance planes" to "I touch your face, your breasts, the flower/holding a world in focus." By juxtaposing two lovers with the "shock and awe" Iraq war discourse, through the linked stanzas of sonnets, Komunyakaa shows how war's catastrophe is bound to all of us— we are all destroyed by war's horror. A book using the collective to represent disaster very differently is Chickasaw writer Linda Hogan's *The Book of Medicines*. There, tension manifests itself through a single speaker or voice who speaks alongside land, plants, and animals, collective voices protesting the disaster of environmental destruction.

The poetry of disaster asks ethical questions about voice. Contemporary poetry does not often ask such questions. Who is speaking and who can speak for whom? How close to a disaster does a poet need to be to write about it? Can one write about a disaster without firsthand experience? Many questions arise in the special fall 2006 issue of the journal *Callaloo* titled *American Tragedy: New Orleans Under Water*. What's most compelling and heartbreaking about the issue is its refusal to privilege one voice of Katrina over any other and its inclusion of oral histories and poems by writers with a range of relationships to New Orleans and the Gulf Coast.

The poetry of disaster challenges our thinking about poetry as a genre. The poetry of disaster not only requires us to question "official history" but also asks that we question poetry as a discourse, as a way of using language to represent the world. Tory Dent's collection, *HIV, Mon Amour*, shows that there may be no other way of documenting the tragedy of the AIDS crisis than through rethinking poetry as form of representation. The book is a compendium of nontraditional texts on which the poems are built: songs, an AIDS quilt text panel, magnetic poetry kits, diary entries kept in quarantine. Over and over, Dent's poems return to visual modes as the body suffering within catastrophic illness becomes an object. Photography and film recur not just as tropes but also as modes of writing the disaster, as in the long poem "Fourteen Days in Quarantine," in which Dent writes, "I'm forced/to witness my own participation in the clinical process, a kind of snuff film/culturally condoned." By referencing the genre of the pornographic film that culminates in the murder of the female "star," Dent highlights the continual bodily violations the poem and the body undergo, and encourages us to ask a different set of questions about the roles of the poet and the reader.

Finally, by revealing the voices, testimonies, and experiences not visible in mainstream representations of disaster, the poetry of disaster, according to Rukeyser, "demands" that we answer. This demand is crucial to our thinking about disaster. Finally, poetry becomes a place where we gather, a space for grief. In 2010 May, the group Poets for Living Waters formed as "a poetry action in response to the BP Gulf oil disaster." This idea echoes Rukeyser's belief in poetry as action. I believe we again see a space that only poetry can make for our understanding of disaster—not just a place where we might find solace, although I hope we can—but also a place where we remember how important it is to speak, to give voice to the experience of disaster: a place like the fence taped with poems in Union Square, like a classroom where we sit together in a circle at a table.

8.3 Some Thoughts

In his interesting essay, "An Interpretation of Disaster in Terms of Changes in Culture, Society and International Relations,"[2] David Alexander comments upon the importance of disasters as symbolic events, as people endow these horrific events with meaning; the notion that tragedy is meaningless is abhorrent to most people and so we grasp at ways to understand it in terms of God, morality, justice, or life's lessons. Symbolically, disasters can be milestones in one's life or community, a metaphor, allegory or parable, can "graphically demonstrate the apparent arbitrariness of fate," or can be a reminder of life's fleeting presence and our mortality.

Art provides a way of understanding and coping that is not present in other venues, including sharing, catharsis, and expressing pain. After the terrorist attack of September 11, 2001, poems appeared almost immediately—on pages tacked to trees, on sidewalks and on bulletin boards around the country.[3] "Prose was not enough. There was something more to be said that only poetry could say."[4] *The New York City Manual for Spiritual Care and Mental Health for Disaster Response and Recovery*[5] suggests writing poetry as part of the spiritual approach to coping, and along similar lines art therapy and storytelling are used as therapeutic techniques in postdisaster recovery.[6] Through the use of imagery and metaphor, poetry offers different perspectives[7] and creates opportunities for empathy. Literature offers a vocabulary of disaster not present in the academic disciplines that traditionally study these events, and provides for narrative and phenomenological approaches to gaining insight into how people cope with hardship.[8]

An ethical dilemma expressed by some artists is that catastrophe should not be represented because it exploits tragedy at the expense of victims, and disaster voyeurism is certainly an issue of concern. But writings of disaster also console; and poetry, because of its often fragmented verse that mirrors the fragmenting of people and places by disaster, can be better suited to this task than traditional narrative approaches. Poetry, especially lyric poetry, has been said to obey "the logic of dreams, of the unconscious."[9]

There is a long history of poetry describing disaster, dating back to the earliest writing of man. Some examples follow.

A Babylonian tale of the great flood dates back to 1626 BC, and possibly back to the nineteenth century BC[10]:

Anzu rent the sky with his talons, He…the land.
[iii.10] and broke its clamor like a pot.[11]

The Roman poet Ovid (43 BCE–17 CE) described the great flood in his poem "Metamorphoses"[12] in following terms:

Mankind's a monster, and th' ungodly times
confed'rate into guilt, are sworn to crimes.
All are alike involv'd in ill, and all
must by the same relentless fury fall.

His dire artill'ry thus dismist, he bent
His thoughts to some securer punishment:
Concludes to pour a watry deluge down;
And what he durst not burn, resolves to drown.

Nor from his patrimonial Heaven alone
is Jove content to pour his vengeance down.
Aid from his brother of the seas he craves,
to help him with auxiliary waves.
The wat'ry tyrant calls his brooks and floods,
who rowl from mossy caves (their moist abodes)
and with perpetual urns his palace fill,
to whom in brief, he thus imparts his will.

Shakespeare wrote in The Tragedy of King Lear:

Blow, winds, and crack your cheeks! rage! blow!
You cataracts and hurricanoes, spout
Till you have drench'd our steeples, drown'd the cocks!
You sulph'rous and thought-executing fires,
Vaunt-couriers to oak-cleaving thunderbolts,
Singe my white head! And thou, all-shaking thunder,
Strike flat the thick rotundity o' th' world,
Crack Nature's moulds…

Somewhat more recently, after the Lisbon earthquake/tsunami/fire (Figure 8.2)[14] that occurred on All Saints Day in 1755 (see the case study at the end of this chapter), Voltaire wrote what was probably the most influential poem of his time that was written about a

FIGURE 8.2 Lisbon.

disaster. This event, sometimes referred to as the first modern disaster, made many Europeans question their faith, as shown in the segment below:

Unhappy mortals! Dark and mourning earth!
Affrighted gathering of human kind!
Eternal lingering of useless pain!
Come, ye philosophers, who cry, "All's well,"
And contemplate this ruin of a world.
…
Confess it freely—evil stalks the land
Its secret principle unknown to us.
Can it be from the author of all good?

Following the Japan tsunami disaster of 2011, many poems were written about it as part of the healing process of the victims[13] and in support of aid.

No longer water
But a powerful demon
Bent on destruction.[15]

> **STUDENT EXERCISE**
>
> Compose a haiku about a disaster. In English, haiku is written in three lines, with each line having an exact number of syllables. The first line contains five syllables, line two contains seven syllables, and line three contains five syllables. Here is my effort:
> *A dread rising tide*
> *Despair, but also courage*
> *Lessons not yet learned.*

Some of the above-quoted poetry reflects common historical themes; disaster as a punishment for the evil-doing of man is a common one. This theme has historical roots going back at least to stories of the great biblical flood, and has been a theme in much literature, one example being *Paradise Lost*[16] by John Milton. There are other interesting themes present in *Paradise Lost* that can be viewed as metaphors for disaster, such as hubris leading to downfall (Satan's fall from heaven), which can be compared to man's overuse of technology in an attempt to control nature. Although now out of favor in secular society, the notion of disaster as punishment is still present as a thread in the modern disaster discourse, as shown in the following example; "At the outset it has to be said that the Christian Churches play a central part in the daily life of the 15 widely scattered Cook Island communities…They were united in attributing the event to an angry God who had applied it to His flock as a punishment for their unspecified moral transgressions."[17] Another recent high-profile example is the comment on the Haiti earthquake disaster in 2010 made by Evangelist Pat Robertson, who said "They were under the heel of the French, you know Napoleon the third and whatever. And they got together and swore a pact to the devil. They said 'We will serve you if you will get us free from the prince.' True story. And so the devil said, 'Ok it's a deal.' And they kicked the French out. The Haitians revolted and got something themselves free. But ever since they have been cursed by one thing after another."[18]

Poetry creates important ways for people to understand and cope with tragedy, and is probably underappreciated within disaster and emergency management studies. Students with a background in literature might find this a promising area to explore.

8.4 Case Study: Burning of the Library at Alexandria

hapless holocaust where ink is offered up instead of blood…where the devouring flames consumed so many thousands of innocents in whose mouth was no guile, where the unsparing fire turned into stinking ashes so many shrines of eternal truth.

Richard de Bury, 1345

…a philosopher may allow, with a smile that the Alexandrian burning was ultimately devoted to the benefit of mankind.

Edward Gibbon, The History of the Decline and Fall of the Roman Empire, *1788*

Caesar: I am an author myself; and I tell you it is better that the Egyptians should live their lives than dream them away with the help of books. The scholar says, What is burning there is the memory of mankind. Caesar replies, A shameful memory. Let it burn.

Bernard Shaw, Caesar and Cleopatra

In ancient Alexandria there were two famous libraries, one in the Brucheion, the palace area, and another smaller one in the Serapaeion, built about a century later; the former is the one most often referred to as the Library of Alexandria (Figure 8.3). It was originally designed by Demetrius of Phaleron, who in the third century BC carried out the instructions of Ptolemy I to build this institution, following the wishes of Alexander the Great, after whom the city was named. The library continued to be supported by later rulers in the Ptolemy reign. The librarians working there attempted to collect all of the world's knowledge, and filled the library with scrolls by buying, copying, or stealing them. Boats visiting the city were required to lend the library any books on board, which were then copied; sometimes the copies were returned and the originals kept.

Alexandria was a geographical focal point of civilization at that time. The largest city in the world, with a thriving papyrus industry, it operated as a trade center between east and

FIGURE 8.3 Ancient Alexandria.

west. As a result, the library became a great center of cultural development. The library included lecture areas, gardens, a zoo, and shrines for each of the nine muses; the word *museum*, in fact, means "a house of the muses." As well as housing scrolls and librarians, there was also a cadre of up to 100 paid in-house scholars, including poets, physicians, mathematicians, historians, astronomers, geographers, and grammarians. The phrase "in house" takes on a very literal meaning, as the resident scholars could not freely leave the grounds. In fact, Alexandria's librarian Aristophanes of Byzantium was imprisoned for life when he was recruited by Alexandria's competitor, the Library of Pergamum.[19] This offers an interesting twist on the notion of tenure!

Famous scholars who worked there include Euclid and Archimedes, and Ptolemy II is reported to have hired 72 Jewish scholars to translate the first five books of the Bible from Hebrew to Greek at the library. Eratosthenes, an astronomer, was another famous scholar resident at the library; among his research was a study concluding that the earth was round, and he successfully calculated its circumference to within 50 miles. Writer and professor Daniel Heller-Roazen comments that "The history of the Alexandrian Museum may well be regarded as the history of the development of classical scholarship as such."[20]

There is a great deal of controversy about the size of the library; different sources suggest that the number of scrolls ranged between 40,000 and 700,000, with good evidence supporting the larger estimate.[21] Factors of 10 were expressed as dots over numerals in

FIGURE 8.4 Theophilus.

ancient manuscripts, which make copying easily subject to errors; it is thought that the 40,000 number may result from a copying error, and that the original number was more likely 400,000. Whatever the numbers, the destruction of the library resulted in the irreplaceable loss of much of our knowledge of ancient times.

There are three popular cultural narratives about how the library was destroyed, although separating reality from myth is difficult. There is much debate over the veracity of the stories and to a large degree the conflicting views might represent cultural biases and political alignments. The three culprits normally identified are Julius Caesar (a Roman), Patriarch Theophilus (Figure 8.4) of Alexandria (a Christian), and Caliph Omar of Damascus (a Moslem); these stories thus provide a multicultural set of villains. It is unlikely that only these three events contributed to the loss of the scrolls; there were numerous other fires or culprits that may have played a factor. Verification through modern rigorous methods is difficult due to the many diverging sources that are poorly documented and because the palace quarter in the northeastern part of the city is now under water.

In *The Civil Wars*,[22] Julius Caesar describes how, in his battle with Ptolemy XIII, he was forced to set the dockyards on fire in 48 BC to ensure his own safety, although some suggest that this only destroyed a warehouse with about 40,000 scrolls and not the main library. Also, the amount of scholarly work that continued at the library after 48 BC, suggests that the number of scrolls lost was not large.[23] Still, Caesar is commonly accused of the first destruction. In 391 AD, 439 years later, Emperor Theodosius I made paganism illegal, and the Temple of Alexandria was subsequently closed by Patriarch Theophilus of Alexandria. According to Socrates of Constantinople, he also demolished the library with the help of a Christian mob, in a fever of religious patriotism. The main source of this narrative is Edward Gibbon.[24] In 640 AD, 251 years after Theophilus, Muslim forces under the command of Caliph Omar conquered Egypt. Omar is reported to have said of the library: "if what is in the library agrees with the contents of the Qur'an, then it is redundant. And, if the contents of the library do not agree with the Qur'an, then such contents are heretic."[25] These latter two events involving Patriarch Theophilus and Caliph Omar are historically unsubstantiated. Writer Uwe Jochum comments that "instead of squeezing the sources in search of absolute truth, we should be aware that both lines are telling a political story reflecting the basic myths of western civilization: Rome is a symbol of power and at the same time one of the main pillars of our own culture, while Alexandria symbolizes deep erudition and, at the same time, oriental luxury and decadence. Both traditions, each of which is included in our own tradition, therefore seek to claim Aristotle's library, because it is through this library that they partake of one of the sources of our western civilization."[26]

Matthew Battles notes the following: "Before the flames, before theft and censorship, the fate of books is bound up in the constant shuffling and transformation. Though Alexandria's libraries were universal in scope, their librarians faced hard choices. Manuscript scrolls were costly and time-consuming to produce, and the scribes' precious labor could not often be lavished on minor texts. Naturally, only the major works were copied in any great quantity. The rest—the secondary, the extra-canonical, and the apocryphal—dropped out of view" and that the library was "moldering slowly through the centuries as people grew indifferent and even hostile to their contents." "What happened to the books of Alexandria? Many, many centuries happened to them—too many for their inevitable dispersal and disappearance to be staved off, no matter whose mobs rioted in the streets, no matter which emperors set fires."[27]

The destruction of books and libraries is a longstanding historical theme. During the crusades, Christians destroyed hundreds of thousands of Greek and Muslim books. The comment by Pope Gregory—who destroyed most of the writings of Livy and forbade the study of the classics—that "Ignorance is the mother of piety" says much about this mindset. The burning of the Library of Alexandria is a story of the duality of greatness and of catastrophe, of the central place books and knowledge have in many societies, and of the longevity of cultural narratives. How disasters are interpreted by cultures is highly dependent upon belief systems and political and cultural affiliations. The destruction of the library has been variously interpreted as an irreplaceable cultural loss, a cultural cleansing, and as a cultural renewal (by releasing people from the cultural burden of the past).[28] In the first burning it was collateral damage in a war zone; in the second it was ideological, aimed at removing pagan idolatry and secular learning from the world, thereby preserving Christian theology; in the third it was again ideological, to support the dominance of the Koran as a source of knowledge. These three cultural narratives are embedded into the story of the destruction of the library, and reinforce political and religious worldviews. The story of the burning of the library is also a metaphor for wanton destruction and the coming of the Dark Ages in Europe.

The loss of the Library of Alexandria is also a good example of problems associated with centralization: Where resources are concentrated, it hurts much more when disaster strikes. A great deal of our knowledge of ancient times survived in small private libraries, where it passed unnoticed from the forces of human destruction. Distributed systems, especially if redundancy is built in, may be less efficient than centralized ones, but they are far more resilient.

Further Reading

Ashrof VAM: Who destroyed Alexandria Library?, *The Milli Gazette* 3(23), 2002. 1–15 December.

Battles, M. *Library: An unquiet history*. WW Norton & Company.

Franz G., "The Ancient Library at Alexandria: Embracing the Excellent, Avoiding its Fate," in *Association of College & Research Libraries Annual Convention* 30, (Philadelphia, March 2011).

Haughton B., "What happened to the Great Library at Alexandria," *Encyclopedia of Ancient History* (2011), http://www.ancient.eu.com/article/207/.

Jochum U: The Alexandrian Library and its Aftermath, *Library History* 15, 1999. S. 5–12.

Philips H., "The Great Library of Alexandria?" *Library Philosophy and Practice* (e-journal) (2010). Paper 417. http://digitalcommons.unl.edu/libphilprac/417.

Thiem J: The great library of Alexandria burnt: Towards the history of a symbol, *Journal of the History of Ideas* 40(4):507–526, 1979.

8.5 A Comment by Joe Scanlon

There are events so significant or dramatic that they stay with us through what is called *flashbulb memory*. For those of us old enough, these include the end of World War II, the assassination of President Kennedy and, perhaps, the death of John Lennon. In Canada, these memories also include disasters such as Hurricane Hazel, the Mississauga evacuation, and the Eastern Ontario ice storm. These events affect so many people that it is no surprise that they inspire literature—and not just poetry but also novels and folk songs; and it is partly through these various forms of literature that we not only shape our perception of these events but also our perception of how people and organizations behave before, during, and after such events.

Canada's best-known novel may well be *Barometer Rising* by Hugh MacLennan, a book about what may be Canada's only catastrophe, the December 6, 1917, Halifax explosion. This munitions ship explosion had one-seventh the power of the Hiroshima atomic bomb and, in a few seconds, killed or injured one-fifth of the 60,000 residents of Halifax, Nova Scotia. Although MacLennan's novel is by far the best known, it is only one of six works of fiction about the explosion. There are three other novels: *A Romance of the Halifax Explosion* (1918) by McKelvey Bell; *The Sixth of December*[29] (1981) by Jim Lotz; and *Burden of Desire*[30] (1992) by former PBS broadcaster Robert MacNeil. There is also a short story—*Winter's Tale* by Thomas Raddall (1979) and a children's book, *Who's a Scaredy-Cat?* by Joan Payzant.

All of these books to some extent mimic reality. Since Bell was for example, Chief Canadian Army Medical Officer in Halifax, and Camp Hill was a 280-bed convalescent hospital for soldiers that admitted 1400 critically injured civilians the day of the explosion, the descriptions of hospitals in Bell's novel mirror his experience: "Maimed and senseless, with blood soaked bandages, with dangling limbs and powder blackened they were carried, or those who could walk, staggered in. Motor cars, carriages, grocery wagons, coal carts and every other conceivable variety of vehicle hastened to the hospital doors with their shattered burdens."

The beds were soon filled to overflowing. Mattresses were dragged out of stores and placed on the floors, in halls, in offices, in every available space. The wounded or dying lay in helpless misery upon bed or mattress, or sat in listless waiting upon the floor. Already dozens of surgeons and nurses were at work staunching blood or stitching up the gaping wounds. Young women from the St. John Ambulance Brigade and other first-aid workers arrived every minute and threw themselves into the breech with consuming zeal and energy. They worked feverishly hour by hour, but still the never-ceasing stream of patients came and still the tramp, tramp of their bearers filled the halls.

Similarly, MacNeil, who had no personal experience of the explosion, draws on an incident that happened to his uncle:

"Abigail frowned but suddenly a mischievous look took over. "Do you know where Michael was when the explosion came to our house?" "Where?" 'Don't tell!' Michael said sternly. 'Don't tell! Don't tell!' With a wicked smile, she whispered through her curled hand, 'He was on the potty. He was on the potty.' She began to sing, 'He was on the potty. He was on the potty.' 'Stop it! Stop it! Stop it!' Michael shouted with rising volume.

MacNeil, 1991, p. 187.

While the above examples reflect what actually happened, the novels also distort. For example, they play down the role of women and play up the role of the military. The explosion's most serious impact was on the city's North end where the only persons at home (it occurred on a weekday morning in wartime) were women, preschool children, and the elderly. Inevitably, women did the initial search and rescue. In the fictional accounts, however, no adult women become involved in search and rescue; however, one woman, Bell's Vera Warrington, does walk through the impact area comforting survivors. The women either play no part at all or get involved in treating the injured (nursing is considered woman's work). Even Hugh MacLennan's Penelope Wain, who is a naval architect—which makes her a very liberated woman in 1917—is blown over by the explosion, cut in the eye and rescued by the two men who love her, Neil Macrae and Angus Murray, a physician. The two put everything else aside, including the needs of other victims, to look for her.

The military, in contrast, was actually preoccupied with its own problems; there were scores of soldiers and their families killed and injured in the explosion. Yet Lotz's hero, Jack Dobney, is with Colonel Thompson when the *SS Mont Blanc* explodes. Thompson immediately starts issuing orders: "Set up first aid posts throughout the city. The Technical College will serve as the Central Medical Depot. Alert the hospitals. Use Camp Hill—move anyone out of it who can walk. Check Rockhead Hospital—see what its status is. The telegraph and telephone lines are probably down so use runners. And motorcycles."

The Canadian Army did establish first aid posts at the Armouries and at Wellington Barracks, but it did so at Wellington Barracks only after the hundreds of injured soldiers and soldiers' families were treated. The Technical University of Nova Scotia was turned into a supply center, although not by the Army but by a committee appointed by the civilian Halifax Relief Committee, and the center was not staffed by soldiers but by pharmacists, Red Cross volunteers, and commercial travelers (who did the deliveries). As for the convalescent soldiers in newly opened Camp Hill, they left on their own. No one had to ask them to clear the way for critically injured disaster victims.

It is not just poems and novels that portray stories of disasters, however. Disasters are also recurring themes in movies and in folk songs. Folk songs seem to provide the most accurate portrayals of what really happens. Researchers at Cape Breton and Carleton universities studied folk songs about mass death incidents in mines in Nova Scotia and on the *Titanic* (there were more than 40 of them). For both events, they found that, for the most part, these songs presented a fairly accurate picture of what happened. They also found that the songs tended to focus on a very short period, ignoring what went on before and afterward. For example, in the case of the *Titanic*, while the songs mention the ship's excessive speed, they do not mention the captain's previous record—he had been in two other collisions—and they do not mention that those in the lifeboats did not rescue people in the water. They also noted an intriguing characteristic of the songs that has yet to be explained: that the songs in Finnish told a somewhat different story than the songs in six other languages. The Finnish songs suggested there was more conflict between passengers and crew than legends would have us believe.

Of course, not all the poems, novels, and folk songs about disaster are meant to be factual. Some are written purely for their entertainment value. Perhaps the most whimsical of these is a song about a polar bear who walked into a pub in Liverpool after the loss of the *Titanic* and asked:

Have you got any news of the iceberg?
My family were on it you see.
Have you got any news of the iceberg?
They mean the whole world to me.

Sources:

Lotz J., *The Sixth of December… is the Halifax Holocaust…*, Markham: PaperJacks, 1981, (1981).
MacNeil R., *Burden of Desire*, (Toronto: Random House, 1993).
Scanlon J., Johnston N., and Vandervalk A., with Sparling H., 101 "Years of Mine Disasters and 101 Years of Song: Truth or Myth in Nova Scotia Mining Songs," *International Journal of Mass Emergencies and Disasters* (March 2012): 34–60.

Scanlon J., Vandervalk A., and Chadwick-Shubat M., "Challenge to the Lord: Folk Songs About the 'Unsinkable' Titanic," *Canadian Folk Music* 45, no. 3 (2012): 3–10.

Scanlon J., "Myths of Male and Military Superiority: Fictional Accounts of the 1917 Halifax Explosion," *English Studies in Canada* 24, (1999): 1001–25.

8.6 A Comment by Robin Cox[31]

Why consider the poetry of disasters in a text on disaster and emergency management? This chapter, a surprise no doubt to most who might be reading this text, is important because it is a surprise. It takes the reader out of the familiar and anticipated knowledge domains of the typical comprehensive text on disasters and in so doing, challenges us as readers. It demands in some ways that we step out of the normative constraints of most academic writing on the topic.

Marshall McLuhan[32] and Margaret Nussbaum[33] remind us that the medium and the message are inseparable. "Writing matters" says Richardson.[34] It is not merely a reflective act but a constitutive one, a means of knowing and constructing meaning. Poetry demands that we, as readers and writers, open up to a different way of thinking to consider with a critical eye and ear the nuance of how truth is constructed. A narrative of fragments, poetry suggests more explicitly the incompleteness of any story. The fragments and the creative spaces in between invite us to adopt a discerning stance to the potential for multiple readings. Poems expose the discourses that in most texts remain unchallenged and invisible. In so doing, they invite us as readers to question these meanings and dare to explore alternatives. This is a discipline dominated by discourses of engineering and management, where disasters are controlled, resisted, and managed, and where the bureaucratic and commercial response to disasters is rarely disputed. The figurative language, the concentrated blend of sound and imagery of poetry, evoke an emotional, embodied reading. In so doing, the poetry of disasters calls us to engage with, rather than simply manage, suffering; to explore the dark side of recovery that Naomi Klein points to, and to examine the benefits and risks of emerging paradigms such as resilience. The poetry of disasters challenges us to read differently, think creatively, and search for new knowledge and meaning in the interstitial spaces of disasters.

Reality Inversion, by R.S. Cox

I watch the hurricane prep,
the hyperbole of the anticipatory horror
leaves me wondering
if we will all be disappointed
secretly, privately,
if it doesn't happen?
The kick of adrenaline,
and fear's slide into horror,
then the cool release of compassion,
until the urgency of the next fix, closer than the last one,
the pulse of our need quickening
as the pace of our lives speeds up.

Reality inversion
where suffering breaks through
yet is contained, commodified, memorialized
and we sense the grace
of witnessing another's suffering
as it stands in for our own.

These televangelistic moments
of buffered anguish, resurrection without loss
and effortless compassion,
allow me to believe
I can give you the finger driving
but still find you a home in the hurricane.

End Notes

1. Cooley N., (n.d.). Poetry of Disaster. Source: This essay originally appeared on poets.org, the website of the Academy of American Poets, http://www.poets.org/viewmedia.php/prmMID/22467. Reprinted with permission. Nicole Cooley grew up in New Orleans and is the author most recently of two collections of poems, Breach (LSU Press 2010) and *Milk Dress* (Alice James Books 2010). She has also published two other collections of poems and a novel. She has received the Walt Whitman Award from the Academy of American Poets, the Emily Dickinson Award from the Poetry Society of America, and a National Endowment for the Arts Grant. Her work has appeared in *The Paris Review, Poetry, American Poet*, and *Callaloo*, among other journals. She directs the MFA Program in Creative Writing and Literary Translation at Queens College-City University of New York where she is a professor of English.

2. In Perry R. W., & Quarantelli E. L., *What is a disaster?: New answers to old questions.* (Xlibris: Corporation, 2005).

3. Alkalay-Gut K., "The Poetry of September 11: The Testimonial Imperative." *Poetics Today* 26, no. 2 (2005): 257–279.

4. Johnson D. L., and Merians V., "Foreword," in *Poetry after 9/11: An Anthology of New York Poets*, ed. Dennis Loy Johnson and Valerie Merians, ix–x (Hoboken, NJ: Melville House, 2002).

5. Harding S., *Nydis Manual For New York City Religious Leaders: Spiritual Care and Mental Health for Disaster Response and Recovery.* (New York: Disaster Interfaith Services, 2007), www.NYDIS.org.

6. Becker S. M., "Psychosocial Care for Adult and Child Survivors of the Tsunami Disaster in India." *Journal of Child and Adolescent Psychiatric Nursing* 20, no. 3 (2007): 148–155.

7. Long S., *Catastrophe and the Arts.* (The Singapore Globalist, January 3, 2013), http://www.tsglobalist.com/2013/01/03/catastrophe-and-the-arts-by-sylvester-long/.

8. Kirmayer L. J., Sehdev M., Whitley R., Dandeneau S. F., and Issac C., "Community Resilience: Models, Metaphors and Measures." *Journal of Aboriginal Health.* (National Aboriginal Health Organization, November 2009).

9. Tanenhaus S., *The Poetry of Catastrophe.* The New York Times. February 9, 2013. (2012), http://artsbeat.blogs.nytimes.com/2011/03/18/the-poetry-of-catastrophe/.

10. …there is a document found in Nippur, which was in ancient Babylonia, in the Fertile Crescent, written in the 19th century before the Common Era. It is more than 4,000 years old and describes a great flood that destroyed civilization. Only a single family survived by building a boat, and from them the rest of civilization rebuilt, http://www.jewishhistory.org/the-great-flood/.

11. *The Great Flood: the Epic of Atrahasis. A Mesopotamian account of the Great Flood*, http://www.livius.org/fa-fn/flood/flood3-t-atrahasis.html.

12. Garth S., and Dryden J., *Metamorphoses*, trans. Ovid. Internet Classics Archive, http://classics.mit.edu//Ovid/metam.html.

13. Los Angeles Times, Capturing Japan's pain in 17 syllables, http://articles.latimes.com/2011/mar/26/world/la-fg-japan-quake-haiku-20110326.

14. Gasperson R. T., (n.d.). *What Can I do to Help Japan?* http://www.squidoo.com/haiku-for-2011-japanese-tsunami-charity. Accessed February 12, 2013.

15. Ibid.

16. Milton J., *Paradise Lost* 2nd ed. (London: S. Simmons, 1674).

17. Taylor A. J. W., "Value Conflict Arising from a Disaster." *The Australasian Journal of Disaster and Trauma Studies* 2 (1999).

18. Smith R., Pat Robertson: Haiti "Cursed" After "Pact to the Devil." CBC News, January 13, 2010, http://www.cbsnews.com/8301-504083_162-12017-504083.html.

19. Franz G., "The Ancient Library at Alexandria: Embracing the Excellent, Avoiding its Fate." In *Association of College & Research Libraries Annual Convention, Philadelphia, March* 30, (2011).

20. Heller-Roazen D., "Tradition's destruction: On the Library of Alexandria." 100 (October 2002): 133–153.

21. Jochum U., "The Alexandrian Library and its Aftermath," *Library History* 15 (1999): S5–S12.

22. Caesar J., *The Civil Wars* 39 (Loeb Classical Library, 1984).

23. See note above 20.

24. Gibbon E., *The History of the Decline and Fall of the Roman Empire* 1 (W. Strahan and T. Cadell, 1776).

25. Joseph I., "Bar Hebraeus and the Alexandrian Library." *The American Journal of Semitic Languages and Literatures* 27 (1911): 335–338.

26. See note above 21.

27. Battles M., *Library: An unquiet history* (WW Norton & Company, 2004).

28. Thiem J., "The great library of Alexandria burnt: Towards the history of a symbol," *Journal of the History of Ideas* 40, no. 4 (1979): 507–526.

29. Lotz Jim., *The Sixth of December* (Markham: PaperJacks, 1981).

30. MacNeil Robert., *Burden of Desire* (Toronto: Doubleday Canada Limited, 1992).

31. Robin Cox is a Professor and Program Head of the Disaster and Emergency Management programs at Royal Roads University. Robin has devoted her research program to understanding disaster resilience and the potential of disasters, as transformative events, to spark social change. Leading Canadian funding agencies, including the Social Sciences and Humanities Research Council of Canada, Canadian Institute for Health Research, Canada's Centre for Security Science, and the Michael Smith Foundation for Health Research, have funded her community- and arts-based research with children, seniors, and those living in rural and remote communities. Robin is an active disaster psychosocial responder and steering committee member with the British Columbia Disaster Psychosocial Services (DPS) network.

32. McLuhan M., *Understanding Media: The Extensions of Man* (London: Routledge, 1994).

33. Naussbaum M., *Loves's Knowledge: Essays on Philosophy and Literature* (New York: Oxford University, 1990).

34. Richardson L., *Writing Strategies: Reaching Diverse Audiences* (Newbury Park, CA: Sage, 1990).

9

Ethics and Disaster

Compassion, in which all ethics must take root, can only attain its full breadth and depth if it embraces all living creatures and does not limit itself to mankind.
Albert Schweitzer (Figure 9.1)

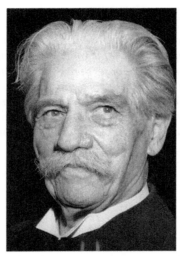

FIGURE 9.1 Albert Schweitzer. *Source: German Federal Archive.*

CHAPTER OUTLINE

9.1 Why This Topic Matters 274
9.2 Recommended Readings 275
9.3 Introduction 277
9.4 Ethics 279
    9.4.1 Utilitarianism 280
    9.4.2 Deontology or Duty-Based Ethics 282
        9.4.2.1 Social Contract Theory   285
        9.4.2.2 Paternalism   288
    9.4.3 Virtue Ethics (Co-authored with Peter Timmerman) 290
    9.4.4 Environmental Ethics 292
    9.4.5 Business Ethics or Corporate Social Responsibility 297

Peter Timmerman is a co-author for this chapter.

9.5 Ethics and the Construction of Risk—a Reflection 298
9.6 Conclusion 299
9.7 Example of an Ethical Dilemma: Temporary Settlement versus Permanent Housing......... 300
9.8 Jean Slick on Ethical Dilemmas 303
9.9 Commentary by Naomi Zack 304
End Notes 306

CHAPTER OVERVIEW

Disasters are studied not only because they are academically interesting but also because of a desire to prevent harm, a desire based upon social values and ethics. Ethics are, therefore, a critical component of disaster studies that encompass all phases of risk, from its initial construction to response and recovery. Sometimes values conflict, and there are a number of ethical theories relevant to disasters, which often leads to dilemmas. The outcomes of these tensions can have enormous consequences for the level of risk taken on by society, and the damage caused by future disasters. Resolving these requires a process that explicitly incorporates ethics as part of the disaster management process.

KEYWORDS

- Corporate social responsibility
- Deontology
- Environmental ethics
- Ethical dilemmas
- Ethics
- Paternalism
- Social contract theory
- Utilitarianism
- Virtue ethics

9.1 Why This Topic Matters

Disasters, by definition, are harmful events that cause suffering. Much of the research related to disasters and all the policies and programs in society are underlain by the social value of reducing the amount of harm they cause. Ethics is therefore fundamental to the study of disasters and has important implications; how research is conducted and how policies and programs are contrived largely depend upon our understanding of the rights and obligations between people, and between government and its citizens. There is no single right answer to the question of what they should be—different values, morals, and ethical theories can lead to different solutions. However, resolving ethical dilemmas transparently and explicitly must be part of the process used to manage disasters. Some areas related to disasters, such as humanitarian response, pay a lot of attention to ethics while others, such as the emergency management community, have largely ignored it.

■ ■ 9.2 Recommended Readings ■

- Beatley, T. (1994). *Ethical Land Use: Principles of Policy and Planning.* John Hopkins University Press. Baltimore and London.
- *Disaster Chaplain Code of Ethics and Guiding Principles,*[1] Nebraska Disaster Chaplain Network of Interchurch Ministries of Nebraska.
- Jenson, E. (1997). *Disaster Management Ethics,*[2] 1st Edition. Disaster Management Training Programme, UNDP.
- Slim, H. (1997). Doing the Right Thing: Relief Agencies, Moral Dilemmas and Moral Responsibility in Political Emergencies and War. *Disasters,* 21(3), 244–257.
- Zack, N. (2009). *Ethics for Disaster.* Rowman & Littlefied Publishers Inc., UK.

■ ■ Questions to Ponder: Ethical Dilemmas ■

- In New Orleans during Hurricane Katrina, a doctor and two nurses at Memorial Medical Center faced an extraordinarily difficult situation. They had been ordered to evacuate, but had severely ill patients who could not be moved. No relief was in sight from response organizations. Their choice was to do nothing and probably let them die in misery or to perform euthanasia. They chose the latter and were subsequently charged with second-degree murder.
  - Which would you choose?
  - Explain the moral grounds for your choice.
- You are a police officer at a disaster site and you observe people breaking into stores and stealing various supplies such as food, water, diapers, baby food, and the occasional luxury item (Figure 9.2).
  - What actions would you take (for example, assisting those obtaining essential supplies and/or arresting those stealing luxury items)? Why?
- You sit on a town council, and a developer has applied to develop a residential community within a local flood plain. This would provide the city with some badly needed tax revenue, and the developer has agreed to build a community center much desired by local residents. His plans include making the houses flood resistant. He claims that people have the right to decide about their own risk. One community group has signed a petition not to allow the development but, rather, to convert the area to parkland and a bird sanctuary. The city is in debt and cannot afford to do this.
  - What are the ethical issues?
  - How would you resolve this?
  - Would you, personally, grant him the licence to build?

FIGURE 9.2 Clapham Junction area is sacked after the third night of riots, on August 9, 2011, in London. Riots start spreading in London after Mark Duggan was shot dead by the police. *Source: Shutterstock/Dutourdumonde Photography.*

- You are a city mayor. During a disaster response you find that you do not have enough heavy equipment to respond to landslide threat. One company (a strong supporter of your campaign for mayor) has some equipment you can use, but refuses to provide it except at an exorbitant cost that you will probably be able to charge to the province. Emergency Management legislation gives you the power to commandeer the equipment, but you are worried about the political fallout if you take such an action. What do you do?
- You are a government employee who has learned of a potentially grave risk to public health,[3] but senior management has forbidden you to follow up this line of research or to speak of it to any person. Disobeying their instructions will almost certainly end your career and probably result in criminal charges due to a confidentiality agreement that you signed. You have a wife and three children, one of whom is disabled and requires expensive care that is mostly paid for by your generous medical plan. Do you speak out?
- A novel and very dangerous bird flu pandemic has begun. A vaccine has been developed, but it will take months for enough to be created to treat all of the population of your country. Some groups are offering to pay substantial sums to purchase it before others, which would greatly help with the cost of this very expensive venture and actually allow you to get more of the vaccine ahead of some other countries. The medical community claims they should get it first since they are the ones helping the sick. The elderly and young children are particularly vulnerable, as are those who are already ill. What process would you use to decide how to allocate the vaccine, and what ethical theories would you use to justify your decisions?
- In the past decade, there have been two destructive floods in your county that caused damage to several communities. You are trying to decide whether or not

to build a dam that would reduce flood risk and generate electricity, both valuable services that seem to far outweigh the cost of building the dam. On the other hand, some natural environment would be flooded, putting a rare and little known species of snail at risk for extinction. Additionally, local First Nations communities (which have also been damaged by flooding) object to the dam because they consider some of the area to be flooded as a sacred ancient burial site. What ethical principles are at play, and how would you resolve them? Is cost-benefit analysis a tool you would use in this case? If yes, what would you include and how would you go about valuation?[4]

> **STUDENT EXERCISES**
> - *Consider the following quotes*:
>   - "What good fortune for governments that the people do not think."—Adolf Hitler.
>   - "Those who can make you believe absurdities can make you commit atrocities"—Voltaire
>   - *How are these quotes relevant to the topic of disasters and ethics?*

9.3 Introduction

In 2008 the military rulers in Myanmar (Burma) initially refused international aid after being hit by Cyclone Nargis, although the death toll and destruction were massive. Even after Myanmar asked for aid from the United Nations, it was difficult for relief agencies to obtain visas. Similarly, in September 2012, North Korea rejected an offer of humanitarian aid from South Korea, which was offered because of storm and flood damage. These two examples highlight the fundamental connection between disaster ethics and political philosophy. Disaster ethics is not an "ivory tower" field of study, but has important implications for how we, as a society, create, prevent, or mitigate human suffering.

This chapter will overview different ethical theories and various arguments regarding the rights and obligations of citizens and government as they relate to planning for and responding to disasters. In ethical discussions there is rarely, if ever, one correct answer. Ethical dilemmas are common, and depending upon underlying ethical values and the ethical theory of choice, different moral judgments can be reached. For example, imagine that as a disaster planner you have to decide how to allocate a limited amount of funding. You are very concerned about the most vulnerable in society (the very poor, the elderly, etc.) and want to develop a program to reduce their vulnerability.[5] On the other hand, the same resources, if put into developing a program for the bulk of the population, would potentially provide protection to a far larger number of people. So—the greatest efficiency (a utilitarian approach) would not target the most vulnerable, and yet you may have a special duty to them. This is an example of an ethical dilemma with no clear answer.

It is important to think about such matters if we are to believe the above quotes by Adolf Hitler and Voltaire. Along similar lines the Milgram experiment (Figure 9.3)[6] (if you are not familiar with this experiment, you really should research it) resulted in the following comment by Prof. Milgram: "Ordinary people, simply doing their jobs, and without any particular hostility on their part, can become agents in a terrible destructive process. Moreover, even when the destructive effects of their work become patently clear, and they are asked to carry out actions incompatible with fundamental standards of morality, relatively few people have the resources needed to resist authority." Judith Boss[7] comments that those most likely to give in to the urging of the authority figure knew it was wrong, but were unable to articulate why it was wrong, whereas those who were able to resist the authority figure were able to provide justifications in the form of moral principles and theory. Most of the work done in disaster risk reduction occurs within bureaucracies in which employees are subject to various levels of authority. If the work that they do is not rooted in ethical considerations, then outcomes can easily reflect other factors such as power, vested interests, or profit—and such criticisms are not uncommon. One example of this is in the foreword in *Another Day in Paradise: International Humanitarian Workers Tell Their Stories* by Bergman, where John le Carré (perhaps overly harshly) refers to "the institutionalized functionaries of global disaster, so integrated with the towering bureaucracy of world aid and so familiar with its weaknesses that they are actually a part of the problem they think they're solving".

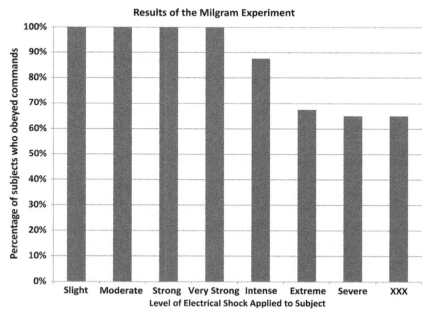

FIGURE 9.3 Compliance by subjects in the Milgram experiment. Shock voltages ranged from 15 V in the slight level, to 450 V in the XXX level. *Source: Milgram, S. (1965). Some conditions of obedience and disobedience to authority. Human Relations, 18(1), 57–76.*

Earlier I mentioned how important it is for ethics and values to be explicit and transparent, and many fields of study and professions go to considerable effort to accomplish this (political theory, as it is actually practised by many politicians, is in my opinion a notable exception). Examples include The Canadian Association of Psychologists, Humanitarian NGOs (Sphere Project) and Medical Codes of Ethics.[8] One exception to this (so far at least) is the field of emergency management, which has devoted little effort to developing ethics as a core element of the profession. A survey of Emergency Management (EM) journals and texts[9] done by some of my graduate assistants revealed only rare references to ethics and human relationships. The International Association of Emergency Managers has developed a set of principles, but they are ad hoc and not firmly rooted in ethical theory. Governments often have codes of conduct or ethical standards for their employees. One example is the Department of Homeland Security Policy #181, which notes that "Public employment is a public trust. DHS employees must not only avoid any potential conflict of interest, but also must avoid any appearance of impropriety. DHS employees must conduct themselves so as to foster public confidence in the integrity of state government", but says nothing about the important ethical issues or processes related to fostering a safer nation. In my opinion, ethics, as applied to disaster, and especially as it relates to the field of emergency management, should be given a great deal more attention than it has thus far.

9.4 Ethics

There are several theories that are particularly relevant to disaster ethics. They include: (1) utilitarianism or consequentialism, (2) deontology or Kantian ethics, (3) virtue ethics, (4) social contract theory, and increasingly (5) environmental ethics. It is not necessary to use only a single ethical theory to analyze a problem; ethical pluralism allows for the notion that many theories may be useful, even when they conflict with each other. Thus we might (and probably should) consider both the net social benefit resulting from some action as well as fundamental rights and duties of the people involved, and balance those often-conflicting approaches.

Extreme examples of moral dilemmas are useful for illustrative and analytical purposes. Consider, for example, the scenario of a lost group of people who have only enough food for some of them to survive. The choice then is for all to die, or for some to be sacrificed. Some forms of equity and distributive justice, depending upon which criteria are used, are satisfied when all die, but most would hardly consider this a desirable outcome. Utilitarian arguments suggest that some be sacrificed so that others may live, thus increasing the social utility of lives saved. A deontological perspective would say that killing people is normally considered immoral if not evil, and is prohibited by most societies except under very particular circumstances such as self-defense. If decided upon, should such actions be voluntary or does the group have the right to select those who will not survive—and if so, how should such a decision be made? Is everybody's life of equal value, or are some people of more worth than others? And even if all have lives of equal worth, are there differences in duty (such as to children)? Questions such as these are not amenable to easy answers, and although fascinating from a hypothetical perspective, may seem an unlikely reality. However, society constantly makes decisions about how to allocate limited or scarce resources to a variety of uses (e.g., health care, education, national defense, scientific research, the arts) and how to manage risk (such as development near hazardous industrial facilities). The consequences of these decisions result in harm to some and good to others, even to the point of life and death. The study of ethics, including lifeboat ethics (Figure 9.4), must inform such decisions.

FIGURE 9.4 What to do when the lifeboat is overloaded?

9.4.1 Utilitarianism

The happiness of individuals, of whom a community is composed, that is their pleasures and their security, is the end and the sole end which the legislator ought to have in view.

Jeremy Bentham, Principles of Morals and Legislation

Actions are right in proportion as they tend to promote happiness; wrong as they tend to produce the reverse of happiness. By happiness is intended pleasure and the absence of pain.

John Stuart Mill

The father of modern utilitarian theory was Jeremy Benthan, who lived in the eighteenth and nineteenth centuries (Figure 9.5). This theory proposes that actions are good if, on balance, they contribute to happiness or some other defined goal or benefit to society. It is sometimes characterized by the notion of the greatest good for the greatest number or the greatest net good for all those affected. This process can accept bad things happening to some people, as long as the end justifies the means. A strength of this theory

FIGURE 9.5 Jeremy Bentham (1748–1832).

is that it forces consideration of the consequences of actions. There are four basic steps in its use:

- Consider who or what will be affected
- Consider what consequences will result
- Maximize utility
- Minimize anti-utility

Many social programs are based upon utilitarian goals where some harm (perhaps in the paying of more taxes) is done to many so that others may benefit. An example is disaster relief,[10] although there are arguments that it is not totally altruistic, since those who pay also benefit as a result of living in a more just and healthier society, or because some common good is restored. A common conundrum is defining what is good; different people or groups often view this in very different ways (such as the value of life or health, or the value of a political philosophy). For example, a nature conservationist and a developer are likely to have very different opinions as to how to weigh the value of the natural environment as compared to a retail development. Another challenge, especially as it relates to the public good, is that it is very difficult to foresee the future (particularly in the complex and rapidly changing world in which we live); this means that outcomes from different decisions made now cannot be easily compared. A high-profile example of this is the environmental and social impact of climate change as a result of increasing greenhouse gas concentrations in the atmosphere, and the debate regarding the reduction of carbon emissions.[11] Another issue is the intensity of harm associated with benefit—for example the harm caused by taxation to support disaster relief. This became an issue during the US presidential race of 2012, when Governor Romney criticized taxation to support disaster relief programs by saying it is immoral.[12]

One argument for mitigation and prevention relates to efficiency, or the maximization of the social utility being considered. Most studies show very favorable cost-benefit ratios for mitigation efforts, although if they are badly done they can increase risk. Therefore actions to prevent or mitigate disaster risk can require much smaller investments than the cost of responding and recovering without such a priori actions, thereby contributing to welfare maximization. Politically, however, it tends to be unattractive because of the difficulty of convincing people that the needs of future disaster victims exceed needs in the present.

Beatley[13] provides an in-depth discussion of utilitarian values as applied to land-use planning. I found his book very good and highly recommend reading it. Many land uses can create significant harm to society or the environment; examples include land and water pollution or an increase in flood risk. Beatley argues that land-use policy makers have the moral and ethical duty to prevent or minimize such harms, particularly if they are public harms. Post-disaster arguments to fund disaster risk reduction based upon a utilitarian framework[14] assume that the marginal value of resources directed toward potential victims exceeds the value of those same resources directed elsewhere because of the acute needs involved, or that the use is more valuable (such as supporting a critical industry or infrastructure).

Various organizations promote the ethical use of floodplains in order to prevent flood disasters. The EU Life-Environment Project proposes the Wise Use of Floodplains, which puts ecology at the heart of integrated water management.[15] The US Association of State Floodplain Managers advocates for a No Adverse Impact approach, which is based upon the principle that the actions of property owners should not adversely affect the rights of each other.[16] Conservation Authorities in Ontario, Canada, have been particularly good at restricting flood plain development in order to protect the public good.[17] Such is not always (or even normally) the case of course; significant development in flood risk areas frequently occurs in many parts of the world.

Zack,[18] after analyzing issues related to limited resources, difficulties in preferential treatment of various groups, and incomplete preparedness, argues that utilitarian ethics result in a best policy of "Fairly Save All Who Can Be Saved with the Best Preparation." What is fair, of course, is subject to some debate. She proposes triage, random selection, or first-come first-served.

Utilitarian arguments are powerful when it comes to disaster ethics and should play an important role in disaster management. However, if unconstrained by ethical thinking related to fundamental rights and obligations, they can lead to morally repugnant actions such as the abandonment of innocents.

9.4.2 Deontology or Duty-Based Ethics

Quotes by Immanuel Kant (Figure 9.6):

- "To be is to do"
- "Do what is right, though the world may perish"

FIGURE 9.6 Emmanual Kant.

- "Morality is not the doctrine of how we may make ourselves happy, but how we may make ourselves worthy of happiness"

Do unto others 20% better than you would expect them to do unto you, to correct for subjective error.

Dr Linus Pauling, Nobel Prize Winner (twice)

Deontology, an ethic based upon duty, argues that there are intrinsic goods and bads that are not situationally dependent but that either should or should not be followed under any circumstances (thus avoiding an important weakness in utilitarian theory), even if the outcomes of the former are bad or the latter are good. Immanuel Kant, one of the founders of this theory, articulated two imperatives in his attempt to ground ethical judgements through rational reflection:

- Universalizability/Reversibility (Golden Rule): Act only according to that maxim by which you can also will that it would become a universal law.
- Respect (no exploitation): Act in such a way that you always treat humanity, whether in your own person or in the person of any other, never simply as a means, but always at the same time as an end.

In this theory, which applies to both duties and rights, morality is a function of the actions themselves and the intent/motive of those who are acting. Seven examples of duties[19] are: beneficence, non-maleficence, fidelity, reparation, gratitude, self-improvement, and justice. There are many works that list people's rights; one was created in 1948 by the United Nations (*The Universal Declaration of Human Rights*[20]), with 30 articles pertaining to rights and duties. Two examples of the articles are: "Everyone has the right to life, liberty and security of person" and "Everyone has the right to a standard of living adequate for the health and well-being of himself and of his family... and the right to security in the event of... circumstances beyond his control."

If followed absolutely, a deontological framework can be critiqued for not allowing gray areas. In addition, one can argue that sometimes doing the right thing can result in horrible consequences—for example, not killing a terrorist about to explode a large bomb in a heavily populated area. In addition, social norms of determining what is right can be confusing; for example, deciding on moral actions often suffers from the influence of vested interests, summarized by the saying "the hat you wear depends upon the seat you sit in." The philosopher John Rawls, in his discussion of societal justice, proposed a thought experiment using a "veil of ignorance" concerning our own interests in order to remove this sort of bias.

Along these lines, Zack[21] lists several values and principles that are commonly accepted in western democratic society:

- "Human life has intrinsic worth.
- Everyone's life is equally valuable.
- Everyone has the same right to freedom from harm by others.
- Everyone is entitled to protection from harm by non-human forces."

The following ethical principles follow:

- "We are obligated to care for ourselves and our dependents.
- We are obligated not to harm one another.
- We are obligated to care for strangers when it doesn't harm us to do so."

There are various implications of these principles for disaster ethics, depending of course upon the values espoused. One is reflected in the principle that preparing for emergencies begins with the individual, who, in Canada, is expected to be self-sufficient for 72 h. Another is that various levels of government have duties toward their citizens to prepare for emergencies through mitigation, prevention, and preparedness, to respond, and to help recovery. The issue of whether policies and actions that are accepted as good are mandated or voluntary is important. Libertarians, for example, hold sacred liberty and property rights, and object to restrictions or costs imposed by government. This has important implications, such as for flood plain development. Libertarians do not like to be told that that they cannot live in a flood plain. Canada and the US have taken very different approaches with respect to managing flood risk. The Canadian government is far more restrictive than the US government, which, unlike Canada, embraced a National Flood Insurance Program (NFIP). The net result of the NFIP might well have been to dramatically increase flood risk in the US[22] as compared to Canada. One interesting study compared flooding in the summer of 1986 in Michigan and Ontario,[23] and found that the difference in costs ($500 million in Michigan and less than $0.5 million in Ontario) was accounted for entirely by the different amount of flood plain development, as a result of different policies and regulations. Libertarians would also object to disaster assistance programs that transfer funds to victims, viewing it as a violation of the rights of non-victims. This is often perceived as a harsh criticism, and one could argue that not providing such aid violates the welfare rights of victims and principles of distributive justice.[24] Other preparedness actions such as purchasing family emergency response kits are clearly voluntary. It is difficult to decide upon the degree to which government should invest in or mandate risk reduction actions, but such decisions are important.

In western societies, egalitarian values are widely accepted and might be considered as a basis for disaster mitigation and relief programs. Dunfee and Strudler[25] critique this approach, noting that equity issues might be applied to the distribution of resources or wealth, the distribution of welfare, or the distribution of opportunity. The first is clearly fraught with difficulties, although in some countries social programs work to reduce the gap between the rich and poor. The second suffers from the problem that those who are worst off in society are often not disaster victims (in fact, disaster victims are often quite wealthy), and therefore it begs the question as to why resources are being diverted in that direction.

The notions of intrinsic rights, wrongs, and duties have a long history in society and fills an important gap in utilitarian theory. Deontological theory underlies much of how society prepares for and responds to disaster. Different values can result in different sets of duties, however, and the inflexibility associated with some actions (such as stealing) can be an issue. For example, looting during a disaster in order to obtain basic goods such as food and water that are needed by your children would be considered acceptable by many, even though it is stealing.

STUDENT EXERCISES

- *Make a list of several policies or actions under the categories of mitigation/prevention, preparedness, response, and recovery, and comment on the obligations of individuals and government for each of them.*

9.4.2.1 Social Contract Theory

"By me kings reign, and princes decree justice."

Proverbs 8.15-16:

Social contract theory contrasts with the medieval belief that kings ruled through divine decree. It is based upon the notion that there is an agreement, implicit or explicit, between citizens and those who govern, that specify rights, freedoms, and liberties. In particular, citizens forego some rights and freedoms in order to live in a state that provides security and safety. It is then the responsibility of government to provide them with a society that is better than that which would have existed without such an agreement. John Locke (Figure 9.7), Thomas Hobbes, and John Rawls are important philosophers who have written on this topic.

FIGURE 9.7 John Locke (1632–1704).

John Rawls' published *A Theory of Justice* in 1972, and posited the interesting notion of decision making behind a *veil of ignorance* to avoid the influence of bias or vested interests. From this he developed Two Principles of Justice, which determine how economic goods and civil liberties are distributed in society. The first principle maximizes the total amount of liberty, ensuring that it is granted to all, whereas the second principle allows inequalities but requires that there be equal opportunity.

The notion of a social contract is fundamental to disaster management, both in planning and response. Alex de Waal, in his book *Famine Crimes,* places it at the center of his analysis, noting that "History is replete with successful methods of preventing famine. Common to them are versions of political contract that impose political obligations on rulers. In the most effective anti-famine contracts, famine is a political scandal." He also says that "An important step in that struggle is for those directly affected by famine to reclaim this moral ownership" (referring to the tendency for humanitarian organizations to co-opt moral ownership of famines during the relief stage).

In western democracies, such a contract does exist, both informally in the minds of citizens and also formally in legislation and policy. The mission of FEMA, for example is "to support our citizens and first responders to ensure that as a nation we work together to build, sustain, and improve our capability to prepare for, protect against, respond to, recover from, and mitigate all hazards".[26] Similarly, the mandate of Public Safety Canada is "to keep Canadians safe from a range of risks such as natural disasters, crime and terrorism".[27] Citizens, as part of the social contract, give up freedoms in exchange for the benefits that government can provide; thus follows the duty of federal, provincial, and municipal governments to engage in mitigation, preparedness, response, and recovery. This is not to say that individuals do not have duties as well. They do (for example, purchasing insurance if it is available). The issue of where individual duties end and collective duties begin is a complex problem that can be resolved only through social discourse. As society becomes increasingly dependent upon critical infrastructure far beyond the locus of control of any individual or small group, the duties of government become greater.

A social contract is the basis for most mitigation, prevention, and preparedness undertaken by different levels of government, yet there is an unclear and worrisome region where government intervention may be excessive. In the pre-disaster state, it especially relates to land-use planning, building codes, and other forms of government regulation designed to make society safer. In the post-disaster state, the issue of when governments can adopt extraordinary powers to respond, and when those powers should be rescinded is critical. Governments do have the responsibility to ensure that they can continue to function (hence the emphasis on Continuity of Operations (COOP) programs) in order to fulfill their side of the contract. At times, however, such powers can become the new normal as rulers use them to bolster their power or restructure society to benefit elites. This is the basis of most of the argument made by Naomi Klein in *The Shock Doctrine*.[28]

Naomi Zack spends a good deal of time in her book discussing the difference between security and safety, and problems that arise when the two are conflated under security. This scenario developed in many western countries, and particularly in the US, after the September 11, 2001 terrorist attacks in order to combat terrorism. She argues that a diminishment of public safety occurs under such a situation "in favor of more dramatic police and military initiatives for security." Other scholars of disaster studies have made similar arguments.[29]

Without trust, a social contract in western democracies cannot function. Unfortunately, trust in politicians tends to be weak in much of the world. Different professions tend to have very different trust levels, as shown in Table 9.1. These have important implications with respect to governance and risk communication.

Table 9.1 Trust among Various Professional Groups

Professional Group	All countries, 2009. % of Respondents who Trust the Various Professional Groups	Professional Group	All countries, 2009. % of Respondents who Trust the Various Professional Groups
Fire service	92	Civil servants	57
Teachers	85	Market researchers	55
Postal workers	81	Lawyers	47
Doctors	81	Trade union representatives	43
Armed forces	81	Journalists	41
Clergy	66	Marketing professionals	39
Environmental protection organizations	64	Banks	37
Police	61	Top managers	33
Charities	60	Advertising professionals	28
Judges	57	Politicians	18

Note: These are average values, and level of trust varies by country. For example, in Greece, trust in politicians is only 6%. GfK Trust Index for spring 2009, GfK Custom Research.

Social contract theory does not necessarily conflict with utilitarianism or Kantian ethical theories—rather, they may be the basis for the type of contract that exists. Social contracts are not static, however, but evolve with changes in environment and culture.

> **STUDENT EXERCISES**
>
> - From a risk perspective, the world is changing rapidly. Trends include globalization, shifting political alliances, environmental destruction, species extinction, greater dependence upon critical infrastructure, etc. How should social contracts change to reflect these realities?

9.4.2.2 Paternalism

The degree to which personal liberties should be considered sacrosanct versus how far government should be allowed to restrict those liberties relates to the notion of paternalism. Paternalistic acts protect individuals from themselves. Such actions are accepted for children by their parents or the state, or for people who are clearly unable to make good decisions, such as those with mental illness. For others, however, to what degree is the state justified in making paternalistic decisions? Examples of state regulations on risk include bike helmets, seat belts (Figure 9.8), air bags, building codes and building inspectors, and conservation authorities limiting the ability of individuals to build how and where they want.

FIGURE 9.8 "Click it or ticket" is a campaign to enforce state seat belt laws. *Source: Shutterstock/Susan Montgomery.*

Paternalism as a duty of government is one argument used in protecting homes in the urban–rural interface against wildfire (for example, the FireSmart Canada Program[30]). Those who wish to live in or near forests argue that they are they have the right to decide what risks they take on as individuals, and should be allowed to do so. This argument is flawed, however, since there are larger social costs such as during a wildfire response. As well, once a development exists it affects future generations, who may not be fully aware of all the risks. Additionally, there may well be others such as children, who do not knowingly or willingly undertake the risk; thus a paternalistic attitude may be justified. Another justification for paternalism relates to the notion of bounded rationality and biases in risk perception. It is simply not always possible for an individual to have access to all the risk-related information that they need, or to be able to process it even if they obtain it. Therefore the need for experts to assess risk can be a justification for paternalism. An interesting aspect of risk theory called risk homeostasis is relevant to paternalism (Chapter 3.5.3).

■ ■ Question to Ponder ■

Over time, what is a public good has changed. Society's dependence on critical infrastructure is very different from what it was in the past, and yet a large majority of critical infrastructure is owned by the private sector. Private companies do not have obligations to the public good the way that a democratic government does, since their main duty is to their stockholders. How should the ethical obligations that governments owe to their citizens take into account this shift of what is a public good?

PRINCIPLES OF CONDUCT FOR THE INTERNATIONAL RED CROSS AND RED CRESCENT MOVEMENT AND NGOS IN DISASTER RESPONSE PROGRAMMES[58]

1. The humanitarian imperative comes first.

2. Aid is given regardless of the race, creed, or nationality of the recipients and without adverse distinction of any kind. Aid priorities are calculated on the basis of need alone.
3. Aid will not be used to further a particular political or religious standpoint.
4. We shall endeavor not to act as instruments of government foreign policy.
5. We shall respect culture and custom.
6. We shall attempt to build disaster response on local capacities.
7. Ways shall be found to involve program beneficiaries in the management of relief aid.
8. Relief aid must strive to reduce future vulnerabilities to disaster as well as meeting basic needs.
9. We hold ourselves accountable to both those we seek to assist and those from whom we accept resources.
10. In our information, publicity, and advertising activities, we shall recognize disaster victims as dignified human beings, not hopeless objects.

9.4.3 Virtue Ethics (Co-authored with Peter Timmerman)

Courage is the most important of all the virtues, because without courage you can't practice any other virtue consistently. You can practice any virtue erratically, but nothing consistently without courage.

Maya Angelou

Heaven and hell suppose two distinct species of men, the good and the bad. But the greatest part of mankind float betwixt vice and virtue.

David Hume

All virtue is summed up in dealing justly.

Aristotle

Virtue ethics emphasizes right being over right action. This form of ethics is different from utilitarianism and deontological theories in that it does not consider actions explicitly; rather, it considers what character traits and virtues a person should adopt, which in turn will help them to live a moral life and to choose moral actions. Its founders go far back in time and include the Greek philosophers Plato and Aristotle.

In terms of how we plan for and respond to disasters, there are many character traits of significance (this is my personal list), including honesty, caring, compassion, generosity, empathy, impartiality, integrity (which refers to acting consistently according to one's stated values or principles), diligence, kindness, openness, reliability, resoluteness, respectfulness, sensitivity, tolerance, toughness, trustworthiness, and truthfulness. Naomi Zak, in particular, emphasizes the importance of integrity and diligence, and notes that "In morally ambiguous extreme cases, we do well to rely on the character or virtues of those in positions to make decisions...." Personal networks and relationships are critical to effective disaster management, and unless there are trusting relationships in place, any process is likely to become dysfunctional. Co-workers and victims alike must have trust in the competence and character of those managing disaster.

Within the list of important character traits, we should include how people view others in terms of their moral worth. The philosophy of Martin Buber[31] provides an interesting and useful perspective on this issue. According to Buber (Figure 9.9), human beings may adopt two attitudes toward the world, "I–Thou" or "I–It." Within an "I–It" relationship, objects or beings are viewed by their functions. Inevitably, in the large and complex world in which we live, this represents the majority of their relationships. In an "I–Thou" relationship, one engages in a mutual dialogue that goes beyond function and acknowledges the fundamental worth of the other. Unlike "I–It" relationships, "I–Thou" ones are imbued with rights, duties, and moral worth. I believe, as suggested by Buber,[32] that this disconnect takes form primarily as a lack of empathy.

This theme is also echoed in the works of others. For example, Tad Homer Dixon,[33] in writing about challenges in dealing with future risk, says "I believe this will be the central challenge—as ingenuity gaps widen the gulfs of wealth and power among us, we need imagination, metaphor and empathy more than ever, to help us remember

FIGURE 9.9 Martin Buber.

each other's essential humanity." At the core of empathy must be an acknowledgment of "I–Thou" relationships. Vanderberg[34] makes this point as well, when he says "technological and economic growth are guided primarily by performance values rather than by human and social values," and that it is critical to understand the larger social context and to ensure that technological and economic systems are compatible with it.

In disaster management, human beings are often treated as means toward the greater good of the greater number—including themselves! In that sense, they are turned into objects/obstacles. A familiar example to everyone is when you are running to get somewhere, and someone is standing in your way; they have ceased for the moment to be a human being and are now a thing in your path. In an emergency, people can very quickly be turned into obstacles to the greater good, and are thus transformed into an "It" as opposed to a "Thou." The most famous expression of this situation is to be found in Simone Weil's work on violence (see *War and the Iliad* by Simone Weil and Rachel Bespaloff). She argues that part of the essence of violence is the turning of one's opponent into a thing (e.g., "dead meat") as one inflicts violence. In order to get to one's goal, one "cuts through" the opposition. The goal is what matters; the dead or ruined that one creates as one drives toward that goal are merely "collateral damage."

A bureaucracy or a dedicated force attempting to resolve a situation can be all too prone to considering people as inconvenient things, even though saving people is what the activity is about. The experiences of Bradshaw and Slansky,[35] two paramedics caught in New Orleans during Hurricane Katrina, reflect this as they observe the following about their treatment: "This official treatment was in sharp contrast to the warm, heart-felt reception given to us by the ordinary Texans. … Throughout, the official relief effort was callous, inept, and racist."

Examples of how "I–It" relationships have resulted in disastrous outcomes in human terms abound. One classic case involves explosions of the Ford Pinto in the 1970s due to a defective fuel system design in which gas tanks often ruptured and exploded in crashes over 25 mph. This may have cost the lives of between 500 and 900 people.[36] The cost of repair was estimated to be larger than the cost of lawsuits from deaths, injuries, and car damages resulting from the explosions; therefore Ford Motor Co. did not feel that a recall was justified. The result of this decision is that many people were injured or killed unnecessarily. In fact, during the period of Ford's successful delaying tactics against Federal Motor Vehicle Safety Standard 301 between 1968 and 1972, some 9000 people were burned to death in all car accidents in the US, and more than 10 million new unsafe cars were sold. A cost–benefit analysis such as used by Ford, placed people in the category of things. Another example of a disaster that resulted from placing people in the category of things is the Johnstown flood of May 3, 1889, which resulted from the catastrophic failure of a dam developed by a group of rich speculators who used it to create an exclusive resort.[37] The dam was inadequately maintained, primarily because the dam owners did not want to spend the money needed to do so, in spite of the risk to downstream communities. It was an instance of the elite abusing their power, with risk being transferred to other less powerful and less fortunate groups.

Both of these examples illustrate a fundamental barrier to effective disaster management. **When the institutions and people who construct risk are disconnected from those who bear its negative consequences, then relationships become of an "I–It" kind, thus removing the issue of moral/ethical values from the risk management equation and reducing it to simple economics.** As noted by Jonathan Glover, "It is more difficult to commit an atrocity against others if they have dignity and respect".[38] One expression of this humanistic approach has been the development of care ethics, a gendered ethic emerging from feminist philosophy that emphasizes the importance of feeling and sentiment in morality.[39] Caring ethics is distinguished from ethical behavior based upon duty by striving to maintain a caring attitude.[40]

There is one final note that I would like to mention about virtues. Over the years, I have often been asked to provide a reference for students applying for a job. Not once has an employer asked me about their marks or their adherence to a utilitarian or Kantian philosophy (even implicitly), but I have frequently been asked about their virtues and character flaws.

Real and fictional case studies in disasters seem to favor egalitarian or deontological moral principles over those of efficiency or a simple utilitarianism that saves the greatest number. In morally ambiguous extreme cases, we do well to rely on the character or virtues of those in positions to make decisions

Naomi Zack

9.4.4 Environmental Ethics

Nature is not a place to visit, it is our home…

Gary Snyder, poet

What deep ecology directs us toward is neither an environmental ethic nor a minor reform of existing practices. It directs us to develop our own sense of self until it becomes Self, that is, until we realize through deepening ecological sensibilities that each of us forms a union with the natural world, and that protection of the natural world is protection of ourselves.

Alan Drengson

Nature does not hurry, yet everything is accomplished.

Lao Tzu

We are not outside the rest of nature and therefore cannot do with it as we please without changing ourselves ... we are a part of the ecosphere just as intimately as we are a part of our own society ...

Arne Naess, Norwegian philosopher and the inspiration behind the Deep Ecology movement

For a good summary of environmental ethics as it relates to land use planning I recommend reading Chapter 7 of the book on land use ethics by Timothy Beatley (*Ethical Duties to the Environment*).[41] The basis of a biocentric or deep ecology approach to environmental ethics is that people owe duties to non-human life, because there exists an intrinsic worth not related to human beings. A more homocentric or shallow ecology approach focuses on the worth of the environment to people, acknowledging the degree to which we depend upon it. Traditionally, environmental ethics has played a weak role in the disaster field but it is becoming much stronger for two reasons. The first is an increasing recognition of the role that the degradation of natural systems plays in exacerbating natural hazards. Examples include the protection provided by mangrove swamps and wetlands to storm surge and tsunamis (case studies would include New Orleans and Hurricane Katrina, and the 2004 tsunami in Indonesia), and the effect of deforestation on landslides and floods (Haiti is one example). The second reason is a growing awareness and disenchantment with the extreme level of impact that the human species is having on the rest of the natural world. This awareness affects many people at a fundamental level, as they are appalled at the massive loss of natural systems. This overuse of the environment is setting the stage for future catastrophes.[42]

> **STUDENT RESEARCH PROJECT**
>
> *Compare the effect of deforestation in Haiti to that of the Dominican Republic, in terms of flooding and landslide risk. In your research, explore the social and political root causes of the different environment paths taken by the two countries.*[43]

Although it was not true to a significant degree before the Industrial Revolution, economic development and population growth can now occur only along with the destruction of the natural environment. This is discussed in the *Millennium Ecosystem Assessment*,[44] which noted that "Many people have benefited over the last century from the conversion

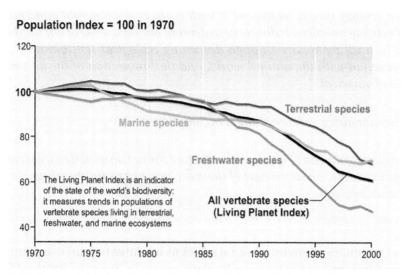

FIGURE 9.10 Changes in biodiversity from 1970 to 2000. *Source: Millennium Ecosystem Assessment.*

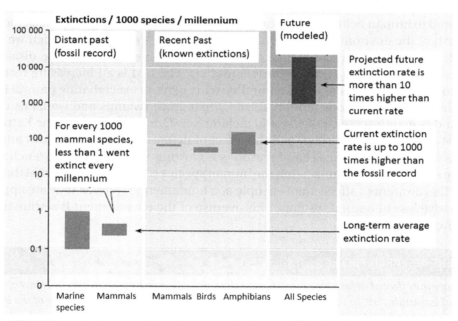

FIGURE 9.11 Extinction rates of non-human species. *Source: Millennium Ecosystem Assessment.*

of natural ecosystems to human-dominated ecosystems and from the exploitation of biodiversity. At the same time, however, these gains have been achieved at growing costs in the form of losses in biodiversity, degradation of many ecosystem services, and the exacerbation of poverty for other groups of people." Figures 9.10 and 9.11 illustrate the impact of economic and population growth on other species.

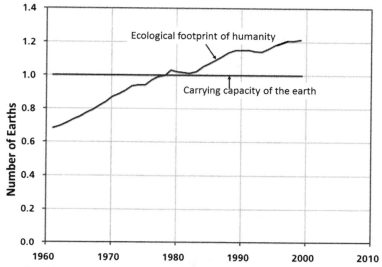

FIGURE 9.12 The number of Earths required to sustain humanity, in terms of renewable resources. *Source: Wackernagel, M., Schulz, N. B., Deumling, D., Linares, A. C., Jenkins, M., Kapos, V., Monfreda, C., Loh, J., Myers, N., Norgaard, R. and Randers, J. (2002). Tracking the ecological overshoot of the human economy. Proceedings of the National Academy of Sciences, 99(14), 9266–9271.*

The change in the ecological footprint of humanity in recent decades is shown in Figure 9.12 which shows the number of Earths required to provide the resources used by humanity and to absorb their emissions for each year since 1960. This human demand is compared with the available supply: our one planet Earth. Human demand has exceeded nature's supply from the 1980s onward, overshooting it by some 20% in 1999. According to Nadeau,[45] non-renewable resources aside, it would take two additional Earths to support the world's current population if developing nations were to develop to the same level as the US. Economic systems rooted in exponential growth, which are embedded within a finite resource supply, must eventually reach limits to growth. This has the potential to be a catastrophic experience if it is not managed by humanity but is imposed upon us by nature instead. In *Limits to Growth: The 30 Year Update*,[46] it was found that "In every realistic scenario, we found that these limits force an end to physical growth...sometime during the twenty-first century. ...In our scenarios the expansion of population and physical capital gradually forces humanity to divert more and more capital to cope with the problems arising from a combination of constraints. Eventually so much capital is diverted to solving these problems that it becomes impossible to sustain further growth in industrial output. When industry declines, society can no longer sustain greater and greater output in the other economic sectors: food, services, and other consumption. When those sectors quit growing, population growth also ceases."

This theme has been explored in numerous books and papers, including *Limits to Growth, Our Final Hour*,[47] *The Gaia Hypothesis*,[48] *Right Relationship: Building a Whole Earth Economy*,[49] and others. Daly and Cobb,[50] in their support of ecological economics, say that "we human beings are being led to a dead end—all too literally. We are living by an ideology of death and accordingly we are destroying our own humanity and killing the planet." The importance of a resilient ecology was the first of six principles of sustainable hazards mitigation proposed by Mileti[51] that "the first guiding principle of sustainability is that human activities in a particular locale should not reduce the carrying capacity of the ecosystem for any of its inhabitants." The future is uncertain, but the risk is great. Environmental ethics must be part of an integrated framework to manage this risk (Figure 9.13).

FIGURE 9.13 Earth Day.

One interesting route to sustainability that has been discussed relates to environmental ethics. In *The Environmental Endgame: Mainstream Economics, Ecological Disaster and Human Survival,* Nadeau suggests that the most probable pathway is through the development of a global environmental ethic, akin to a religious movement. There is empirical evidence for this adaptation strategy. Studies have found that "pre-scientific societies can and do adopt conservation practices on the basis of their experience; practices that in the past were implemented through the medium of religious beliefs."[52] Nadeau's argument emphasizes the importance of embracing environmental ethics as part of a global ethos. Daly and Cobb echo this by noting that, in their judgment, changes at a deep religious level are needed to avoid the trap of ad hoc and insufficient responses to crisis. In their books, Brown and Garver[53] and Parfit[54] also talk about a spiritual awakening and the moral responsibility that we have to future generations as part of the solution to the environmental crisis.

Taylor, in his book *Respect for Nature: A Theory of Environmental Ethics,*[55] outlines several rules and principles of environmentally ethical behavior:

Rules:

- "The rule of non-maleficence: a duty not to harm creatures in the natural environment: particularly those that do not harm human beings." An example would be hunting for pleasure, as opposed to sustinance.
- "The rule of noninterference: the duty to refrain from denying freedom to organisms and a general hands-off stance for ecosystems and organisms."

- "The rule of fidelity: retractions to deceiving or betraying wild creatures."
- "The rule of restitutive (or compensatory) justice: requirements to restore or compensate for previous injustices done to organisms and ecosystems."

Principles:

- "Self-defense: individuals can protect themselves."
- "Proportionality: in conflicts between humans and other species, non-basic human interests cannot justify overriding a basic interest of other species."
- "Minimum wrong: people must choose alternatives which do the least damage to the natural world and harm or destroy the fewest number of organisms."
- "Distributive justice: where conflicting interests are all basic, fair shares are equal shares, and certain adjustments are required to ensure all species are treated fairly."
- "Restitutive justice: compensation is required for injustices done in the past; the greater the harm done, the greater is the compensation required."

> **STUDENT EXERCISES**
>
> *Give one example for each of the above rules and principles.*

The rules of the environmental game are changing, and, as students of disaster and citizens of our planet, we need to give serious consideration to ethical issues that underlie the choices made by society. If, indeed, we are in an environmental end-game (and it is my belief that we are), then the decisions that humanity makes now will largely determine future catastrophes. Sadly, important social and political paradigms reject or minimize the inclusion of environmental ethics in society. It is common for interests vested in economic development to see environmental concerns as contrary to their purposes. In North America, legislation put to Parliament by the Conservative Government of Canada under Prime Minister Harper have rolled back decades of environmental progress,[56] and the word "environment" seemed to be entirely absent in the political dialogue of the 2012 US presidential race between President Obama and Governor Romney; only jobs and economics were deemed to be important enough to discuss. Similarly, but to a greater degree, the environment has suffered greatly in the name of economic development in many developing countries, such as China and Brazil. As of 2010, one study ranked the 12 countries with the worst absolute composite environmental impact as Brazil, US, China Indonesia, Japan, Mexico, India, Russia, Australia, Peru, Argentina, and Canada.[57] As long as environmental protection is perceived as being a barrier to economic growth, the future will be bleak from an environmental perspective.

9.4.5 Business Ethics or Corporate Social Responsibility

Business has a responsibility beyond its basic responsibility to its shareholders; a responsibility to a broader constituency that includes its key stakeholders: customers, employee, NGOs, government—the people of the communities in which it operates.

Courtney Pratt, Former CEO Toronto Hydro

We know that the profitable growth of our company depends on the economic, environmental, and social sustainability of our communities across the world. And we know it is in our best interests to contribute to the sustainability of those communities.
Travis Engen, CEO, Alcan

It takes 20 years to build a reputation and five minutes to ruin it.
Warren Buffet

In the past several decades, discussions around Corporate social responsibility (CSR) have become increasingly prevalent and important. Professor Mark Schwartz, a colleague of mine at York University (for more on this subject, I recommend reading his book *Corporate Social Responsibility: An Ethical Approach*[59]), raises the following questions in the introduction to his book:

- What should a firm's responsibilities be toward society?
- Should firms merely maximize profit while obeying the law?
  - Or do firms possess additional ethical, or even philanthropic (i.e., charitable), obligations toward society?
- What are the key arguments supporting a more narrow (i.e., profit-based) approach to CSR, as opposed to a broader (i.e., beyond profit) approach?

These are important questions, especially since most of the critical infrastructure in many countries belongs to the private sector; in the US and Canada, it is over 85%.[60] Citizens depend upon these systems for their health and welfare, and therefore they should be considered public goods. What are the ethical implications for private sector companies that own public goods, in terms of the obligations of companies and the government that regulates them? Trends toward privatization, as well as our increasing dependence on the systems that support us, make this issue increasingly important.

CSR focuses on the notion that corporations have obligations beyond creating profit, although certainly some disagree. These obligations integrate social, environmental, and governance issues into the practice of business.

STUDENT EXERCISES

Consider the questions above, for the case of a hydro-electric company, privately owned, supplying the only source of power to an isolated community that is totally dependent upon it for energy.

9.5 Ethics and the Construction of Risk—a Reflection

The discussion of ethics often centers on decisions that individuals make; however, these decisions are almost always made within a larger context that is a function of culture, geography, and institutions, among other things. People can be moral and make ethically based decisions, but is it reasonable to evaluate an institution as being either moral or ethical, when it is simply a thing? I am not sure of the answer to this, but what is clear to me is that there are important structural issues related to institutions when it comes to ethical decision making.

One of the most important structural issues that lies at the root of the disaster problem is the social, geographic, and economic gap between those (both people and institutions) that benefit from risk construction and those that suffer from it. There are innumerable examples, such as child labor or pollution-creating industries (the disaster at Bhopal, India in 1984 is a perfect example of this). My observation is that those who walk the halls of power rarely live in dangerous polluted locations or have children who work in factories. Their perception of risk is fundamentally different than other, less fortunate segments of society.[61] In his book *Collapse*,[62] Jared Diamond gives an interesting discussion of how flood management in the Netherlands has *not* fallen prey to this social trap, because they are all down in the polder together.

In most places in the world, this risk gap has been increasing as a result of globalization; one sign of this is the change in the difference in wealth between the upper and lower classes. Ethical disaster management means explicitly incorporating ethical decision making by individuals and emergency/disaster management organizations, but unless the larger-scale structural issues are also addressed, progress in this direction will be limited. Of course it is important to have ethically informed preparedness and response, but it is more critical to have ethical risk construction. Structural problems create social traps[63] such as the tragedy of the commons,[64] extreme poverty, environmental degradation, and rigid institutions, all of which impede ethical decision making and destroy resilience.

> **STUDENT EXERCISES**
>
> *Consider the case of a community working to increase its resilience. How might ethical decision making be of relevance?*

9.6 Conclusion

Understanding ethical considerations is fundamental to understanding disasters. Ethics underlies every aspect of disaster management, from the construction of risk to response and recovery. There is not one right answer when it comes to morals and ethics, but what is necessary is that ethical considerations be a part of the process used to manage disaster risk. It is unfortunate that ethics has thus far been insufficiently present.

> **STUDENT RESEARCH PAPER**
>
> - Research the case of nuclear contamination at Hanford, Washington, USA, and analyze it from an ethical perspective (Figure 9.14).

FIGURE 9.14 Nuclear Reactors at Hanford, along the Columbia River in January 1960. *Source: U.S. Department of Energy.*

9.7 Example of an Ethical Dilemma: Temporary Settlement versus Permanent Housing

Esra Bektas[65] discusses the ethical dilemma faced by government after an earthquake in Turkey devastated the city of Duzce in 1999. There was an urgent need for shelter (Figure 9.15); at the same time, providing shelter in the short term created long term negative impacts in the city, such as increasing informal settlements and slum areas. Experience has shown that most temporary housing settlements exist far beyond their expected life and can become permanent in everything but name. This is a specific example of the Samaritan's dilemma,[66] which is about the possible adverse effects of altruism by sabotaging positive adaptive behavior. A psychologist might call this "enabling." It is important to remember that altruism also has positive effects (which, of course, is why we do it), by providing tools for disaster victims to recover and become productive and self-supporting.

FIGURE 9.15 Earthquake Damage in Duzce, November 12, 1999. *Source: Ansal, A., Bardet, J. P., Barka, A., Baturay, M. B., Berilgen, M., Bray, J., Cetin, L. Cluff, T. Durgunoglu, D. Erten, M. Erdik, I. M. Idriss, T. Karadayilar, A. Kaya, W. Lettis, G. Olgun, W. Paige, E. Rathje, C. Roblee, J. Stewart & Ural, D. (1999). Initial geotechnical observations of the November 12, 1999, Düzce earthquake.* A Report of the Turkey–US Geotechnical Earthquake Engineering Reconnaissance Team. *NSF-sponsored GEER Report #003, November,* http://www.geerassociation.org/GEER_Post%20EQ%20Reports/Duzce_1999/Report-Duzce.htm.

The provision of short-term aid can save lives and alleviate suffering for victims immediately after a disaster, and satisfies a Kantian ethical imperative of helping those in need. Providing shelter for those who have lost their homes in a disaster can be viewed as an intrinsic good, and is supported by Article 25 of the UN Universal Declaration of Human Rights.[67] There can be no doubt as to the needs of disaster victims; consider the words of Mrs Roseline Ogwe, a widow with three children who lives in Nigeria and suffered through a flood disaster:

I lived in an old house left for us by my father and that house has been washed away; the little money I realised from the sales of my cassava and cocoyam had been spent here in the camp because the aid is not regular,

Roseline Ogwe

Another widow, Mrs. Ellami Philip, and her six-year-old grandson said she managed to escape from being trapped by floodwaters in her town of Abua through the help of some youths who ferried them across to safety.

I did not have any time to harvest the little crops I have in the farm; everything has been washed away. I have not gone back to see what has happened to my house, box of clothes, and other belongings.
I am seeing an entirely different lifestyle here but my worry is what happens after now. We will not stay in this camp forever; how do we start again?'

Ellami Philip[68]

Needs can be overwhelming, as can be the scale of disaster. The following description from the CARE web page about a humanitarian crisis in the Horn of Africa is an example (Figure 9.16):

FIGURE 9.16 Refugee camp for Rwandans in Kimbumba, eastern Zaire (current Democratic Republic of the Congo), following the Rwandan genocide. *Source: Center for Disease Control and Prevention.* http://www.cdc.gov/nceh/ierh/Gallery/Zaire%201Lg.jpg.

Chronic food insecurity has spiraled into a massive humanitarian crisis in the Horn of Africa, where today more than 10 million people are in acute need of assistance. The situation, affecting large parts of Djibouti, Ethiopia, Kenya, Somalia and Uganda, is only expected to deteriorate at least until next year, with some areas experiencing the worst drought in 60 years and no sign of resumed rains in sight.
Some 1300 refugees a day, the vast majority from war-torn Somalia, are pouring into the Dadaab refugee camps in Kenya— now the world's largest such site, with almost 400,000 displaced people in three camps originally designed for 90,000.[69]

It has also been shown, however, that aid (depending upon how the aid is delivered) can increase vulnerability by creating cultures of dependency through the creation of disincentives to autonomy and sustainable development. The anticipation of charity and aid can "crowd out" preventive actions by local governments and peoples. This argument underlies much of the critique by Alex de Waal of famine relief in Africa.[70] He argues that long-term international relief efforts have undermined the political processes of governments in Africa, which have abdicated responsibility for preventing famine in favor of foreign experts who tend to adopt technological solutions that do not address underlying social causes.

From the Samaritan's point of view (s)he must commit to help, which provides an opportunity for both passive and active exploitation by recipients, whether they be individuals or governments. This is an example of a moral hazard. In an extreme case, this can be catastrophic for a recipient country, since it can lead to a spiral of disempowerment and dependency. Unless the capacity and commitment of the Samaritan is endless, the outcome may ultimately be very harmful. It is always a question of degree, however, and in this sense the Samaritan's dilemma is a poor representation of the true choices with which donors are faced. The Good Samaritan cannot ignore victims, but can make decisions regarding how much aid to provide, and in what form. This is what Buchanan[71] called strategic courage.

Kantian ethics suggest that short-term aid must be provided; otherwise potential donors are allowing death and suffering when they have the ability to prevent or mitigate them. In this case, the ethical question is not whether aid should be provided but, rather, how much aid and in what form. Utilitarian ethics suggest that aid (to a greater or lesser degree) should be withheld in order to force adaptive responses on the part of recipient nations and peoples, which leads to a greater good—eventually. Of course, none of us can predict the future precisely, and it may well be that withholding aid would create more harm than its provision. Some level of capacity and resilience must be present in the affected nation and population in order for the withholding of aid to be a reasonable ethical strategy. A corollary of the utilitarian position might mandate supporting the development of local capacities. One risk of the utilitarian argument is the withholding of aid for the supposed good of the victims, but which may really serve selfish ends of potential donors.

Of course, real-life situations are generally not so black-and-white. As noted previously, much depends on how aid is delivered. Aid that supports local development is helpful, whereas aid that increases debt, harms local industry, is not culturally appropriate, or serves as a mechanism to increase the wealth of companies from donor countries is probably harmful. In practice, this issue can be largely addressed not by the reduction of aid but, rather, by the rethinking of strategies and the creation of more effective aid programs. Such strategies should incorporate procedural ethics by using a transparent and fair system of decision making that involves recipients in deciding their fates. A paternalistic approach is more likely to fall into the trap that this dilemma addresses.

9.8 Jean Slick on Ethical Dilemmas

Jean Slick, who teaches disaster management at Royal Roads University and spent many years working for the International Red Cross, discussed, in an interview with me, three cases of ethical dilemmas that she faced during her operational years. The first two were during the Red River floods in Manitoba, Canada of 1997 and the Kelowna, British Columbia fires of 2009, and revolved around the issue of how to apply a formula for disaster financial assistance. The formula included the variables of loss, need, debt, income, and age. The dilemma involved the conflict between socialist and libertarian ethics. Aid given to impoverished victims creates moral hazard by sometimes rewarding those who have not worked hard to accumulate financial resources, while creating a relative disadvantage for those who did. Should those who were living in substandard housing be given an equivalent dwelling as a result of disaster aid, or should it be upgraded to what Canadians consider a minimum acceptable level? How should squatters be dealt with in terms of shelter assistance? Eventually it was agreed that the government had a moral obligation to provide improved dwellings when it was previously substandard, and would do so.

One of the tensions in this decision making existed because of the merging of two ethical norms related to the aid that was being provided. Disaster aid is based upon a principle of helping those who have suffered, through no fault of their own, as a result of disaster, and returning them to their pre-disaster state as much as possible. This does not necessarily provide aid to those who are at most need (victims may be quite wealthy), such as the urban homeless. The ethical norm of charity is based on helping those in need who suffer in normal times. When some victims fell into both categories, the ethical basis for providing aid became muddled. If funds came from a disaster financial assistance source, is it ethical to provide them to people based on the norm of charity?

After the Indonesian tsunami of 2004, the Red Cross was involved in developing criteria for rebuilding. Issues of whether to rebuild in the previously occupied at risk locations or whether to move to new locations involved some difficult trade-offs between future risk versus current needs. A complicating factor was that if communities were moved, livelihoods were affected, and no government compensation was provided. Should those decisions rest with the community, or should a paternalistic approach be taken? The Red Cross does have an ethical duty to their donors to use the funds wisely, and a community-based approach could recreate morally repugnant risk for future generations.

9.9 Commentary by Naomi Zack[72]

I agree with Professor Etkin that ethical reasoning is of primary importance in disaster. A situation, question, or decision is an ethical matter if human well-being or life is at stake, and disaster certainly has that aspect. I also agree that we need a plurality of ethical theories to draw on in thinking about disaster. What this means is that even experts will disagree, and events may prove that a decision was mistaken. However, in practical matters, people are judged to behave ethically if, at the time they chose a course of action, the available evidence supported their beliefs about the facts of the situation, and their choice of action was ethical. For example, if the evidence suggests that an earthquake has or will soon occur at sea near coastal development, the right thing to do is to sound alarms for a tsunami and immediately activate evacuation plans. That is an ethical choice of action, because it will save lives. However, there may be factual disagreement about the likely force of an imminent earthquake, and, in that case, given evidence and beliefs about the cost and dangers of evacuation, those who believe the earthquake will be minor and not result in a tsunami may make an ethical choice in not activating evacuation plans.

Behaving ethically is something that each individual does voluntarily and some are more motivated to do the right thing than others. However, people in positions of authority, especially government officials in democratic societies, are accountable for their actions involving human well-being. So even if ethics does not matter to them personally, it is in their professional self-interest to behave ethically. This requires that they have some knowledge of the ethics of the situation they are in and have deliberated on their choices of action, including not acting. Officials in positions of authority need to do the right thing, which should appear to others to be the right thing to do—and for which they can give good reasons for their choice of action, both before and after a disaster event. The right thing to do in one set of circumstances may be the wrong thing in another. For example, it would be unethical to give children the same medical doses as adults, because it is commonly known that children need smaller doses. The ethical rule of helping or treating people is not thereby relative, but merely applied differently as those receiving help or treatment differ.

All too often in disasters, as well as emergencies, members of the public experience the actions of government authorities as either too overbearing and intrusive, or neglectful in failing to provide prompt and effective aid, both of which occurred in the aftermath of Hurricane Katrina. That government may disappoint disaster victims means that individuals, families, private and social organizations, as well as community groups need to take responsibility on their own for disaster preparation and response. In fact, there is a strong history of private individual aid and rescue efforts; for example, immediately after the Boston Marathon bombing in April 2013, runners and others rushed to the aid of victims, with little concern for their own safety.

Members of the public are concerned mainly with their own and others' safety in times of sudden upheaval, while the government may be more concerned with preserving its own continuity and imposing order. Professor Etkin mentions my discussion in *Ethics for Disaster* of the difference between safety and security. The precedence of security concerns over safety concerns following the September 11, 2001 terrorist attacks (9/11) in the United States, evident in an All-Hazards response to disaster under the Department of Homeland Security, suggests that the ethics of keeping people safe now takes a back seat to the logistics of apprehending wrong doers.

It may be a broad truth about human nature that people get far more excited about deliberate acts of violence than other kinds of risk. Thus, while fewer than 3000 people died in the 9/11 collapse of the twin towers of the World Trade Center, more than 10 times that number die each year in American vehicular accidents. The risk of automobile accidents is factored into the institution of insurance and considered acceptable in contemporary society. But the risk of death by terrorist attack, which is a comparably small risk, is far more motivating for garnering resources and engaging in extensive preparation, mitigation, and prevention.

Theories and practices of safety, which take into account the safety of vulnerable populations such as the poor and elderly, women and children, and some racial and ethnic minority groups, have not yet been fully developed and applied. Neither has there been transparent public discussion of the ethics of favoring the rescue of those deemed more important to the community, or the enrichment of some government contractors in disaster events. Democracy demands such development and transparency. Instead, what prevails is a crude form of utilitarianism, expressed by the slogan, "Save the Greatest Number!" But that slogan, which masquerades as an ethical principle, can be a threat to the safety of many if it is applied before the fact of emergency, by planning to sacrifice some members of a community. On the ground, in the heat of the moment, saving the greatest number may be the best ethical decision, given limited personnel and resources. However, planning to save the greatest number beforehand is a poor ethical principle when that "greatest number" would be far greater if there were more adequate preparation, better budgeting, and more foresight in building and development (Professor Etkin has aptly considered mitigation in terms of zoning policies and risks of flooding), It is also unethical to decide beforehand who will be "sacrificed" without informing or consulting those victims.

Government officials cannot be forced to behave ethically, which requires virtues of loyalty, integrity, benevolence, and compassion. But these virtues of safety, which are calmer, quieter, and more modest than the currently lauded heroic security virtues of bravery, ferocity, and pride, are like those more spectacular virtues in what it takes to acquire any virtue. Aristotle has left us with the enduring wisdom that a person acquires a virtue by performing acts that are acts of the virtue in question, by acting for the right reasons, and by knowing what one is doing as the end result of a process of deliberation and reflection. For example, one becomes courageous by doing courageous acts, because they are courageous, in full awareness of the risk involved. The practice of virtues that ethically preserve life and respect all individuals in disaster situations could strengthen the safety virtues in some individuals and gradually change the values of an entire society, resulting in new heroic ideals and role models—to take the place of the action figures and covert operators who are so admired in contemporary American entertainment. Getting it right in how we deal with disaster is an ongoing ethical process, a collective struggle with the highest stakes.

End Notes

1. *Disaster Chaplains – Providing Spiritual Care in Times of Disaster.* http://cretscmhd.psych.ucla.edu/nola/Video/Clergy/Articles/Ecumenical/Practice_Scope.pdf.
2. Jenson E., Disaster Management Ethics. UNDP Disaster Management Training Programme, (1997), http://iaemeuropa.terapad.com/resources/8959/assets/documents/UN%20DMTP%20-%20Ethics.pdf.
3. Note the Health Canada scandal, Whistleblowers go public on threat of bovine growth hormones, http://www.cbc.ca/thecurrent/episode/2011/11/22/whistleblowers-go-public-on-threat-of-bovine-growth-hormones/.
4. See Cragg W., and Schwartz M., "Historical Injustice: Lessons From the Moose River Basin," *Journal of Canadian Studies* 31, no. 1 (1996): 60–81 for a case study of a similar situation.
5. An example is the Family Refugee Program, funded by Canadian provinces, to assist refugee families with young children. http://www2.news.gov.bc.ca/news_releases_2009-2013/2012JTST0019-001554.htm.
6. Milgram S., "Behavioral study of obedience," *Journal of Abnormal and Social Psychology* 67, no. 4 (1963): 371–378. Interesting Audio clips of participants in the Milgram experiment can be found at http://www.simplypsychology.org/milgram.html.

7. Boss J. A.,.*Analyzing Moral Issues* (NY, NY: McGraw-Hill, 2005).
8. Canadian Medical Assocation, CMA Code of Ethics (2004), http://policybase.cma.ca/dbtw-wpd/PolicyPDF/PD04-06.pdf.
9. This is one example of the usefulness of graduate assistants!.
10. Canada, for example, both the federal government and provincial governments have disaster relief programs.
11. Issues around climate change are addressed in more detail in Chapter 3. Authoritative information on this topic can be obtained from a number of sources, including the Intergovernmental Panel on Climate Change (IPCC) website. Beware the many sites that provide biased and untrue information!.
12. Blodget B., "Mitt Romney: It's Immoral To Borrow Money for Disaster Relief," Business Insider: Politics (2012), http://www.businessinsider.com/romney-disaster-relief-immoral-2012-10.
13. Beatley T., Ethical Land Use: Principles of Policy and Planning (Baltimore and London: John Hopkins University Press, 1994).
14. Dunfee T. W., and Strudler A., "Moral Dimensions of Risk Transfer and Reduction Strategies," in *Managing Disaster Risk in Emerging Economies* (Washington, D.C: The World Bank, Disaster Management Facility, 2000).
15. The EU Life-Environment Project has an interesting set of case studies on the Wise Use of Floodplains. http://www.floodplains.org.
16. In the United States, the No Adverse Impact (NAI) floodplain management approach also supports the use of floodplains in such a way as to avoid flood damage. For more details visit the webpage of the Association of State Floodplain Managers. http://www.floods.org/index.asp?menuID=460.
17. Brown D. W., Moin S. M., and Nicolson M. L., "A comparison of flooding in Michigan and Ontario:'soft'data to support'soft'water management approaches," *Canadian Water Resources Journal* 22, no. 2 (1997): 125–139.
18. Zack N., Ethics for Disaster, (U.K.: Rowman & Littlefied Publishers Inc., 2009).
19. Ross W. D., The Right and the Good Edited, with an Introduction, by Philip Stratton-Lake (New York: Oxford University Press, 2002), rpt. of original 1930 edition.
20. *The Universal Declaration of Human Rights: United Nations.* http://www.un.org/en/documents/udhr/index.shtml.
21. See note 18 above.
22. Burb R. J., "Flood insurance and floodplain management: the US experience," *Environmental Hazards* 3, (2001): 111–122.
23. See note 17 above.
24. See note 14 above.
25. Ibid
26. FEMA. http://www.fema.gov/about-fema.
27. Public Safety Canada. https://www.publicsafety.gc.ca/cnt/bt/index-eng.aspx.
28. Klein N., *The Shock Doctrine. The Rise of Disaster Capitalism* (Vintage Canada, 2007).
29. Mitchell J. K., "The Fox and The Hedgehog: Myopia about Homeland Security in U.S. Policies On Terrorism," in *Terrorism and Disaster: New Threats, New Ideas*, ed. Lee Clarke (Research in Social Problems and Public Policy, Volume 11) (Emerald Group Publishing Limited, 2003), 53–72.
30. FireSmart Canada. https://www.firesmartcanada.ca/.
31. Buber M., I and Thou, translated by Ronald Gregor Smith (New York: Charles Scribner's Sons, 1958).
32. Buber believed that the expansion of a purely analytic, material view of existence was at heart an advocation of 'I-It'relations - even between human beings.

33. Homer Dixon T., The Ingenuity Gap: Facing the Economic, Environmental and other Challenges of an Increasingly Complex and Unpredictable World (Knopf, 1995).
34. Vanderburg W. H., The Labyrinth of Technology: A Preventive Technology and Economic Strategy as a Way Out (Toronto: University of Toronto Press, 2000).
35. Bradshaw L., and Slonsky L. B., The Real Heroes and Sheroes of New Orleans, SocialistWorker.org (2005) http://socialistworker.org/2005-2/556/556_04_RealHeroes.shtml.
36. Birsch D., and Fielder J. H., The Ford Pinto Case: A Study in Applied Ethics, Business and Technology (Albany: State University of New York Press, 1994), 315.
37. McCullough D., The Johnstown Flood, (Rockefeller Center, NY: Simon and Schuster Paperbacks, 1968).
38. Glover J., Humanity: A Moral History of the Twentieth Century, (Yale: Yale University Press, 2000).
39. Noddings N., *Caring: A Relational Approach to Ethics and Moral Education* (Univ of California Press, 2013).
40. Boss J. A., *Analyzing moral issues* (Mayfield Publishing Company, 1999).
41. See note 13 above.
42. The case of Easter Island is often used as a metaphor for how environmental degradation might result in collapse.
43. There is a nice case study of this in chapter 11 of Diamond, J. (2005). Collapse: how societies choose to fail or succeed: revised edition. Penguin.
44. *Ecosystems and human well-being: biodiversity synthesis* (Island Press, 2005).
45. Nadeau R. L., The Environmental Endgame: Mainstream Economics, Ecological Disaster and Human Survival (Rutgers University Press, 2006).
46. Meadows D., Randers J., and Meadows D., Limits to Growth: The 30 Year Update (Vermont, U.S.A.: Chelsea Green Publishers, 2004) 368.
47. Rees Sir M., *Our Final Hour: A Scientist's Warning: How Terror, Error, and Environmental Disaster Threaten Humankind's Future In This Century–On Earth and Beyond* (New York: Basic Books, 2003).
48. Lovelock J. E., and Cayley D., *The GAIA hypothesis* (CBC RadioWorks, 1995).
49. Brown P., and Garver G., Right relationship: Building a whole earth economy (Berrett-Koehler Publishers, 2009).
50. Daly H. E., For the common good: redirecting the economy toward community, the environment, and a sustainable future (No. 73), (Beacon Press, 1994).
51. Mileti D., Disasters by Design - a reassessment of natural hazards in the United States (Washington: Joseph Henry Press, 1999).
52. Gadgil M., Hemam N. S., and Reddy B. M., "People, refugia and resilience," in *Linking social and ecological systems: Management practices and social mechanisms for building resilience*, ed. Berkes F., and Folke C. (Cambridge University Press, 1998), 30–47.
53. Brown P.G., and Garver G., Right Relationship: Building a Whole Earth Economy and Restoring the Public Trust (San Francisco: Berrett-Koehler Publishers, 2009).
54. Parfit, Reasons and Persons. (UK: Oxford University Press, 1987).
55. Taylor P. W., Respect for nature: A theory of environmental ethics (Princeton University Press, 2011).
56. For example, the omnibus budget legislation before parliament proposes changes to the Navigation Protection Act, the Fisheries Act and the Environmental Assessment Act that weaken environmental assessments and allows development with fewer constraints. As well, the Hazardous Materials Information Review Commission is being eliminated, with its authority transferred to the Minister of Health; thus what was previously a civil service function is now political.

57. Corey J. A., Bradshaw, Giam X., Sodhi N. S., and Willis S., "Evaluating the Relative Environmental Impact of Countries" *PLoS ONE* 5, no. 5 (2010): e10440, doi:10.1371/journal.pone.0010440.
58. Code of conduct for the International Federation of Red Cross and Red Crescent Societies. http://www.ifrc.org/en/publications-and-reports/code-of-conduct/.
59. Schwartz M. S., Corporate social responsibility: An ethical approach (Broadview Press, 2011).
60. Public Safety Canada (2004). Government of Canada Position Paper on a National Strategy for Critical Infrastructure Protection, http://www.acpa-ports.net/advocacy/pdfs/nscip_e.pdf Gao (2006). Critical Infrastructure Protection Progress Coordinating Government and Private Sector Efforts Varies by Sectors' Characteristics. Report to Congressional Requesters, GAO-08-39.
61. This is discussed more in the chapter 3 on Risk. Of particular interest is the "White Male Effect" in risk perception.
62. Diamond J., Collapse: How Societies Choose to Fail or Succeed (N.Y.: Penguin Group, 2005) 575.
63. Platt J., "Social Traps," *American Psychologist* 28, no. 8 (1973): 641–651.
64. Hardin G., "The Tragedy of the Commons," *Science* 162, no. 3859 (1968): 1243–1248.
65. Bektaş E., A Post-Disaster Dilemma: Temporary Settlements In Düzce City, Turkey (Erasmus University Rotterdam, HIS, 2005).
66. Buchanan J. M., "The Samaritan's dilemma," in Altruism, morality and economic theory, ed. Phelps E. S. (Russell Sage Foundation, 1975), 71–85.
67. The Universal Declaration of Human Rights. United Nations. "Everyone has the right to a standard of living adequate for the health and well-being of himself and of his family, including food, clothing, housing and medical care and necessary social services, and the right to security in the event of unemployment, sickness, disability, widowhood, old age or other lack of livelihood in circumstances beyond his control." (1948), Source: http://www.un.org/en/documents/udhr/.
68. Ikey A., "What Next for Flood Victims after Relief Camps?" The Nation (November 14, 2012), http://thenationonlineng.net/new/featured/what-next-for-flood-victims-after-relief-camps/.
69. CARE (n.d.) Dadaab Refugee Camps: Update on the Crisis in the Horn of Africa. http://www.care.org/careswork/emergencies/dadaab/. accessed April 25, 2013.
70. De Waal A., Famine crimes: politics & the disaster relief industry in Africa (Indiana University Press, 1997).
71. Buchanan J. M., (1972): "The Samaritan's Dilemma," reprinted in: J. M. Buchanan, (1977): Freedom in Constitutional Contract, Texas A&M University Press: 169–185.
72. Naomi Zack received her PhD in Philosophy from Columbia University. Zack has taught at the University at Albany, SUNY, and has been Professor of Philosophy at the University of Oregon since 2001. Her latest book is The Ethics and Mores of Race: Equality after the History of Philosophy (Rowman and Littlefield, 2011). Zack's recent books are Ethics for Disaster (Rowman and Littlefield, 2009 and 2010), Inclusive Feminism: A Third Wave Theory of Women's Commonality and The Handy Answer Philosophy Book (Visible Ink Press, 2010). Zack's earlier books include: Race and Mixed Race (Temple, 1993); Bachelors of Science (Temple, 1996); Philosophy of Science and Race (Routledge, 2002); Inclusive Feminism (Rowman and Littlefield, 2005) and the short textbook, Thinking About Race (Thomson Wadsworth, 2nd edition 2006). Zack has also published a number of articles and book chapters and spoken widely about race and feminism. Her work on disaster ethics has been received internationally, including at UNESCO headquarters in Paris in 2011 and the United Nations University in Tokyo in 2012. She was invited to present the keynote address at a conference on Disaster Justice at the Faculty of Law and COST, University of Copenhagen, in February 2014. Zack's book in progress is A Theory of Applicative Justice.

10 Workshop on Principles of Disaster Management

It is easier to produce ten volumes of philosophical writing than to put one principle into practice.
Leo Nikolaevich Tolstoy (Figure 10.1)

FIGURE 10.1 Portrait of Leo Tolstoy (1887) by Ilya Efimovich Repin. *Source: Tretyakov Gallery, Moscow,* http://simple.wikipedia.org/wiki/File:Ilya_Efimovich_Repin_(1844-1930)_-_Portrait_of_Leo_Tolstoy_(1887).jpg.

CHAPTER OUTLINE

- 10.1 Why This Topic Matters 312
- 10.2 Recommended Readings 313
- 10.3 Why Are Principles Needed for Disaster Management? 314
- 10.4 The Complexity of Current Principles 318
- 10.5 Two Models: Clarifying Principles 319
  - 10.5.1 Principles Pyramid 319
  - 10.5.2 Principles Matrix 322

Ian Davis is a co-author for this chapter.

10.6 Tasks for Breakout Groups	323
End Notes	326

CHAPTER OVERVIEW

One of the most important applications of disaster theory is to articulate principles upon which management practices are based. Many sets of principles, which are often quite different from each other, have been published. It is not necessarily that they conflict, but rather that they emerge from different institutional and cultural contexts and are meant to serve different purposes. This workshop is an opportunity for students to define a context that is relevant to them and engage in a thoughtful discussion of what principles should underlie practice. The principles that are developed should not be ad hoc, but rather grounded in theory and social values.

KEYWORDS

- Disaster management
- Principles
- Workshop

10.1 Why This Topic Matters

A people that values its privileges above its principles soon loses both.

Dwight D. Eisenhower (Figure 10.2)

FIGURE 10.2 Dwight D. Eisenhower, 42nd President of the United States and war hero.

Principles are important for a number of reasons:

- *First, they allow organizations to create more coherent sets of policies of procedures.* These policies would assist institutions with different values and mandates to better understand and talk to each other. Beyond such discourse, however, if clearly defined principles are accepted and agreed upon between different organizations, then it is possible for genuine cooperation and coordination to occur on the basis of consensus.
- *Second, principles can provide an agreed upon and ethical basis for action.* It is essential to emphasize the ethical dimension in all aspects of disaster risk management since the lives of people and the viability of communities are at stake. Principles can assist in enabling decision makers to distinguish between relative and universal ethical issues. Ethical principles form the bedrock or platform to assist decision makers as they seek (or are reluctantly pushed) to become more accountable to beneficiaries, as well as increasingly transparent in handling their operations and managing their finances.
- *Third, principles are needed to guide various elements of disaster planning and implementation.* They can assist in the development of policy, strategy, planning, tactics, and actions on the ground as well as learning and adapting after the disaster. It is essential to undertake disaster planning in all countries; without guiding principles, disaster risk management can be little more than a directionless formality or knee-jerk response. There are an abundance of principles to guide disaster managers, and each of these relative or locally applicable principles can be tailored to suit an organization and its role. It is important to recognize that while some principles may be explicit and consciously followed, others may be implicit or subconsciously recognized and applied.

■ ■ 10.2 Recommended Readings ■

- Davis I., and Murshed Z., *Community-Based Disaster Risk Management*[1] (Bangkok: Asian Disaster Preparedness Center, 2006).
- IAEM Principles of Emergency Management[2]

■ ■ Question to Ponder ■

- *I have been a selfish being all my life, in practice, though not in principle.*
  Mr Darcy in Jane Austen's Pride and Prejudice.
    - Is it possible to be selfish in practice, but not in principle?

10.3 Why Are Principles Needed for Disaster Management?

> **STUDENT EXERCISE**
>
> - *Pick a partner and debate the following two quotes.*
>   - *Important principles may, and must, be inflexible.* Abraham Lincoln (Figure 10.3)
>   - *I am a man of fixed and unbending principles, the first of which is to be flexible at all times.* Everett Dirksen
>
>
>
> **FIGURE 10.3** Photograph of Abraham Lincoln, 16th President of the United States.

The *Oxford Dictionary* defines *principle* as a "fundamental truth as (a) basis of reasoning." Principles guide people's decisions and actions, policies and procedures developed by organizations, and laws and doctrines of political entities. The *Collins English Language Dictionary* further defines *principle* as "A general rule that you try to obey in the way that you try to achieve something." These definitions place emphasis on the implicit authority contained in a principle as a fundamental truth or general rule. Their purpose concerns practical action; thus principles exist to guide actions, achieve something, or define a way of acting.

The statement "Everyone has the right to life, liberty and security of the person and the right not to be deprived thereof except in accordance with the principles of fundamental justice," from the Canadian Charter of Rights and Freedoms, is one of principles (Figure 10.4).[3] The same is true for the following, from the United States Declaration of Independence: "We hold these truths to be self-evident, that all men are created equal, that they are endowed by their Creator with certain unalienable Rights, that among these are Life, Liberty and the pursuit of Happiness"[4] (Figure 10.5). If there is not a clear understanding and statement of principles, then there cannot be a consistent, cohesive, and embracing disaster management strategy or effective communication between different organizations. A further incentive to develop principles guiding disaster management[5] comes from external pressures exerted by donor governments and international financial institutions. In return for their support to developing countries needing grants and loans following disasters, they are

FIGURE 10.4 Canadian Charter of Rights and Freedoms.

FIGURE 10.5 John Trumbull's 1819 painting *Declaration of Independence*, depicting the five-man drafting committee of the Declaration of Independence presenting their work to Congress.

increasingly demanding from beneficiaries improved accountability for assistance and overall transparency of operations, especially in financial management. For these demands to be satisfied, shared ethical principles are needed to support policies and practice.[6]

Within the field of emergency and disaster management there are a plethora of principles[7] described in various books and organizational websites.[8] These principles purport to provide a guiding and enduring basis for how the practice of disaster management is pursued. Yet, a perusal of the various sets of principles reveals little convergence. Why is this? What are the implications of this diversity?

We suggest that divergence emerges because of three basic reasons:

- *The first relates to differences in fundamental values and organizational mandates.* For example, a nongovernmental organization (NGO), such as the Red Cross or CARE, with a strong focus on disaster assistance at the community level will not share all of the same values or purposes as, for example, the World Bank, which tends to work at international and national governmental levels, although disaster management is important to both. Their cultures are quite different—one rooted in humanitarian assistance and the other in a highly politicized economic environment where development has traditionally been viewed through the perspective of neoclassical economics. Other differences may relate to discipline. A meteorological agency may focus on technology and advance warning, whereas a development agency might focus on community sustainability.
- *Second, divergence exists because different people or organizations address disaster management from different operational perspectives.* An academic might be philosophical, a government agency strategic, and a relief-based operation tactical. As such, their principles, which should reflect their personal or organizational purpose, would look quite different though they might not be in conflict with each other. For example, the first of the eight principles from a philanthropy website[9] is "Do no harm," whereas the first principle from Erik Auf der Heide's[10] book *Disaster Response: Principles of Preparation and Coordination* is "Because of the limited resources available, disaster preparedness proposals need to take cost-effectiveness into consideration." These two principles bear little relationship to each other, though it is quite likely that the proponents of one would not object to the assertion of the other.
- *Finally, people or organizations may work in different parts of the disaster management spectrum* (e.g., mitigation, preparedness, response, or recovery). Each of these pillars has its own requirements that would result in varying concerns and strategies.

Beyond the more idealistic aspects of organizational mandates lies the often unstated tendency of organizations to ensure their own survival and growth, even at the expense of optimally assisting disaster victims. Numerous examples of this self-interest can be detected. For example, after the 2004 Asian tsunami, national and international agencies poured into the affected countries and embarked on energetic funding campaigns, often in competition with other agencies even though it rapidly became apparent to everyone in the relief system that there was a plethora of agencies present, well beyond local needs. It also was apparent that far more money had been collected than could possibly be managed given limited local capacities or available funding channels. In addition, there was a marked lack of cooperation between many of the hundreds of NGOs working to assist the disaster victims. Successive evaluations of this chaotic situation have highlighted the urgent need for some consensus to be reached with regard to agreed-upon guiding principles. This would enable agencies to better align their organizations. Without such cooperation one can expect more scenarios like the Sri Lanka NGO circus of the uncoordinated actions of hundreds of international and national NGOs, where each pursues their individual goals. The risk is of this pattern being repeated in future megadisasters that attract the attention of vast numbers of agencies. This competition between agencies exists not only for funding but also for media attention.[11] The tension between short-term gain and holding to principles is ubiquitous and often corrupting, as reflected in the quote by W. Somerset Maugham, "The most useful thing about a principle is that it can always be sacrificed to expediency."

The quote above by Maugham is highly cynical, and I much prefer the approaches of Dwight Eisenhower, who said "A people that values its privileges above its principles soon loses both," and Lester Pearson (Figure 10.6),[12] who said in his Nobel Laureate speech[13] that "one must never betray the principles on which the United Nations Charter is based" (referring to the work of negotiating on international problems). Other barriers to principled actions relate to political turf wars, such as during and after the Hurricane Katrina disaster in the United States. Few, if any, organizations are monolithic enterprises—competing agendas and internal priorities inevitably exist even in disaster situations.[14] These issues of agency self-interest highlight the continual need for guiding principles that assert the priority or primary mission of humanitarian agencies to be based exclusively on the needs of the affected community, rather than any other internal consideration. This was the motivation of the Good Humanitarian Donorship Initiative[15] and the Red Cross when they first promoted the International Code of Conduct[16] in 1995. Over 400 agencies have signed the code, meaning that they will seek to abide by its conditions or principles. The following two codes give a flavor of the overall focus:

FIGURE 10.6 Lester B. Pearson, August 11, 1944, 14th Prime Minister of Canada and winner of the 1957 Nobel Peace Prize. *Source: Toronto Star, The Ottawa Journal/Library and Archives Canada/e002505448.*

Code of Conduct No. 1.

The Humanitarian imperative comes first. *The right to receive humanitarian assistance, and to offer it, is a fundamental humanitarian principle which should be enjoyed by all citizens of all countries....*

Code of Conduct No. 2.

Aid is given regardless of the race, creed or nationality of the recipients and without adverse distinction of any kind. Aid priorities are calculated on the basis of need alone. *Wherever possible, we will base the provision of relief aid upon a thorough assessment of the needs of the disaster victims and the local capacities already in place to meet those needs. Within the entirety of our programmes, we will reflect considerations of proportionality. Human suffering must be alleviated whenever it is found; life is as precious in one part of a country as another. Thus, our provision of aid will reflect the degree of suffering it seeks to alleviate. In implementing this approach, we recognize the crucial role played by women in disaster-prone communities and will ensure that this role is supported, not diminished, by our aid programmes. The implementation of such a universal, impartial and independent policy, can only be effective if we and our partners have access to the necessary resources to provide for such equitable relief, and have equal access to all disaster victims.*

Given the high levels of agency staff turnover in international NGOs it is possible that initiatives such as the Good Humanitarian Donorship or the Code of Conduct may be totally unknown to new staff. In 2007, Ian Davis, coauthor of this chapter, then a consultant to one of the largest global NGOs (that was developing an international strategy to guide their global humanitarian program), noted a total absence of any reference to the Code of Conduct within a document on ethical concerns. This absence was despite the fact that this agency was one of the early signatories, agreeing to abide by the requirements of the code. Subsequent enquiries indicated that this was because key staff were totally unaware of the existence of the code and their own agencies' agreement to abide by its contents.

Thomas Drabek,[17] a professor at the Disaster Research Center at the University of Delaware,[18] presents another reason why the field of disaster management does not have a well-defined set of principles: there is no general theory that underlies it. He argues that there are aspects of theories, such as those coming from social constructionism, sustainable development, and vulnerability theory, that can and are used as a foundation of an emergency management theory, but it is still very much in a stage of development. Along a similar vein, David Alexander[19] notes that "Models and interpretations of disaster abound, but the phenomenon is so multi-faceted that a general theory of universal explanatory power is unlikely ever to be formulated."

10.4 The Complexity of Current Principles

The Parliament of Canada, affirming that the Canadian Nation is founded upon principles that acknowledge the supremacy of God, the dignity and worth of the human person and the position of the family in a society of free men and free institutions...

Preamble to the Canadian Bill of Rights

A 2012 Google Internet search using the phrase "disaster management" resulted in 8.7 million hits, whereas "principles of disaster management" resulted in 78,700 hits. Clearly, the phrase "disaster management" is much in use, though its principles are somewhat neglected in comparison.[20] To get a sense of the variance of stated principles, we analyzed 15 sources including various government and NGO websites and books. The stated principles varied greatly in number, perspective, and depth. Some were comprised of a few short statements, sometimes embedded in much longer documents (for example, the Republic of South Africa Disaster Management Bill[21]), whereas others went into considerable depth and were multitiered ("The Wingspread Principles: A Community Vision for Sustainability"[22] and "Gujarat State Disaster Management Policy"[23]). Some statements emphasized values and ethics ("South Asia: Livelihood Centered Approach to Disaster Management: A Policy Framework"[24]), whereas others were more management oriented (*Disaster Response: Principles of Preparation and Coordination*[25]). These examples support the notion that, in terms of principles, the field of disaster management lacks a cohesive approach.

The three examples shown in Table 10.1 illustrate some of these points. The first, taken from the Government of Canada, is managerial in context, reflecting responsibilities at different levels of society. There is nothing in this list that reflects normative values or ethics or how disasters should be coped with in terms of types of actions. The second, taken from the SPHERE Humanitarian Charter, is very different, emphasizing how people should live and act and the fundamental values that drive organizations. The third example, taken from Auf der Heide, is much more practically oriented, focusing on implementation strategies and error avoidance.

10.5 Two Models: Clarifying Principles

In view of the somewhat chaotic state of existing principles, we propose that the following two models be used to provide structure to the discussion.

10.5.1 Principles Pyramid

First, we propose a four-level hierarchy of principles (Figure 10.7). Level 1, the broadest, reflects the fundamental values and ethics that motivate behaviors. Level 2 is strategic, whereas level 3 is tactical. Level 4 deals with implementation. Levels 1 and 2 are broad enough that they should be generally applicable over a large range of possibilities. However, levels 3 and 4 become increasingly sensitive to local culture and legislation and are very difficult or impossible to generalize.

Level 1: Ethical, Core Value Principles, which relate to the underlying shared beliefs and concerns of organizations and of their mandate. Using a food metaphor, level 1 would relate to the ethics of food production (such as a human rights–based approach). An example would be the SPHERE principle in Table 10.1: "A right to a life with dignity."

Level 2: Strategic Principles are informed and based on the ethical principles articulated in level 1 (such as what actions to consider taking; why, where, and with what expected

Table 10.1 Examples of Principles of Disaster Management from Three Sources

Source	Principles
(1) Fact Sheet: Canada's Emergency Management System[26]	Emergency management in Canada is based on the following principles: 1. It is up to the individual to know what to do in an emergency. 2. If the individual is unable to cope, governments respond progressively, as their capabilities and resources are needed. 3. Most local emergencies are managed by local response organizations, which are normally the first to respond. 4. Every province and territory also has an Emergency Management Organization (EMO), which manages any large-scale emergencies (prevention, preparedness, response, and recovery) and provides assistance and support to municipal or community response teams as required. 5. Government of Canada departments and agencies support the provincial or territorial EMOs as requested or manage emergencies affecting areas of federal jurisdiction. From policing, nuclear safety, national defense, and border security to the protection of our environment and health, many federal departments and agencies also work to prevent emergencies from happening or are involved in some way in a response and recovery effort.
(2) Sphere Humanitarian Charter and Minimum Standards in Disaster Response[27]	We reaffirm our belief in the humanitarian imperative and its primacy. By this we mean the belief that all possible steps should be taken to prevent or alleviate human suffering arising out of conflict or calamity, and that civilians so affected have a right to protection and assistance. It is on the basis of this belief, reflected in international humanitarian law and based on the principle of humanity, that we offer our services as humanitarian agencies. We will act in accordance with the principles of humanity and impartiality, and with the other principles set out in the code of conduct for the international Red Cross and Red Crescent Movement and non-governmental organisations (NGOs) in disaster relief (1994). *1.1 The right to life with dignity* This right is reflected in the legal measures concerning the right to life, to an adequate standard of living and to freedom from cruel, inhuman or degrading treatment or punishment. We understand an individual's right to life to entail the right to have steps taken to preserve life where it is threatened, and a corresponding duty on others to take such steps. Implicit in this is the duty not to withhold or frustrate the provision of life-saving assistance. In addition, international humanitarian law makes specific provision for assistance to civilian populations during conflict, obliging states and other parties to agree to the provision of humanitarian and impartial assistance when the civilian population lacks essential supplies. *1.2 The distinction between combatants and non-combatants* This is the distinction which underpins the 1949 Geneva Conventions and their Additional Protocols of 1977. This fundamental principle has been increasingly eroded, as reflected in the enormously increased proportion of civilian casualties during the second half of the twentieth century. That internal conflict is often referred to as "civil war" must not blind us to the need to distinguish between those actively engaged in hostilities, and civilians and others (including the sick, wounded and prisoners) who play no direct part. Non-combatants are protected under international humanitarian law and are entitled to immunity from attack. *1.3 The principle of non-refoulement* This is the principle that no refugee shall be sent (back) to a country in which his or her life or freedom would be threatened on account of race, religion, nationality, membership of a particular social group or political opinion; or where there are substantial grounds for believing that s/he would be in danger of being subjected to torture.

(3) Erik Auf der Heide: Disaster Response: Principles of Preparation and Coordination[28]

1. Because of the limited resources available, disaster preparedness proposals need to take cost-effectiveness into consideration.
2. Planning should be for disasters of moderate size (about 120 casualties); disasters of this size will present the typical inter-organizational coordination problems also applicable to larger events.
3. Interest in disaster preparedness is proportional to the recency and magnitude of the last disaster.
4. The best time to submit disaster preparedness programs for funding is right after a disaster (even if it has occurred elsewhere).
5. Disaster planning is an illusion unless: it is based on valid assumptions about human behavior, incorporates an inter-organizational perspective, is tied to resources, and is known and accepted by the participants.
6. Base disaster plans on what people are "likely" to do, rather than what they "should" do
7. For disaster planning to be effective, it must be inter-organizational.
8. The process of planning is more important than the written document that results.
9. Good disaster management is not merely an extension of good everyday emergency procedures. It is more than just the mobilization of additional personnel, facilities, and supplies. Disasters often pose unique problems rarely faced in daily emergencies.
10. In contrast to most routine emergencies, disasters introduce the need for multi-organizational and multi-disciplinary coordination.
11. In disasters, what are thought to be "communications problems" are often coordination problems in disguise.
12. Those who work together well on a daily basis tend to work together well in disasters.
13. Disasters create the need for coordination among fire departments, law enforcement agencies, hospitals, ambulances, military units, utility crews, and other organizations. This requires inter-agency communication networks utilizing compatible radio frequencies.
14. Procedures for ongoing needs assessment are a prerequisite to efficient resource management in disasters.
15. A basic concept of triage is to do the greatest good for the greatest number of casualties.
16. Triage implies making the most efficient use of available resources.
17. Good casualty distribution is particularly difficult to achieve in "diffuse" disasters, such as earthquakes and tornadoes, which cover large geographic areas.
18. Effective triage requires coordination among medical and non-medical organizations at the disaster site and between the site and local hospitals.
19. Panic is not a common problem in disasters; getting people to evacuate is.
20. Inquiries about loved ones thought to be in the impact zone are not likely to be discouraged, but can be reduced or channeled in less disruptive ways, if the needed information is provided at a location away from the disaster area.
21. Many of the questions that will be asked by reporters are predictable, and procedures can be established in advance for collecting the desired information.
22. Newsworthy information will rapidly spread among news organizations and from one type of media to another.
23. The media will often withhold newsworthy disaster stories it feels would be detrimental to the public.
24. Local officials will have to deal with different news media in times of disaster than those with which they interface on a routine basis.
25. Adequate disaster preparedness requires planning with rather than for the media.
26. The propensity for the media to share information and to assume "command post" perspective facilitates the establishment of a central source of disaster information.

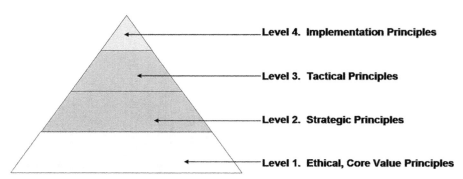

FIGURE 10.7 A four-level hierarchy of principles.

consequences?). Using a food metaphor, level 2 would be a nutrition guide. An example of this level of principle would be the principle in Table 10.1, "If the individual is unable to cope, governments respond progressively, as their capabilities and resources are needed."

Level 3: Tactical Principles deal with the practical outworking of the strategic principles. Using a food metaphor, level 3 would be a cookbook (such as how to adopt the agreed-upon strategy considering staffing, financial implications, etc.). An example of this might be the development of mutual aid agreements or the use of post audits to evaluate disaster response (such as occurred with Federal Emergency Management Agency after Hurricane Katrina).

Level 4: Implementation Principles are related to all the preceding levels: core values, strategies, and tactics (such as actions taken, as well as their monitoring and evaluation). Using a food metaphor, level 4 would be eating the meal as well as congratulating the cook or writing a letter of complaint to the restaurant. An example might the exchange of vulnerability and victim information between NGOs.

It is important to note that the process of creating principles is not a linear unidirectional one, but rather a process that requires continual feedback between ethical principles and how they are implemented. It is not just that theory informs practice—it is also the reverse. As a person or organization develops strategies, fundamental principles would have to be revisited on an ongoing basis, and how changes to values might affect higher levels of the pyramid must be considered. It is not just about creating documentation, but more about developing a process that engages staff and incorporates ethics and values in an ongoing way.

10.5.2 Principles Matrix

The practice and theory of disaster management depends on various factors, such as (if the comprehensive emergency management model is being used) which pillar of disaster management is being considered (mitigation, preparedness, response, or recovery) and the type, capacity, scale, and complexity of the disaster. Although underlying values are likely to be fairly robust, strategies, tactics, and implementation increasingly depend on these factors. For example, the mitigation of drought might include several levels of government working together to develop strategies to conserve water, develop crop insurance plans, and create incentives to switch to drought-resistant crops. In contrast, responding to terrorism might emphasize a command and control first responders approach. At larger scales of mitigation (for natural hazards in particular), environmental stewardship and sustainable development would be important to include, though not in the case of response to smaller-scale technological emergencies. A matrix methodology can be helpful in distinguishing between these factors (Figure 10.8). Although Figure 10.8 is based on the comprehensive emergency management model, others could be used.

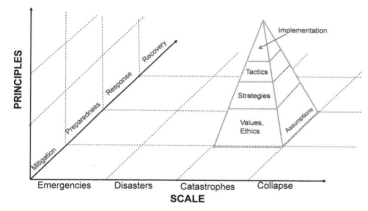

FIGURE 10.8 An example of how the pyramid discussed above might be slotted into a matrix model to help focus the development of principles. Similar figures could be constructed using different variables; type of disaster is the most obvious. For example, disasters that are rapid onset, well defined and understood, of natural origin, and of short duration would require a very different set of coping strategies than those that are slow onset, diffuse, ill defined, poorly understood, and of technological/human origin.

STUDENT EXERCISE

Develop a principles frame based on the pressure and release model (Chapter 6, Section 6.5.2).

10.6 Tasks for Breakout Groups

Reverence for life affords me my fundamental principle of morality, namely that good consists in maintaining, assisting, and enhancing life, and that to destroy, to harm, or to hinder life is evil.

Albert Schweitzer

The purpose of this workshop is to give students an opportunity to think about how ethics and principles relate to each other, to engage in a dialogue with other students and faculty, and to participate in an exercise to develop of a set of principles of disaster management for different scenarios. Each breakout group should select a chair, whose job it is to ensure that each member of the group has an equal opportunity to contribute to the discussion and that a consensus is reached. As well, each group should select a recorder, who will report back to the class at the conclusion of the workshop.

Constructing principles of disaster/disaster risk management is a complex task that should—if it is to be effective—involve an entire organization. A useful process must allow for discussion and should begin at a very fundamental level, one that defines world view and then moves increasingly toward a more detailed perspective. A three-step process can be used.

- *Step 1: Define a Frame of Reference.* This refers to a person's or institution's role as it relates to disaster management and both reflects and partly determines their values, moral code, and world view. Examples of roles are managing a government agency that provides disaster assistance, a business continuity manager for IBM, a victim without access to resources who cannot recover without help, or a Red Cross volunteer who responds to disasters. Of course, people in different frames of reference might share the same values, but it is not uncommon for them to approach disasters from a very different set of needs and perspectives; hence, the sort of postdisaster conflict that can arise between recovering victims and insurance companies.[29] In cases such as this the values associated with disaster relief can conflict with other important institutional values, such as profitability.
- *Step 2: Define a Purpose of Disaster Management.* Depending on philosophy, ethic, and job, different purposes seem possible. Three possible ones are listed below; more can certainly be constructed (the third item in the list is unlikely to be explicitly acknowledged but can play an important role within an organization's culture).
  - Minimize the loss, pain, and damage caused by disasters within the larger social context.
  - Minimize the damage caused by disasters while maintaining the structures of rights, power, and wealth within society, as well as the institutions that support them.
  - Provide jobs, careers, and pensions to people who work in organizations related to disaster management and ensure that these organizations are well funded.[30]
- *Step 3: Construct a Set of Principles.* Linked to the above, the principles should reflect the hierarchical structure and matrix models discussed above.

The discussion should begin with explicit statements of the nature of the social contract and moral theories that are chosen. Clear distinctions need to be drawn between descriptive ethics (what is) and normative ethics (what ought to be). In cases where rights and duties conflict with each other, a resolution process should be provided.

It is clear that different organizations will arrive at different results using the above process. There is no "correct" answer; in fact, engaging in the process[31] is probably more important than any specific set of results, as noted by President Dwight D. Eisenhower, who said "I have always found that plans are useless but planning is indispensable."

This workshop is more about process than content. The purpose of the breakout groups is to provide workshop participants with an opportunity to engage in process and to have a discussion about the various aspects of developing sets of disaster management principles. Below are four suggested frames of reference that can be used.

- *Save-the-Victim:* This is a nonprofit international NGO dedicated to disaster relief. It has been in existence for 32 years but has recently run into difficulties obtaining funding. Several years ago it was kicked out of a country in which it was running relief operations because of perceived bias in terms of who received aid. Senior management is rather upset with the Red Cross at the moment because of some conflicts with them during a couple of recent disaster responses.
- *Hall Mart:* This is a huge multinational company that sells a variety of goods to the general consumer. Hall Mart is very successful financially but has a poor reputation as a global citizen. It would like to enhance its image if it does not cost too much, and it is concerned about business continuity issues, particularly after Hurricane Katrina. Some senior managers are wondering whether there is a business opportunity in this field.
- *The Province of Riskario:* This is a medium-sized province in the country Scario. This is a very poor country that is deeply in debt to the World Bank. It is also a military dictatorship, and nobody in the government seems very concerned about disasters unless they themselves are affected. The few people who run the country are very rich, but there is almost no middle class. A group of revolutionaries is making it difficult for the government and army, and police corruption is rampant. Poor people are mostly afraid of the police and army and increasingly of the revolutionaries. Other than the army, there is no governmental organization that does disaster management. NGOs are reluctant to work there because several of their members were killed last year.
- *Canadian Disaster Development Division:* This is a federal government organization devoted to development issues related to reducing vulnerability to disasters in less developed countries. It is very bureaucratic (you just wouldn't believe the amount of paperwork!), somewhat underfunded, and subject to strong political pressures. It tends to be viewed by NGOs as being too politicized, given its position within the government. The professionals who work there are committed and idealistic but overworked and getting rather jaded.

When the working groups have completed their work and have reached a consensus, a reporter from each group will present to the class, outlining their principles and the arguments used to reach them.

The first principle is that you must not fool yourself—and you are the easiest person to fool.

Richard Feynman

End Notes

1. Davis I., and Murshed Z., *Community-Based Disaster Risk Management* (Bangkok: Asian Disaster Preparedness Centre Klong Luang, Thailand, 2006), http://www.docstoc.com/docs/134379040/Community-Based-Disaster-Risk-Management-_CBDRM_.

2. International Association of Emergency Managers, Principles of Emergency Management (2007), http://www.iaem.com/publications/documents/PrinciplesofEmergencyManagement.pdf.

3. *Section 7 of the Canadian Charter of Rights and Freedoms*, http://laws-lois.justice.gc.ca/eng/Const/page-15.html.

4. Jefferson Thomas, ed., Declaration of Independence: Jefferson's draft as amended and accepted by Congress (United States Library of Congress, June 1776).

5. In recent years these terms have been widely accepted. Disaster management refers to the postdisaster management of emergencies, whereas disaster risk management describes the proactive processes of risk assessment and risk reduction. Another term, disaster recovery management, referring to longer-term disaster recovery, is gradually being added to these descriptions as recovery secures belated recognition among policy makers and funding institutions.

6. ADB, OECD and Transparency International (2005). Curbing Corruption in Tsunami Relief Operations Manila: Proceedings of the Jakarta Expert Meeting organized by the ADB/OECD Anti-Corruption Initiative for Asia and the Pacific and Transparency International, and hosted by the Government of Indonesia Asian Development Bank, Jakarta, Indonesia, April 7–8, 2005, Organisation for Economic Co-operation and Development (OECD) and Transparency International.

7. CRHNet (2005). "The Principles for Disaster Management as a Key for Successful Management," in *Reducing Risk Through Partnerships*. Proceedings of the 2nd Annual Canadian Risk and Hazards Network (CRHNet) Symposium, Toronto, Canada, November 17–19, 2005, www.crhnet.ca.

8. For example: Alexander D., Principles of Emergency Planning and Management (Oxford University Press, 2002) or Eight Principles of Disaster Management, http://www.onphilanthropy.com/bestpract/bp2002-08-16.html.

9. *Eight Principles of Disaster Management*, http://www.onphilanthropy.com/bestpract/bp2002-08-16.html.

10. Auf der Heide Erik, Disaster Response: Principles of Preparation and Coordination (1989), http://orgmail2.coe-dmha.org/dr/Images/Main.swf.

11. Clinton B., Lessons for a Safer Future: Drawing on the experience of the Indian Ocean tsunami disaster. Eleven key actions for building nations' and communities' resilience to disasters (New York and Geneva: ISDR, 2006); Scheper E, et al., Impact of the Tsunami Response on Local and National Capacities (London: Tsunami Evaluation Coalition, 2006); Telford J., and Cosgrave J., "The international humanitarian system and the 2004 Indian Ocean earthquake and tsunamis," *Disasters* 31, no. 1, (March 2007): 1–28.

12. Prime Minister of Canada from 1963 to 1968, and winner of the Nobel Peace Prize in 1957.

13. Award Ceremony Speech for Lester Bowles Pearson by Gunnar Jahn, Chairman of the Nobel Committee, http://www.nobelprize.org/nobel_prizes/peace/laureates/1957/press.html.

14. Ian Davis was on the management board of an international relief agency during the 1970s and witnessed the forces of self-interest in action. He recalls some alarming boardroom discussions where the agency financial director would express the "need" for a major disaster to occur within a given financial year to produce the consequent influx of funds from agency supporters to ensure that staff redundancies would not occur. This was on account of a 14–20% allocation of administrative and handling charges that the agency deducted from every contribution and disasters provided the major "financial surges" needed to fill the agencies' administrative budget. Therefore, we as board members were faced with the blunt reality that if there were few disasters in a given year the agency had to loose staff and cut back on essential administrative requirements. However, some "creative accounting"

mechanisms were introduced by certain agencies to offset this risk by dubiously charging the salaries of home- or overseas-based aid administration staff as a project or donation item as a way to boost the administration top slice." Needless to say, loyal supporters of agencies were never informed about such subtleties as the agency "adjusted" (or manipulated) their supporters' contributions to meet the agency's internal requirements.

15. Good Humanitarian Donorship (2003), Principles and Good Practice of Humanitarian Donorship June 17, 2008, http://www.reliefweb.int/ghd/a%2023%20Principles%20EN-GHD19.10.04%20RED.doc.

16. International Federation of Red Cross and Red Crescent Societies: Code of Conduct, http://www.ifrc.org/en/publications-and-reports/code-of-conduct/.

17. Drabek T. E., "Theories Relevant to Emergency Management versus a Theory of Emergency Management", Journal of Emergency Management 3, no. 4 (2005): 49–54.

18. Academics working at this center, such as Quarantelli E. L., Russell Dynes, Thomas Drabek, Eugene Haas, and others, have been primary contributors to the sociological understanding of disaster. I highly recommend going to their websites and reading some of the papers they have published.

19. Alexander D., Principles of Emergency Planning and Management (Oxford University Press, 2002).

20. For comparison's sake, a search for lipstick through the same search engine resulted in 65 million hits.

21. Disaster Management Act (2002). Republic of South Africa, http://www.info.gov.za/view/DownloadFileAction?id=68094.

22. The Wingspread Principles: A Community Vision for Sustainability. Disaster-Preparedness: Building for a Sustainable Future, http://www.smartcommunities.ncat.org/wingspread2/wingprin.shtml.

23. Gujarat State Disaster Management Policy (GSDMP). Gujarat State Disaster Management Authority, http://www.gsdma.org/pdf/GSDMPolicy%20_revised%20on%2025th%20October_.pdf.

24. Ariyabandu M. M., and Bhatti A., A Livelihood Centred Approach to Disaster Management: A Policy Framework for South Asia Practical Action & Rural Development Policy Institute (RDPI), Kirulapone, Sri Lanka (2005), 101.

25. Der Heide E. A., and Irwin R. L., Disaster Response: Principles of Preparation and Coordination (Mosby, 1989), 363, http://www.coe-dmha.org/Media/Disaster_Response_Principals.pdf.

26. Emergency Management in Canada. Public Safety Canada, http://209.87.31.75/Community/Emergency_Management_in_Canada.pdf.

27. *The Sphere Handbook: Humanitarian Charter and Minimum Standards in Humanitarian Response*, http://www.sphereproject.org/handbook/index.htm.

28. See note 25 above.

29. For example, after the Hurricane Katrina disaster, victims launched a class-action suit against State Farm Insurance and American International Group regarding denial of claims or lack of response. A settlement was reached in January 2007 and amounted to hundreds of millions of dollars. http://www.msnbc.msn.com/id/16774511/ns/business-us_business/t/state-farm-settles-hundreds-katrina-lawsuits/#.UHxIJ1F7pTs.

30. Our thanks to Dennis Mileti for this suggestion.

31. This perspective was popularized by Dwight D. Eisenhower, who said, The plan is useless; it's the planning that's important.

11

Final Reflections

What it lies in our power to do, it lies in our power not to do.
Aristotle

Justice and power must be brought together, so that whatever is just may be powerful, and whatever is powerful may be just.
Blaise Pascal

CHAPTER OVERVIEW

Disasters need to be studied and understood in order to mitigate their impacts as much as is reasonably possible. They encompass all human dimensions, and are intricately bound up in the structures of society that determine the everyday flows of power and resources. We have a moral obligation to reduce the impact of such disasters, especially as those impacts become more devastating with the influence of climate change, urbanization, environmental degradation, system complexity, and globalization. There are a variety of models to use to improve risk reduction and resilience; each model has both strengths and weaknesses that must be considered, depending upon what problem is being addressed. It is unclear what the future holds, but an effort must be made to understand disaster so as to successfully combat and reduce its impact.

KEYWORDS

- Comprehensive emergency management
- Critical infrastructure
- Emergency plans
- Ethics
- Resilience
- Risk reduction
- Society
- Vulnerability

■ ■ Question to Ponder ■

Why, with all the knowledge and technology that has been developed over past decades, are disaster losses continuing to increase?

Disasters need to be studied and understood, because they are devastating events that shatter people, places, and communities. I believe that we have a moral obligation, both as individuals and as a society, to mitigate their impacts as much as is reasonably possible. This is becoming more urgent as climate change, urbanization, environmental degradation, system complexity, and increasing connectedness are creating a globalized system more prone to catastrophe. However, we are far from meeting this moral obligation.

Disasters are often thought of in too narrow a way; in fact they encompass all human dimensions. They become part of our social fabric and have a presence in our myths, stories, histories, and social structures. This presence is composed of three parts. The first and most common part is recognition. We hear, we see, and we experience; the result is an acknowledgment of the reality of disaster. This acknowledgment is especially virulent when we are ourselves affected, but is easily put aside when victims are distant in geography or personal connection. The second, and perhaps most difficult, part to resolve is cognitive dissonance. Disasters challenge our deeply held beliefs about the safeness of our world and our place in it, and can require us to shift our worldviews. Such shifts can be extraordinarily difficult; often avoidance and denial are used as psychological strategies to maintain our conceptual paradigms. Hence, the tendency is for people to rebuild in hazardous regions or to continue to adhere to failed risk reduction strategies. The third and most important part is the reflections of these disasters in terms of how we arrange our lives and communities, our policies, and the decisions made by societies to reduce disaster risk. Although disasters are not always learning moments, they do create windows of opportunity for change that are sometimes used to create a safer world.

Mankind has been dealing with disasters since time immemorial. How we have perceived and understood these events has changed according to cultural norms, which have varied from fatalism to a belief in the arbitrariness of a random universe. From simply being victims (in ancient times), we have increasingly changed our viewpoint to one that bears responsibility for outcomes.

Sometimes disasters happen because of unknown hazards such as an earthquake associated with an undiscovered fault line, a new virus, or a meteor strike. Others happen because we live in exposed and vulnerable communities, even when the circumstances that create them are well recognized. Often there are good reasons for this, for example, in order to have access to natural resources such as rivers, or because of a lack of choice. It is common, though, for people to voluntarily increase their risk beyond what might seem reasonable for reasons such as bounded rationality, denial, social traps, or the pursuit of short-term gain.

However, we also live in a society that is strongly differentiated, particularly in terms of power and wealth—and current trends are increasing disparities. The people and institutions that pursue gains can often achieve them only at the expense of the less gifted or unfortunate, or of future generations. A byproduct of this process is the creation of communities that are vulnerable and lack resilience, and in this sense disasters flow out of human activities in a very fundamental way. This is the reason why corruption and disasters are so closely linked.

Disasters are intricately bound up in the structures of society that determine the everyday flows of power and resources. Ken Hewitt described this in the following terms: natural disasters "depend upon the ongoing social order, its everyday relations to the habitat and the larger historical circumstances that shape or frustrate these matters."[1] Because of this, as technology, society, and the natural environment change and evolve, so do disasters. In addition to the traditional disaster landscape, new layers are being added that reflect novel threats that once did not exist; new pathways are being created that allow their destructive nature to affect more people ever further away; and greater levels of complexity frustrate our attempts at control.

Ultimately, any methodology or descriptive model of disaster must be embedded within a theoretical or conceptual meta-framework that organizes the thinking processes that select and categorize information. For this, risk (Chapter 3) is an extraordinarily useful framework, which is the reason why the UN ISDR has adopted disaster risk reduction as their paradigm. Risk is defined in a variety of ways by different disciplines and organizations. In particular, the empirical definition Risk = Hazard × Vulnerability has proved to be a useful model, and has assisted the shift away from the dominant hazards paradigm toward vulnerability and resilience perspectives. Creating effective risk reduction strategies requires understanding that much of risk is socially constructed, of recognizing the importance of biases in risk perception, of social traps, and the principle of risk homeostasis, as well as other theoretical considerations. Increasingly, as Ulrich Beck describes, in *The Risk Society*, society has organized itself around the concept of risk, although this is certainly not the only organizing concept. Power and wealth, for example, remain important organizing forces that often act to increase risk, and this process seems likely to continue as various aspects of hazard and vulnerability increase in magnitude over time.

Another extremely useful concept is that of resilience, previously used within the fields of engineering, ecology, and psychology (among others). There is a tendency, however, for organizations to "jump on the bandwagon" of resilience, to the neglect of other important aspects of disaster risk. Resistance and vulnerability are still essential concepts that should not be neglected as organizations embrace the latest "flavor of the day." Hazard, vulnerability, and resilience are all connected via a complicated set of dynamics that are incompletely understood. In particular, considering vulnerability and resilience as opposite ends of a spectrum leads to too simplistic a picture of how they relate to each other. Our goal should not be to minimize vulnerability and maximize resilience, but to search for optimal relationships between them.

The emergence of complexity theory is leading to a different understanding of the relationships among hazard, vulnerability, and resilience. It addresses the question, To what extent can disaster risk be managed? Chaotic systems exhibit the properties of emergence and self-organization, which are often seen in disasters and catastrophes. Management approaches must embrace these characteristics. In particular, Normal Accident Theory has been an important contribution to our understanding of risk assessment and what levels of risk may be acceptable. The more traditional notion that complex systems can be effectively managed in a top-down fashion must be discarded in favor of more sophisticated approaches that recognize its limitations. Big things have small beginnings, and the pathways that create them can be, and often are, unpredictable.

If our goal is to better understand and manage disasters, then the use of models is inevitable. There are a variety of disaster and disaster management models used by academics and practitioners, and every disaster management program is based on one of them whether or not it is realized. The two most common ones are variants of Comprehensive Emergency Management (prevalent in North America and Britain) and Disaster Risk Reduction (promoted by the UN ISDR). Not long ago I heard a discussion between two emergency managers on whether the comprehensive emergency management model really had four pillars or five (the fifth being prevention). It was a specious argument because they were confusing models with reality. No model is perfect—the better model is the one that serves some purpose to a greater degree. Practitioners and researchers are best served by having a set of models in their analytical toolbox, and the ability to choose ones that are most useful. Each model has its strengths and weaknesses, but one not commonly used—the First Nations Wheel—has useful strengths that are not present in the others, particularly the wisdom of elders and the moral basis upon which the model rests. This is one example of how the field of disaster management can benefit from cross-fertilization between different cultures.

Choosing an appropriate model is just a starting point. There are other barriers to having a good understanding of disaster and engaging in effective management strategies. Vested interests are one barrier that can actively work against risk reduction. Another is disaster myths, which are beliefs that empirical research have shown to be false. Many of these myths are propagated in the media and are extraordinarily persistent. Some are innocuous, but others can create great harm by delaying response efforts or leading to poor decision making. The myth that disasters are always accompanied by widespread looting and violence is particularly damaging, since it leads to an overemphasis on security issues as compared to search and rescue. This was evident in New Orleans after Hurricane Katrina. Lee Clarke's discussion of fantasy documents is also enlightening. Emergency plans can serve many purposes, which are not always for the purpose of effective emergency management—they may address other organizational, social, or political needs. In addition, there are other myths that are more like world views, which can bias risk estimations and lead to unduly optimistic behaviors. Some myths are easily dispelled through the use of empirical research, whereas others are more fundamental and extraordinarily persistent.

Academics in the field of disaster studies mainly use scientific approaches to understanding disaster. However, other ways of understanding are also important—hence the inclusion of the chapter on disaster poetry. Art has been used by people throughout history as a way to understand and cope with trauma and disastrous events, and provides opportunities for people to gain meaning not available in scientific approaches. Should we see a disaster as an impersonal damaging event, or as a "a powerful demon, Bent on destruction"?[2] As a component of disaster studies, poetry and other art forms have been largely ignored, but are deserving of far more attention.

It is difficult to underestimate the importance of ethics and values. Largely absent from emergency management literature, it is more prevalent in other areas of disaster studies, particularly ones dealing with humanitarian and health issues. Several ethical theories are particularly relevant to disaster studies, including utilitarianism, deontology, virtue ethics, social contract theory, and environmental ethics. There is no doubt that the practice of disaster management rests upon a foundation of ethics and values, even when they are not explicitly stated, and I expect ethics to become increasingly visible as an area of discussion within emergency management.

As disasters continue to morph because of the changing context within which they occur, the application of ethical theory will need to reflect these changes. One example of this is our increasing dependence upon critical infrastructures, which are mostly privately owned but can also be considered a public good. How should the social contract between government and citizens reflect this shift? Understanding disasters requires a multidisciplinary approach; they are political, cultural, geographical, ecological, physical, environmental, social, psychological, ethical, religious, and spiritual. They reflect the values and moral choices made by society, and I believe that it is in this realm that our greatest power lies.

I wonder what the future holds? There are many pathways available to us. Some seem likely to lead to global catastrophe, with a hostile climate, dying oceans, degraded ecosystems, and increased political strife as different regional societies compete for diminishing resources. Other pathways are sustainable. We certainly have the knowledge that we need to maintain a healthy and safer planet. All it takes is the choice to do so.

End Notes

1. K. Hewitt, ed., *Interpretations of Calamity: From the Viewpoint of Human Ecology*, vol. 1 (Unwin Hyman, 1983).
2. R. T. Gasperson, *What Can I do to Help Japan?* http://www.squidoo.com/haiku-for-2011-japanese-tsunami-charity. Accessed February 12, 2013.

Appendix 1
Selected Disaster Data

Disaster	Deaths	Deaths Per Million	GDP in Billions (USD)	Total Cost in Millions	Cost as % of GDP	Population in Millions	Total Affected Population	Affected (per million)
2000 Iceland earthquake	0	0.0	$8.7	$12.0	0.14%	0.3	91	324
2000 Botswana flood	3	1.8	$5.6	$5.0	0.1%	1.7	138,776	83,954
2005 Alberta (Canada) floods	4	0.1	$1133.8	$357.0	0.0%	32.3	5000	155
2005 Alberta floods	4	1.2	$201.2	$357.0	0.2%	3.3	5000	1512
2006 Namibia flood	5	2.4	$8.0	$8.5	0.1%	2.1	2100	992
2007 Gloucestershire/ Worcestershire UK flood	7	0.1	$2814.0	$4000.0	0.1%	61.0	340,000	5575
2007 Los Angeles (US) wildfire	8	0.0	$14,287.0	$2500.0	0.0%	301.2	640,064	2125
2007 Los Angeles (CA) wildfire	8	0.2	$1801.0	$2500.0	0.1%	36.6		17,512
2004 Bahamas cyclone Jeanne	9	28.6	$7.1	$550.0	7.8%	0.3	1000	3177
2008 Jamaica Hurricane Gustav	12	5.3	$13.9	$66.2	0.5%	2.3	4000	1764
2001 Nicaragua cyclone Michelle	16	3.1	$4.1	$1.0	0.0%	5.1	24,866	4830
2005 Cuba hurricane Dennis	16	1.4	$42.6	$1400.0	3.3%	11.3	2,500,000	22,214
2002 Morocco flood	80	2.7	$40.4	$200.0	0.5%	29.5	15,017	510
2001 Chad flash flood	100	11.5	$1.7	$1.0	0.1%	8.7	175,763	20,203
2001 Russian cold wave	145	1.0	$306.6	$0.1	0.0%	144.8	6120	42
2009 Saudi Arabia flood	161	5.8	$377.2	$900.0	0.2%	27.9	10,000	358

Continued

—Cont'd

Disaster	Deaths	Deaths Per Million	GDP in Billions (USD)	Total Cost in Millions	Cost as % of GDP	Population in Millions	Total Affected Population	Affected (per million)
2003 Turkey earthquake	177	2.5	$303.3	$155.0	0.1%	70.2	290,520	4137
2009 Australia wildfire	180	8.2	$991.9	$1300.0	0.1%	21.9	9954	455
2006 Chinese drought	184	0.1	$2712.9	$2910.0	0.1%	1311.0	180,000,000	137,299
2007 Sichuan (China) flood	535	0.4	$3494.2	$4425.7	0.1%	1317.0	105,004,000	79,730
2004 Phillippenes cyclone Willie	1619	18.8	$90.4	$78.2	0.1%	86.2	881,023	10,215
2005 Hurricane Katrina (US)	1833	6.4	$12,623.0	$125,000.0	1.0%	288.4	500,000	1734
2005 Hurricane Katrina (LA)	1833	451.0	$183.0	$125,000.0	68.3%	4.1	500,000	123,153
2003 Algerian earthquake	2266	71.0	$67.9	$5000.0	7.4%	31.8	210,261	6612
2003 French heat wave	19,490	324.0	$1796.7	$4400.0	0.2%	60.1		324
2011 Japanese tsunami	19,846	157.0	$5869.5	$210,000.0	3.6%	126.2	368,820	2921
2008 Wenchuan (China) earthquake	87,476	66.0	$4520.0	$8500.0	0.2%	1324.0	45,976,596	34,725
2010 Haiti earthquake	222,570	22,945.0	$6.6	$8000.0	121.2%	9.7	3,700,000	381,443

Appendix 2
Statistics Canada: Factors and Measures Related to Community Resilience

Statistics Canada has identified the following holdings as being relevant to the measures of community resilience.

Factor 1: Economic Health of the Community
- economic diversity
- household income and expenditures
- incidence of low income
- employment rates
- housing affordability
- home ownership
- median share of income
- share or employment by industry
- percentage of population with insurance

Factor 2: Community Access to Communication
- telephone (landline) coverage and usage
- cell phone access and usage
- internet access and usage
- television usage
- radio usage
- print media usage
- emergency information access/access to community announcements
- perceptions of trust in media/access to other means of communication

Factor 3: Sufficient Response and Recovery Capacity among Community Services
- number of police per capita
- number of fire fighters per capita

- number of medical personnel per capita
- shelters per capita
- hotel rooms per capita
- volunteers per capita

Factor 4: Community Socio-Demographic Characteristics

- sex
- education level
- age structure
- population density
- immigrant/recent immigrant status
- visible minority status
- Aboriginal status
- religion
- living arrangements
- languages (spoken, written)
- crime and social disorder

Factor 5: Transportation and Evacuation Capacity

- valid driver's licences per capita
- public transit availability and usage
- number of vehicles per household
- access to a household vehicle or public transit
- access to roads
- perception of reliance on others for transportation needs

Factor 6: Availability of Social, Civic, and Religious Organizations

- number of religious organizations per capita
- number of volunteer/nonprofit organizations per capita
- number of organizations with emergency response mandates per capita
- number of community sports, recreation or arts centers per capita
- number of community centers, resource centers, libraries etc., per capita

Factor 7: Health Status

- self-rated health and well-being
- prevalence of mental health issues

- obesity
- high blood pressure
- smoking
- alcohol use per 100,000 population

Factor 8: Activity Limitation/Degree of Assistance Required

- number of people with limitations/restrictions due to a mental or physical disability or condition, or age
- activities limited by physical or psychological conditions
- disability rates
- severity of disability

Factor 9: Neighborhood/Community Belonging and Social Cohesion

- sense of belonging to community/neighborhood
- trust in neighbors
- willingness to help neighbors

Factor 10: Social Resources/Social Capital

- proximity to family
- proximity to close friends
- frequency in communication with family
- frequency in communication with friends
- membership or participation in religious/social recreational groups

Factor 11: Sense of Civic Empowerment/Self-Efficacy among Community Members

- individuals sense of self-efficacy/ability to problem-solve/manage change

Factor 12: Generalized Trust

- trust in strangers
- willingness to help strangers
- overall perception of community safety and security

Factor 13: Civic Engagement/Participation

- involvement in local/community events/activities
- volunteering
- local political engagement
- voting

Factor 14: Household Composition

- household living arrangements
- relationship between household members
- average number of residents per household

Factor 15: Official Language Proficiency

- mother tongue
- knowledge of official languages

Factor 16: Literacy

- proportion of adult population with high school diploma or equivalent
- proportion of adult population who are functionally illiterate/low literacy levels

Appendix 3
Interviews with Ian Burton and Ken Hewitt

I am fortunate to have been mentored in my career by Professor Ian Burton and through him to know Professor Ken Hewitt. Both of these brilliant academics are giants in the field of hazard and disaster studies, and I am indebted to them for much of what I have learned. Both Ian and Ken, who are now retired, live in Toronto and agreed to meet with me to be interviewed about their perspectives on disasters. I provided them with a list of tentative questions before the interview; an edited transcript of their responses follows.

Interview with Ian Burton, January 2014

Etkin
Looking back over your research career in several fields related to disasters, what do you think were your most important insights?

Burton
Most of my research has been collaborative, working in a team with one or more academic colleagues and graduate students. A measure of the success of such collaboration is that it becomes impossible to identify the specific person or source of the idea. I also owe a debt for their help in stimulating new insights of many people in government, in international organizations and (nongovernmental organizations), and in the private sector, as well as a few political friends.

Etkin

Well, let's modify the question, then, to include collective insights.

Burton

I have been working on hazards, risks, and disasters for more than 50 years and continue to do so in my semiretirement. During that time there have been many new ideas and insights that seemed powerful in their day but which are now commonplace. For example, some of us at Chicago were able to demonstrate and explain how it is that the provision of flood protection through dams and dykes (levees) can often lead to an increase in flood damage. Engineering works can instill a false sense of security and encourage more development in flood plain lands, leading to more losses when the magnitude of the design flood is exceeded. Related are issues of risk taking and risk perception—the tendency to underestimate the likelihood of extreme events and a lack of effective management and governance.

Another insight concerns the benefit of finding and assessing all alternative courses of action. The disaster field has long been dominated by the natural sciences (studies of the physical processes that give rise to extreme events) and humanitarian response. To a remarkably large extent this social construction still exists. Follow any breaking news story about a disaster in the mass media and you can observe the unfolding of a familiar narrative. First we hear about the extreme event itself—the earthquake or tropical storm and its magnitude—and then we hear about the losses and damage, and the help that is being provided and promised from countries and places beyond the immediate area of the disaster. Rarely, if ever, is a question asked about the reasons for the magnitude of the losses and the suffering or why events of similar magnitude can result in heavy losses in some places and much lower losses elsewhere. It is commonly assumed that the natural event was to blame: There was an earthquake offshore and then a tsunami, and so many lives were lost in the "natural disaster." So disasters are "natural" or acts of God. A change in the media reporting of disasters is long overdue. In fact, it has been well understood in the research community for many years now that disasters are in fact caused by the everyday decisions of individuals, groups, and organizations over many years. As we say in a phrase attributed to Tony Oliver-Smith, "first the earthquake and then the disaster."

In the past few decades a lot has been learned (new insights, if you like) about how and why disaster losses are continuing to increase at a rapid rate in both rich and poorer countries. Broadly speaking, the increase in property losses can be attributed to the growth of the human population and the growth of the global economy. There are just many more people and more property in the path of danger. But this explanation is insufficient and misleading. We now have a much greater understanding of the natural trigger events (their magnitude, frequency, spatial distribution, and so forth) and a much better capacity to build hazard-resistant infrastructure and buildings (factories, hospitals, schools, residences), the availability of stronger building materials, and the capacity to adopt building codes and standards that minimize risk. We also understand well how to use land use planning to guide human settlements away from exposed places on steep slopes, coastal sites, and other higher-risk zones. Given all this increase in knowledge and capacity, why have disaster losses continued to grow at an increasing rate? A simplified answer is that the knowledge available to decision makers at all levels is not used properly or not used at all.

Almost every week there are news stories about disasters. When you dig into the backstory it is common to find that the available knowledge and technology was rejected as too expensive or not in the best "interests" of the planners, decision makers, or investors. In some instances a disaster can be attributed to corruption or to blind and short-sighted self-interest. Perhaps more often it can be explained by a decision-making process that gives more priority to growth and economic expansion than to safety and the less material values in the quality of life. In this way disasters are but one more symptom of unsustainable development or a rush to achieve short-term gain at longer-term risk.

Recognition of these circumstances has led to a growing movement in the research community to seek out "root causes" of disaster. Why does humanity persist in practicing unsustainable development? How might these practices be understood better and exposed more clearly in the media? And how might people become better educated about the consequences of today's decisions?

This leads me to one of the most fundamental insights. An unavoidable element in our human condition and circumstances is *uncertainty*. An elaborate analysis of disaster risk is now standard practice in many collective public and private decisions, but this does not and cannot eliminate uncertainty. It is quite impossible to factor uncertainty into a risk analysis in such a way that a clear and unambiguous result emerges. Ultimately, we face value choices in an ocean of uncertainty.

Let me add a short postscript to this rant about insights. The good news is that human beings are blessed with a capacity to learn from experience. We can record and analyze experience and share it. The bad news is that we easily forget the lessons of experience. Some institutional ways of correcting this could help—and here is a good result. Despite the increase in magnitude of disasters and disaster losses, the actual loss of life and rate of injury are going down. This may be attributed to greatly improved forecasting and warning systems and the capacity to evacuate people more quickly and effectively.

Etkin

The next question is about who has played an important role in this evolution of ideas. Which disaster researcher do you most respect?

Burton

I respect a very large number of them, and I hesitate to name some names and not others. I will name a few and will no doubt offend many other colleagues who should be included in this response to your question. My choice is guided largely by experience—people I have worked with. First and foremost are Gilbert White (sadly no longer with us) and Bob Kates, with whom I have worked for many years since we were together at the University of Chicago in the late 1950s and early 1960s.

I also have very much appreciated colleagues at the University of Toronto (in the Geography and Environment departments) and in Environment Canada (in the Corporate Policy Group and in the various climate research organizations). To name a few names I would like to mention Ken Hewitt, a highly respected and innovative scholar; likewise, not in any order, Terry Cannon, Ian Davis, Maarten van Aalst, Joel Smith, Kris Ebi, Allan Lavell, Ian Noble, Barry Smit, Olga Pilifosova, Andris Auliciems, Roger Kasperson, Anne V. Whyte, Derrick Sewell, Amos Tversky, Monirul Mirza, Saleemul Huq, Tony Oliver-Smith, and so many others. Perhaps, David, you would like a shorter list of people who might be considered the "giants" in the field. But I am not persuaded that we stand "on the shoulders of giants." Disaster research is a very closely knit, collaborative field of colleagues, and long may it remain so.

Of course there are many other players in the disaster field, and the area of research itself is expanding as new dimensions are added. One of the things we have to ask is why—considering that there is so much more knowledge now and a better understanding of the physical and natural sciences of the hazards and of the social and economic processes that lead to exposure and vulnerability—do losses continue to grow? Why has so much excellent research apparently had so little effect?

Etkin

Is that an assumption? If they had not done all that research, would we be in a much worse state?

Burton

That is a hypothetical question. There is no doubt in my mind that things could have been much worse without the increase in knowledge and understanding. The problem, in my view, has more to do with the lack of sufficient application of what we know.

Etkin

This leads me into the next question, about your degree of optimism or pessimism about the world's future over the next 60 years or so.

Burton

Of course it is very hard to make reliable predictions, especially, as they say, about the future! I see one big reason to be quite optimistic and many reasons for concern. I am optimistic about the human power and capacity for ingenuity and creativeness. The transformation in the human conditions of life for many people over the last 250 years is quite astounding. But I think that we are, perhaps, only at the beginning of this transformation process. The power of science and technology to change our lives is greater than ever. Major advances are to be expected, especially, perhaps, in the biological sciences, which hold out hopes for radical improvements in human health and disease control and make possible the massive expansion of food production that the world will certainly need in the next century.

All these past and future changes bring new risks as well as great benefits. My concerns have most to do with the requirement to apply enough wisdom in the management of the change. Our organizational capacity and governance, especially at the international level, seem to me to have fallen way behind our scientific and technical capacity. We live in a period of considerable pessimism, and there are many dire predictions of calamity. There is a widespread view that the present human path of development is unsustainable. We need a new vision for the future of humanity and a new paradigm for development. I believe that some needed features of the new direction are now becoming apparent. We need to change the idea of growth away from more and more consumption and use and pollution of the earth's resources toward nonmaterial growth. More and more evidence is showing that beyond a certain level of material comfort and security our sense of happiness or well-being does not increase. Happiness and sense of well-being do not depend on absolute value but on comparative value. This means that the new less material growth paradigm must be combined with a high level of equity. The gap between rich and poor, within and between countries, has to be drastically reduced. The pessimism of our times stems partly from the fact that we seem to be on a treadmill of addiction to material growth and a continuing increase in inequity. My reason for some optimism is that these errors and misdirection are now increasingly recognized. We have not yet reached the point of action, but a swing toward the new paradigm is surely coming. My concern is that it will not be strong enough nor come soon enough.

The growth in disaster losses can be seen as one symptom of our global situation. While there is much that can be achieved to contain the growth in losses, it cannot be accomplished within the confines of disaster risk management by itself but has to be part of a much larger change. The narrative of this book makes an important contribution to this goal.

Interview with Ken Hewitt, December 2013
I presented the following set of questions to Professor Hewitt

- If you look back over your career as it relates to disasters, what do you think was your most important insight?
- Who do you most respect as a disaster researcher? Why? If (s)he were with us now, and you could ask one question, what would that question be?
- What is your degree of optimism or pessimism about the world's future over the next 50 years? Why?
- Does it make sense to look for understanding of disaster in terms of grand narratives or universal underlying themes, or is the human experience too splintered for that? If yes, what are those grand narratives? If no, what are the implications of that for disaster managers?
- What advice would you give to a young, idealistic graduate student eager to study disaster management in order to better the world?
- If you were given the power to make one law in Canada, what would it be?
- How successful do you think international efforts in DRR [disaster risk reduction] have been?
- I once heard you comment, "What if this is the way it is supposed to be?" Can you explain what you meant?
- If you were me, what question would you ask Dr. Kenneth Hewitt?

Hewitt

My sense is I would have a dozen different answers to each question depending on context. It would range from whether the context is academic or institutional, whether it is in Canada or some other country, or whether it is global. This is an answer to the questions in the sense that I think that's more important than anything else—perspective, circumstances, who we are talking about, where we are operating, and what purpose we have in answering these questions. So, if you ask me who or what I consider to be the most effective if I was in the Himalaya, then I can think of a dozen men and women whose awareness of risk, whose actions to limit or avoid dangers, whose courage and effectiveness when damaging events occur seem to me to outshine any academic whom I know, including myself. And yet I know that none of them would have the foggiest idea of what to do if they were here, least of all would they be listened to even if they did have a good idea what to do in southern Ontario.

So context is really huge. No one said more about the importance of complexity than you have, so you know that part of complexity is that it is not just complex but it is very different depending on where you are coming from. So yes, I couldn't give you one person or one answer to most of these questions, but a spectrum of responses would suit me. Does that seem all right?

Etkin

Yes, that makes total sense. Well, let's pick a context, for instance, the North American context. You have had a big influence on this field transitioning from the hazard paradigm to a vulnerability paradigm. I think it probably would have happened in any case, but you certainly gave it a huge push. As well, the importance of the normal day-to-day processes in society as constructing vulnerability largely came from you. So would you say that is your greatest contribution? Do you think that is the insight that you think is most useful in terms of understanding disasters and how to manage them?

Hewitt

It certainly had a huge effect on my approach of doing things because I came into this field as an applied geomorphologist working with floods and glacial hazards. Those kinds of things still interest me and I do most of my work on them, but this sense that disasters are mostly about vulnerability (though not just about that) that is socially constructed became the theme. That society is actually the biggest actor in the creation of risk, rather than the planetary environment, is one of the important things I got from my early work with the natural hazards group, especially Gilbert White, Ian Burton, Bob Kates, and Myra Schiff. They forced me to think about these issues and it was, in many ways, their inspiration that led to the "Interpretations of Calamity" workshop and book. To the extent that I may have contributed to ideas related to social vulnerability, they came especially from other members of the multidisciplinary group who contributed to that book.

It is important to emphasize that the second and more profound relevance of "social construction" is in its stricter meaning of how thought and actions derive from and are shaped by "interests" and, usually, the more influential groups within the social order. I do not see vulnerability as something sitting there, some inevitable natural or social condition that puts some people more at risk than others. It is manufactured, maintained, magnified, or reduced and mainly by more or less continuous human actions.

Now this is an aspect many disaster specialists in North America are not too comfortable with. I mean, they see it as a sort of left-wing something or other. I don't, although some of the more forthright analyses have come from people who place themselves on the left of the political spectrum. However, I think well-informed people generally understand the fact and the intentional nature of political intervention in such matters (especially the "realists" in Ottawa, Washington, and Calgary or wherever risk is socially constructed). They and others go to great lengths to redistribute risk—to protect some and put it on others. This view does differ [from] the dominant view in North America prior to the emergence of the social vulnerability paradigm. To my mind we are again returning to the old "hazards" and "patterns of war" view, in Claude Gilbert's terms. With respect to our work in the so-called vulnerability paradigm, I feel that we've largely failed to get this idea through to the most influential institutions in society. In academia we talk about it a lot (those who are in disaster reduction and related fields), but it is much like the way high schools foster environmental and sustainability concerns. Almost every class and student seems really concerned about it, but society, or recent actions of the state in Canada, suggest the opposite. Likewise, it is my sense that, in the last 20 years or so, the disaster community, or the most effective institutions and well-funded actions, have gone back to a 1950s view: a hazards mind-set and a civil defense style of response focusing on emergency measures, funding, and threat management. In this sense I'm not feeling that we have been particularly successful.

Etkin

What would it take to be more successful than we have been so far?

Hewitt

Well, I think if, in fact, the Hyogo Framework for Action [HFA] for DRR, and the principles that the Red Cross and Red Crescent societies, Doctors Without Borders, and Oxfam follow, and that increasingly agencies like the British Overseas Development Agency subscribe to – *if* they were leading the way, then this would have a huge impact. Instead, what I find is that the principles are all in reports, conferences, and classes. Conversely, the major institutions and especially the major funding is poorly allocated (you know we talked about the recent [Overseas Development Institute] report on funding of DRR, showing singularly low levels and limited amounts going to the most vulnerable populations). To my mind HFA was the logical outcome of the vulnerability risk reduction perspective, which comes from institutions and meetings that have been strongly influenced by the sort of thinking I would like to be identified with. However, in the first decade of this century DRR has received about 20% of all the funding, usually divided up into little tiny packages. That being the case, we can't really say that things are going in a direction that is promising. Similarly with climate change. In fact, a lot of DRR funding has moved into the climate change adaptation arena but again [is] going to small local projects, islands, costal zones, and assessments for cities and so on.

Few people are talking about the only promising way for humans, which is, of course, to substantially reduce fossil fuel use and reorganize many activities that go along with massive industrialized, climate-impacting activities. And if your question is about what's actually happened to date, one must say there is much talk but little action that is productive.

Etkin

To what extent do you think the barriers or what hinders the more successful approach result from fundamental structures in society? By this I mean the way we organize power and wealth and the way risk is constructed in boardrooms and by major corporations and small groups of elite people. There are structural issues that create barriers, and as well there are corruption issues, too. This is different from the biased risk perceptions sort of issues that could presumably be corrected through more research, education, and this sort of thing. Do you think we can be at all optimistic about making progress in the face of those sorts of barriers, or would it require some sort of evolution, or revolution in society, to do it?

Hewitt

In every respect we have to talk about the emergence of the state and of large corporate enterprises as dominant organizations and how they administer their affairs in those structures. It is very difficult to change them, certainly to change them quickly. But of course that can happen—witness the almost complete turnaround of smoking in Canada in a much shorter time span than key changes in emissions could be managed. This question of optimism and pessimism is difficult. No one can grasp enough of what is happening to really know what is the overall balance of developments going on, not least of all because our awareness is so powerfully influenced by the media, by access to information, [by] our own main interests, and by the ability of these same organizations to control what is known. Some of us may know more than most people, but it is still not enough. We can certainly point to developments that show these things are not monolithic. Even the fact that we are talking about these things is because there are also laws, principles, institutions, values, and arrangements that actually go profoundly against what is happening. I mean, given the scientific consensus, how come there is no massive effort in conservation and fuel use efficiencies, eminently feasible for almost every area of fossil fuel use? Instead, *after* Kyoto, after Rio, things got worse. Is it really true that the structures are so inflexible? Or is there massive, deliberate resistance to the changes required? The effectiveness of climate change denial suggests the latter. We only have to look at what happens with wars to know that they can be changed practically overnight, and then we can look at a few things [such as] urban conflagrations, famine, and, again, smoking in wealthier nations to see that dramatic social changes affecting risk can be brought about quite quickly—and even against what elites and dominant spokespersons have said could happen.

So, optimism or pessimism is a very difficult area in which to operate, but as scientists or public servants, like most university teachers, it is our job to point out what has happened on both sides and, clearly, it seems irresponsible not to highlight the worst outcomes and scenarios or bask, as some seem to, in an optimistic fog: "If I'm alright, the world's alright." Then again, you can point to some areas where modern changes have been amazingly effective, such as public health or urban fire hazards. The history of these developments also testifies to how progressive ideas were blocked and subverted. As long as we've got this sense that disasters are not being driven by the planet, as long as we have the sense that climate change is not actually something out there that is being done to, not by, society, by social, technological, economic, political processes, then we know it can be changed. It may not be changing, but it *can be* changed—for better or worse.

Etkin

In the last election in the United States, Romney made the comment that—I won't get this exactly right—but that disaster financial assistance was immoral. Do you remember that? He said he didn't believe in the federal government giving disaster aid and he considered it immoral, which is a kind of libertarian perspective. Do you think that mind-set or the values that underlie that statement are a strong thread that runs through groups of people in western countries, maybe eastern countries too?

Hewitt

Well, I don't remember it, but my guess is it arose from a religious perspective; the history of religion includes leaders who blame disaster on sin, while success, including economic success, is due to virtue. Those unaffected by disaster have, supposedly, done the right things. I am guessing that is where Mr. Romney comes from. I have never subscribed to that view because I can see absolutely no evidence that those who suffer are the ones who are unusually sinful and survivors unusually virtuous. However, suppose we recognize that disaster is not some random, natural, out-of-the-blue, unstoppable force—though Romney may see it as an "Act of God"—but, in most cases, a consequence of the way society functions and organizes itself, then it is certainly valid to in some sense blame it on society's misdeeds. What is odd is to find Romney in the same school as Lenin. Lenin, at least prior to the October revolution, was against humanitarian relief after disasters on the grounds it is just to wallpaper that ultimately reinforces the economic conditions that brought the disasters about, while making charitable people feel better about it. I don't agree with that either, but it also seems to be where Mr. Romney is coming from.

However, all such comments seem beside the point when we see what is happening in corrupt, failed, other-directed humanitarian actions. So many reconstruction and rehabilitation projects involve funding for projects that state and other powers already wanted. Opportunism is a huge part of it. Anyway, I do think that a necessary—not necessarily "sufficient"—condition for successful humanitarian relief and reconstruction is where those at risk or surviving victims have a major say in what is done.

Etkin

We need people like you in positions of authority to make these decisions. How would you feel about taking on such a role?

Hewitt

No. While I appreciate that talk may be cheap, given the structures and divisions of labor in modern society, it is a gratuitous form of correctness to say we must always put our money where our mouth is! It is unreasonable to expect that the administrative and professional skills necessary to effect political or operational changes will develop from a life spent in research, writing, and teaching. We may have to recognize that our institutions are not set up to deal well with public, popular safety measures, least of all for those already marginalized within the world community, in any given society, city, or family. Successful long-term disaster reduction needs different, largely absent institutions working within, or in direct cooperation with, communities, in families and in neighborhoods as popular civil actions, with "experts"—let alone ministers or cabinets—listening and suspending judgment until balanced assessments are in. Responsible institutions must set the principles, the transparency and responsibilities. It would be a far cry from what now prevails, notably in homeland security and the like.

Etkin

If you look at it at that local level, I see what you are saying, but at the same time much of the risk that we see is constructed at much higher levels of society. Critical infrastructure spans countries and regions, so would we need different sorts of institutions to address those risks that are constructed at the larger scale?

Hewitt

My sense is that countries like Canada and the US, even Japan and China, have laws, principles, and so on in their constitution that should prevent this. The fact that they don't work is a problem. It is not so much a case of changing the way things are done as making sure that what is supposed to happen does happen. We hear this a lot—that recommendations from disaster inquiries and accident inquiries are not implemented—and that had those recommendations been in place most of the later disasters would not have happened. The Walkerton, [severe acute respiratory syndrome], and 2003 power outage inquiries all came to such conclusions. In most cases governments say yes, they will implement the recommendations, and yes, those issues really have to be addressed, but what then happens? At best a select few items are addressed, in the cases just mentioned, not the more critical ones. That is why we have high hopes of a free press instead of one that is listening to and mouthing what leaderships are saying, one that is doing inquiry and in-depth analysis, reporting all of the news and not just that which seems to be convenient.

Etkin

It seems like you would be in favor of an auditor general to monitor disaster risk reduction measures.

Hewitt

Given the setup that we have, yes, absolutely—an ombudsman and auditor general with powers and a high degree of independence. Yes, if they are credible. At least they would give the public the sort of awareness that other existing ones do, although it is by no means clear the auditor general gets the desired action. Bill Clinton's office did something like this on Haiti earthquake funding. It acted like an audit. And I believe the EU has changed its funding principles for Haiti and elsewhere as a result—in this case more cash and direct development assistance to Haitians themselves.

Etkin

When I was having lunch with you and Ian Burton, you commented, "What if this is the way it is supposed to be?" I think I know what you meant by it, but I wondered if you could explain it a bit more.

Hewitt

It is obviously about context. I suspect that what I meant was that many of those actions or inactions that we regard as failures are the result of deliberate decisions and are not necessarily "bad" for those making the decisions. An academic might say, "if only the government had been properly informed they would have done this or that." But is it not more likely they will usually do what they want to do or to please their more important constituencies?

Etkin

Your description of vulnerability as a kind of by-product generated from social processes, not something that is intended but rather something that just happens as a result of normal processes, such as the generation of wealth and power is powerful. From that point of view—I just have to think my way through what I'm trying to ask—that kind of perspective seems to lead to a very fatalistic view of disasters because they are so deeply embedded in the fundamental social fabric of who we are a species, that it leads to a kind of acceptance from an intellectual point of view, although you could be morally outraged. How would you respond to that?

Hewitt

I would respond in two ways. In one way you are right. Disaster is not only, or mainly, about disaster, and if your focus is strictly "disaster management" you are likely solving the wrong problems, which is always a prescription for fatalism. On the other hand, if this is the case, the only way to change it is to change those structures, and far from the argument being that you cannot change them—which would be fatalistic—it turns upon the assertion that they can be changed and, in key cases and for those least impacted by disasters, usually do change. Of course, you can only change a structure if you recognize that it exists. The hazards/patterns of war approach, mentioned above, does not even recognize the issue exists. The "emergency first" approach to disaster management never gets around to the DRR problem and so remains locked into what I once called "sustainable disaster." One example: Modernization, enforced migration, and land use changes in India have driven people into cities on a huge scale, where there are no real facilities or investment in the means to settle them safely. That was a major factor in the exposure and vulnerability of so many in the Indian ocean tsunami of 2004: flight to and concentration of people, labor, and urban slums in coastal zones. The processes that put people at risk and failed to offer them protections may seem a long way from what we usually think of as disaster reduction. That doesn't mean they cannot be changed, once you identify the risk relationship.

Etkin

So do you think the truth, if I can use that word, do you think it is not said?

Hewitt

I think mainstream media commentary, political statements, and many other high-profile views do not draw this connection with these typical at-risk conditions, which I argue disasters *reveal* rather than cause. The sense is that we are not dealing [with] something that arises from economics, or social, political, and legal decisions, but misfortunes to do with the planetary environment or with the human condition, not something that can be changed. Whether they are misled, or engaged in misleading the public, is another question.

Etkin

We've evolved biologically from a tribal creature, and it seems to me that we developed strategies, coping strategies and rationalities, that worked well for us in that kind of an environment. But when you look at what has happened in recent history as we have transitioned into a global species, we are scaling up these strategies to regional and global scales. Authors [such as] Vanderberg, who wrote the Labyrinth of Technology, have talked about how microscale rationality becomes irrationality when you aggregate it—a tragedy of the commons type of thing—if you look at it that way it's a huge barrier for us to overcome, that scaling up of strategies kind of issue. If you agree with that argument, how should we try to deal with that?

Hewitt

I find all of those arguments dubious, specious pleadings. I don't think we should talk about "we" in this context. I mean, rather than one *Homo sapiens*, humans are like hundreds of distinct species that adapt to and find so many distinct niches in the biosphere and change over time. I support the notion of "one humanity" in [an] ethical and humanitarian sense. I don't think the situations you mention make sense of a "we" doing anything. Aggregate statistics of population may imply a conventional we, but through a great many, many simplifications; I, you, our group, our institutions, these are the ones that generally make decisions or have decisions made for them as distinct entities. Then it is the power of the institutions, their values, and the groups they serve that become decisive. The key developments affecting disaster risks and responses very much occur within particular institutional frameworks, culture and value systems, as well as the influence of powerful individuals within those institutions.

And while it is indeed complex, we can and have identified a lot of groups and places where historical developments and particular powers have made some people much more at risk than others. With respect to your original question, which was that this just happens, well no, it doesn't just happen, but is the work of human hands, tools, trained and orderly groups usually obeying instructions. It is not only or mainly a side effect but more like the decision of a husband or parent not to get life insurance or not to do other things that would actually give better protection to their children, a government to pay down the debt rather than invest in job creation or improved health care, when in fact they have the capacity to do otherwise. People know that converting agricultural land to development is going to have consequences on local food production. They know that mechanization is going to push them off land and destroy jobs. They choose to do that and defend doing so, usually in terms of a taken-for-granted view of [the] "greater good" or progress. It is also not accidental that disaster management has been kept largely separate from the examination of such decisions in terms of what HFA calls "underlying risk factors."

Etkin

What is your reaction to sort of large metanarratives like Beck's risk society? Do you have a degree of comfort with that grand narrative? I am getting a sense that you ascribe a large degree of importance to individual and local narratives in decision making.

Hewitt

I don't see it as an either/or situation. I think we have to be aware of these big-picture ideas and critiques that you have been raising, too. And yes, it is a totality, a matter of contexts, uneven impacts of progressive changes, contexts. DRR can be a quite different thing trying to deal with a village versus a city, in Nepal as against the North American "homeland security" system, and to see the latter in relationship to what is happening in other developed, and developing, countries.

It has always been the role of intellectuals to look at the total audience that you are dealing with, and Beck did that, though I feel his case was overwhelmingly about increasingly popular concerns in West Germany in the 1970s and '80s, and to some extent similar ones that were found in Canada. To my mind it's a powerful metaphor, but it is a mistake to apply it much more widely or even to assume that it would apply to all of Germany, least of all then East Germany, or all of the industrialized countries of that time. And I suspect Beck's metanarrative is fundamentally mistaken. Modern dominant institutions states and alliances and the common markets are not acting in a "risk first" or risk obsessions as against former wealth, investment, and growth. I find the major financial, productive, and governmental organizations continue to have and act upon a "growth first" priority. Beck has been talking about a particular segment of the population—you and I (and people like us) who were becoming conscious of the fact that the benefits of consumer society in the postwar boom, which had seemed so progressive as well as beguiling, came at considerable cost and danger. In this I believe he was right and draw attention to particular and real ways in which we are increasingly at risk. We spend an enormous amount of time buying the right food, avoiding risk, getting insurance, worrying about our kids, not using toys from China, and all of it. That is the risk society and it is a brilliant synthesis. You just have to recognize who it involves, and it is far from global even in German society, where it was coming from.

Etkin

I'll ask one more question and then wrap it up. If you were me, what question would you ask yourself?

Hewitt

I thought you might ask, "What do you think it is really about?" Do you have a bottom line that drives you to continue doing and make sense of this work? You kind of did ask that at the beginning, but it got lost. This is where one is on the shifting ground of having responsibilities both as an academic and as a concerned citizen.

Etkin

That is a fair enough question. How would you answer it?

Hewitt

Bottom line? Disaster is about suffering. The only reason, justification, or excuse to work in this field is if you are reasonably convinced you can say and do something that can reduce it. That should be the first priority in every analysis, everything you teach, everything you advise, and every recommendation. Much like the world of crime, however, there is a strong sense that disaster management is built around official, technical, and statistically evaluated responses. As with policing, detection, and punishment of crime, they may not be without merit, but give all too little real attention to the plight, experiences, voices, needs, and concerns of the victims of disaster.

Index

Note: Page numbers followed by "b", "f" and "t" indicate boxes, figures and tables respectively

A
Aboriginal networks, 135
Adaptive management perspective, 132
A New Species of Trouble (Erikson), 6
Average Annual Loss (AAL), 213

B
Bam earthquake, 27–29, 28f
Bird flocking, 160, 161f
Business ethics, 297–298

C
Canadian Disaster Database, 5–6
CAPRA. *See* Central American Probabilistic Risk Assessment (CAPRA)
Carbon monoxide (CO), 95
Caring ethics, 292
CAS. *See* Complex adaptive systems (CAS)
Catastrophe (CAT) model, 213–215, 213f, 254
CAT models. *See* Catastrophe (CAT) models
CEM. *See* Comprehensive emergency management (CEM)
Central American Probabilistic Risk Assessment (CAPRA), 81
Centre for Research on the Epidemiology of Disasters (CRED), 4–5, 26
Chaotic systems, 156–157, 199
CISD. *See* Critical incident stress debriefing (CISD)
Climate system, 157–159
Climatology, 13–15
Coastal Cities, 90
Community-based approach, 119
Community resilience, 130–134
Complex adaptive systems (CAS), 162
Complexity theory, 155, 171, 331
Complex systems
  Chaos theory, 169
  characteristics of, 155–166
    adaptation, 162–163
    emergence, 159–160
    probabilities, comment on, 163–166
    self-organization, 160–162
  NGOs, 169
  normal accident theory, 166–168
    high reliability theory, 168
Comprehensive emergency management (CEM), 202–205, 203f, 332
Computer simulation models, 159
Constructionist approach, 59–60
Corporate social responsibility. *See* Business ethics
Coupling-complexity chart, 166, 167f
CRED. *See* Centre for Research on the Epidemiology of Disasters (CRED)
Critical incident stress debriefing (CISD), 129–130

D
Data
  availability, 35
  databases, 35–41
    Center for Research on the Epidemiology of Disasters, 36–39
    Munich Reinsurance Database, 40–41
    World Bank, 39–40
  economic data, 35
  Hurricane Hazel and Toronto, 46–50
  measuring loss, 26–30, 27t
  postresponse collection, 33–34
  quality, 30–35
    biases in, 33–35
    sources and methodology, 30–33
  small business enterprises, 35
Deontology/duty-based ethics, 282–289
Deterministic system, 156

355

Disaster deficit index, 81
Disaster management
  complexity of, 318–319
  frame of reference, 324
  fundamental barrier to, 292
  human beings in, 291
  principles, 319–323
    breakout groups, tasks for, 323–326
    constructions on, 324
    examples of, 320t–321t
    matrix, 322–323, 323f
    pyramid, 319–322, 322f
  purpose of, 324
Disaster models
  CARE model, 207–208, 208f
  CAT model, 213–215, 213f
  CEM, 202–205, 203f
  characteristics, 7–8
  climatological perspective, 13
  CRED, 4–5
  disaster risk reduction, 218–219
  ecological models, 215–217, 215f
  emergencies and catastrophes, 8
  Europe, heat wave in, 10–18
  human-caused, 8–9
  ICS, 210–212, 212f
  labeling, 8–9
  linear risk management model, 209–210, 209f
  meaning of, 3–10
  Panarchy: Understanding Transformations in Human and Natural Systems, 6, 7f
  PAR model, 205–207, 206f
Disaster Resiliency/Preparedness Index, 82
Disaster risk reduction (DRR), 132, 218–219
Dynamic nonlinear systems, 157

E
Earthquake hazard, 109, 109f
Ecological footprint, of humanity, 295
Ecological models, 215–217, 215f
Ecological paradigms, 217
Ecological resilience, 126–127
Economic risk, 39, 43f
Ecosystem approach, 182, 242
Emergency Disasters Database (EM-DAT), 80

Engineering approach
  catastrophe (CAT) models, 83–84
  HAZUS, 84, 86f
  limitations, 84
Engineering resilience, 124–126
Environmental ethics, 292–297
Environmental vulnerability index (EVI), 81
Equilibrium models, 217
Ethical behavior, 292, 296
Ethics
  business ethics, 297–298
  construction of risk, 298–299
  deontology/duty-based ethics, 282–289
  environmental ethics, 292–297
  Jean Slick on, 303–304
  temporary settlement *vs.* permanent housing, 300–303
  utilitarianism, 280–282
  virtue ethics, 290–292
EVI. *See* Environmental Vulnerability Index (EVI)
Exceedence probability (EP) analysis, 213, 214f

F
Fail-safe designs, 124
Famine Crimes (Alex de Waal), 286
Fat tails, 163–164, 165f
Federal Emergency Management Agency (FEMA), 57, 109, 182, 186, 256
FEMA. *See* Federal Emergency Management Agency (FEMA)
Flashbulb memory, 266
Floodplain management, 176, 185–186
Fragility curves, 213–215, 214f
Freezing rain
  accumulations of, 94, 94f–95f
  formation of, 92, 92f

G
GDP. *See* Gross domestic product (GDP)
Geographic Information System, 84, 85f–86f
Global Urban Risk Index, 80–81
Gross domestic product (GDP), 27–29, 39, 43f
Groundwork for the Metaphysics of Morals (Kant), 137–138

H

Hazard, 113–122
  event profiles, 107f
  floods cause, 109–111
  mapping, 109
  paradigm, 106–107, 331
  "S"-shaped curve associated with, 115, 116f
Hazards United States (HAZUS), 84, 86f, 213
Heat wave, in Europe
  climatological perspective, 13
  climatology, 13–15
  human responses, 13
  mortality, 11, 12t, 15–16, 18
  political response, 16
  temperature anomalies, 10–11
High Reliability Theory (HRT), 168
Homeostasis, 63
Hooke's law, 124, 125f
HRT. *See* High reliability theory (HRT)
Hurricane Katrina, 27–29, 28f, 256–257
Hurricane Pam, 182–183
Hydro Quebec, 95

I

ICS. *See* Incident Command System (ICS)
IEM. *See* Integrated Emergency Management (IEM)
Incident Command System (ICS), 169, 210–212, 212f
Indices approach
  CAPRA, 81
  examples of, 80–82
  on hazard and vulnerability, 79
  issues in, 83
  limitations, 80
Integrated emergency management (IEM), 204, 205f
International community, 32–33
International Strategy for Disaster Reduction (ISDR), 26

J

Janoff-Bulman's research, 242

L

Laissez-faire management style, 69
Laundry list approach, 119

Local Disaster Index, 81
Lorenz strange attractor, 157, 158f

M

Mapping Vulnerability: Disasters Development and People, 121
Mercury contamination, 139f
Millennium ecosystem assessment, 293–294
Mission Improbable: Using Fantasy Documents to Tame Disaster (Lee Clarke), 244
Mortality, 11, 12t, 15–16, 18
Munich Taxi Experiment, 62
Myths
  at core of internal world, 241
  earthquakes, 233–236, 234f
  of human behavior, 237–240
  human nature*Panarchy: Understanding Transformations in Human and Natural Systems*, 243, 242
  of technological and economic growth, 240–241
  tornadoes, 231–233, 232f

N

National Flood Insurance Program (NFIP), 181, 284–285
National Incident Management System (NIMS), 210, 212
National Oceanic and Atmospheric Administration (NOAA), 33
Natural disasters
  deaths from, 37–38, 39f
  definition, 26
  economic impact of, 37, 38f
  number of, 36–37, 37f
  people affected by, 38, 40f
  by region, 38–39, 42f
Natural hazard system, 108, 108f
NFIP. *See* National Flood Insurance Program (NFIP)
Nonequilibrium models, 217
Nongovernmental organizations, 31
Non-human species
  extinction rates of, 293–294, 294f
Normal Accidents: Living with High Risk Technologies (Charles Perrow), 166
Normal accident theory, 166–168, 331

O

Optimal risk management strategies, 58
Organized anarchy, 160

P

Panarchy model, 217, 218f, 242–243
Panarchy: Understanding Transformations in Human and Natural Systems, 6, 7f
PAR model. *See* Pressure and release (PAR) model
Participatory planning, 144
Paternalism, 288–289
Philosophical approaches
   cause and effect, 198–200
   ethics and values, 200–202
Poetry
   Alexandria, Library at, 262–266
   Nicole Cooley essay, 254–258
Policy decisions, 159
Pressure and release (PAR) model, 205–207, 206f
Prevalent Vulnerability Index, 81
Psychological resilience, 128–130
Psychometric paradigm, 67, 68f, 108
Public Health Agency of Canada (PHAC), 142–143

Q

Quasistationary frontal zone, 92, 93f

R

Rationalist approach, 59–60
Rationalist-constructionist spectrum, 60, 60f
Rationality badges, 244
Regional disaster mortality data, 38–39, 42f
Resilience
   community resilience, 130–134
   ecological resilience, 126–127
   engineering resilience, 124–126
   index, 130
   psychological resilience, 128–130
   religion and spirituality, 134–138
   shift toward resilience thinking, 123–124
   vulnerability, 114

Resistance, 331
Restitutive justice, 297
Reversibility, 283
Risk
   communication strategies, 245
   definition, 331
   as feeling, 73–77
   homeostasis, 62–66
   ice storm disaster, 1998, 92
   management index, 81
   management strategy, 209, 209f–210f
   measuring, 78–87
      case study approach, 85–87
      engineering approach, 83–84
      indices approach, 79–83
   perception, 66–73
   psychometric paradigm of, 67, 67f–68f
   sea level rise and subsidence, 87–91
   as social construct, 59–61
   society, 77–78

S

Sea level rise, 87–91
Sensitivity, 82
"S"-shaped vulnerability curve, 115, 116f
Situational approach, 119
Social capital, 132
Social contract theory, 285–288
Social referencing, 128
Social vulnerability, 112–113, 119
   index, 81–82
Spirituality, 134–138
   providers of, 137
Strange attractors, 157
Structural mitigation, 179–180
Subsidence, 87–91

T

Target risk, 62–63
Taxonomy approach, 119
The Book of the Dead (Muriel Rukeyser), 255
The Crowd (Gustav LeBon), 248
The Oak Tree and the Reeds, 122f, 123
Traditional risk management framework, 61, 61f
Transformability, 127

U

UN International Strategy for Disaster Reduction, 5
United Nations Development Programme (UNDP), 80
Universalizability, 283
US Office for Foreign Disaster Assistance (OFDA), 34
Utilitarianism, 280–282

V

Vested interests, 332
Virtual risk, 70
Virtue ethics, 290–292
Vulnerability, 331
assessment, 119–122
creative tensions and debates, 113–115
definition, 113–122
key spheres of, 115, 116f
resilience, 114, 115f
scales and causation, 115–119
theory, 123

W

White male effect, 73, 74f
Windstorm disasters, 38, 42f
World Wildlife Federation, 82

X

X-rays, risk factors for, 67, 67f

Printed by Printforce, the Netherlands